Hydrogels for 3D Printing

Hydrogels for 3D Printing

Editors

Enrique Aguilar
Helena Herrada-Manchón

Basel • Beijing • Wuhan • Barcelona • Belgrade • Novi Sad • Cluj • Manchester

Editors

Enrique Aguilar
Department of Organic and
Inorganic Chemistry
Universidad de Oviedo
Oviedo
Spain

Helena Herrada-Manchón
Mechatronics and Digital
Industry
Fundación IDONIAL
Gijón
Spain

Editorial Office
MDPI
St. Alban-Anlage 66
4052 Basel, Switzerland

This is a reprint of articles from the Special Issue published online in the open access journal *Gels* (ISSN 2310-2861) (available at: www.mdpi.com/journal/gels/special_issues/085GU2740Y).

For citation purposes, cite each article independently as indicated on the article page online and as indicated below:

Lastname, A.A.; Lastname, B.B. Article Title. *Journal Name* **Year**, *Volume Number*, Page Range.

ISBN 978-3-7258-1240-0 (Hbk)
ISBN 978-3-7258-1239-4 (PDF)
doi.org/10.3390/books978-3-7258-1239-4

© 2024 by the authors. Articles in this book are Open Access and distributed under the Creative Commons Attribution (CC BY) license. The book as a whole is distributed by MDPI under the terms and conditions of the Creative Commons Attribution-NonCommercial-NoDerivs (CC BY-NC-ND) license.

Contents

About the Editors . . . . **vii**

Preface . . . . **ix**

Enrique Aguilar and Helena Herrada-Manchón
Editorial for the Special Issue "Hydrogels for 3D Printing"
Reprinted from: *Gels* **2024**, *10*, 323, doi:10.3390/gels10050323 . . . . **1**

Helena Herrada-Manchón, Manuel Alejandro Fernández and Enrique Aguilar
Essential Guide to Hydrogel Rheology in Extrusion 3D Printing: How to Measure It and Why It Matters?
Reprinted from: *Gels* **2023**, *9*, 517, doi:10.3390/gels9070517 . . . . **4**

Arpana Agrawal and Chaudhery Mustansar Hussain
3D-Printed Hydrogel for Diverse Applications: A Review
Reprinted from: *Gels* **2023**, *9*, 960, doi:10.3390/gels9120960 . . . . **25**

David Patrocinio, Victor Galván-Chacón, J. Carlos Gómez-Blanco, Sonia P. Miguel, Jorge Loureiro and Maximiano P. Ribeiro et al.
Biopolymers for Tissue Engineering: Crosslinking, Printing Techniques, and Applications
Reprinted from: *Gels* **2023**, *9*, 890, doi:10.3390/gels9110890 . . . . **60**

Maria Minodora Marin, Ioana Catalina Gifu, Gratiela Gradisteanu Pircalabioru, Madalina Albu Kaya, Rodica Roxana Constantinescu and Rebeca Leu Alexa et al.
Microbial Polysaccharide-Based Formulation with Silica Nanoparticles; A New Hydrogel Nanocomposite for 3D Printing
Reprinted from: *Gels* **2023**, *9*, 425, doi:10.3390/gels9050425 . . . . **95**

Sónia Oliveira, Isabel Sousa, Anabela Raymundo and Carlos Bengoechea
Three-Dimensional Printing of Red Algae Biopolymers: Effect of Locust Bean Gum on Rheology and Processability
Reprinted from: *Gels* **2024**, *10*, 166, doi:10.3390/gels10030166 . . . . **114**

Jana Matejkova, Denisa Kanokova, Monika Supova and Roman Matejka
A New Method for the Production of High-Concentration Collagen Bioinks with Semiautonomic Preparation
Reprinted from: *Gels* **2024**, *10*, 66, doi:10.3390/gels10010066 . . . . **133**

Christina Kaliampakou, Nefeli Lagopati, Evangelia A. Pavlatou and Costas A. Charitidis
Alginate–Gelatin Hydrogel Scaffolds; An Optimization of Post-Printing Treatment for Enhanced Degradation and Swelling Behavior
Reprinted from: *Gels* **2023**, *9*, 857, doi:10.3390/gels9110857 . . . . **156**

Héctor Sanz-Fraile, Carolina Herranz-Diez, Anna Ulldemolins, Bryan Falcones, Isaac Almendros and Núria Gavara et al.
Characterization of Bioinks Prepared via Gelifying Extracellular Matrix from Decellularized Porcine Myocardia
Reprinted from: *Gels* **2023**, *9*, 745, doi:10.3390/gels9090745 . . . . **187**

Andrea Souza, Matthew Parnell, Brian J. Rodriguez and Emmanuel G. Reynaud
Role of pH and Crosslinking Ions on Cell Viability and Metabolic Activity in Alginate–Gelatin 3D Prints
Reprinted from: *Gels* **2023**, *9*, 853, doi:10.3390/gels9110853 . . . . **199**

Preetham Yerra, Mario Migliario, Sarah Gino, Maurizio Sabbatini, Monica Bignotto and Marco Invernizzi et al.
Polydopamine Blending Increases Human Cell Proliferation in Gelatin–Xanthan Gum 3D-Printed Hydrogel
Reprinted from: *Gels* **2024**, *10*, 145, doi:10.3390/gels10020145 . . . . **212**

About the Editors

Enrique Aguilar

Enrique Aguilar was born in Noreña (Asturias, Spain) and received his Ph.D. from the Universidad de Oviedo under the supervision of Prof. J. Barluenga and Prof. S. Fustero in 1991. After postdoctoral research with Prof. A. I. Meyers at Colorado State University (1991–1994), he returned to the Universidad de Oviedo as a researcher, being promoted to assistant professor in 1996, associate professor in 2002, and full professor in 2019. He has been a visiting scientist at the University of Colorado (1996, with Prof. G. A. Molander). His current interests are centered on the development of synthetic organic methodology, organometallic chemistry, homogeneous metal catalysis, and 3D bioprinting.

Helena Herrada-Manchón

Helena Herrada-Manchón, born in Vilanova i la Geltrú (Barcelona, Spain), holds a Ph.D. in Chemistry from the University of Oviedo. She also earned a master's degree in Chemistry and Sustainable Development as well as a bachelor's degree in Pharmacy from the University of Barcelona. After several years dedicated to community pharmacy, during her master's studies she joined the staff of the IDONIAL Technology Centre, where she currently serves as the technical manager for the development of inks and bioinks for 3D printing in sectors of high interest, such as tissue engineering, regenerative medicine, or the pharmaceutical industry. Her main interests lie in hydrogels, rheology, material characterization, and all 3D printing technologies applied to the healthcare field.

Preface

Dear Colleagues,

Hydrogels are 3D cross-linked networks of flexible polymer chains that contain a large amount of water as the filling solvent. There is a wide variety of polymeric hydrogels that can be classified according to their synthetic methods, physical properties, polymer source (natural, synthetic, or hybrid), ionic charge, and method of cross-linking (via pH or temperature changes, ionic cross-linking, photochemical reactions, enzymatic cross-linking, or guest–host interactions).

Three-dimensional printing, as an emerging versatile manufacturing technology, has been applied in the fabrication of hydrogel constructs with complex structures and potential applications in tissue engineering, regenerative medicine, delivery systems (drugs, proteins, genes, and cells), implantable devices, sensors, and diagnostic devices, among others.

This Special Issue aims to present a collection showcasing the recent progress in hydrogels, including natural polymer hydrogels, synthetic polymer hydrogels, and derivative hydrogels to be used in extrusion printing, inkjet printing, laser or light processing printing, 3D bioprinting, and 4D printing.

Enrique Aguilar and Helena Herrada-Manchón
Editors

Editorial

Editorial for the Special Issue "Hydrogels for 3D Printing"

Enrique Aguilar [1,*] and Helena Herrada-Manchón [2,*]

1. Centro de Innovación en Química Avanzada (ORFEO-CINQA), Departamento de Química Orgánica e Inorgánica, Instituto Universitario de Química Organometálica "Enrique Moles", Universidad de Oviedo, C/Julián Clavería 8, 33006 Oviedo, Spain
2. Fundación Idonial, Parque Científico y Tecnológico de Gijón, Avda, Jardín Botánico 1345, 33203 Gijón, Spain
* Correspondence: eah@uniovi.es (E.A.); helena.herrada@idonial.com (H.H.-M.)

Citation: Aguilar, E.;
Herrada-Manchón, H. Editorial for
the Special Issue "Hydrogels for 3D
Printing". *Gels* **2024**, *10*, 323.
https://doi.org/10.3390/gels10050323

Received: 6 May 2024
Accepted: 8 May 2024
Published: 9 May 2024

Copyright: © 2024 by the authors.
Licensee MDPI, Basel, Switzerland.
This article is an open access article distributed under the terms and conditions of the Creative Commons Attribution (CC BY) license (https://creativecommons.org/licenses/by/4.0/).

Hydrogels, which are three-dimensional networks of hydrophilic polymers capable of absorbing and retaining large amounts of water, have emerged as versatile materials with vast potential in various fields. In the context of 3D printing, hydrogels offer unique advantages due to their tunable mechanical properties, biocompatibility, and ability to encapsulate cells or bioactive molecules. The process of 3D printing with hydrogels comprises several crucial steps, each of which plays a critical role in determining the final structure and properties of the printed object.

A key aspect in this process is the formulation of hydrogel materials used for printing. The selection and optimization of raw materials or components are crucial to ensure the desired rheological properties and printability of the biomaterial ink or bioink. During printing, the hydrogel material is deposited layer by layer, following the path determined by slicing software. The rheological properties of the hydrogel formulation are essential at this stage to ensure the proper deposition of the material and maintain the structural integrity of the printed object.

In addition to rheological considerations, the choice of hydrogel formulation must consider the specific requirements of the intended application. For instance, in tissue engineering applications, hydrogels must provide a supportive matrix for cell growth and tissue regeneration. Therefore, factors such as biocompatibility, cell adhesion, and degradation kinetics are of utmost importance.

Furthermore, post-processing steps such as crosslinking are often necessary to stabilize the printed structure and enhance its mechanical properties. Crosslinking agents, such as chemical or physical crosslinkers, are used to form covalent or physical bonds between polymer chains, resulting in a more robust and stable hydrogel network.

This Special Issue showcases a diverse array of seven articles and three reviews that explore the multifaceted aspects of hydrogels for 3D printing, encompassing all the aforementioned steps, as follows:

1. The review "Essential Guide to Hydrogel Rheology in Extrusion 3D Printing: How to Measure It and Why It Matters?" by Herrada-Manchón et al. examines the crucial role of rheology in extrusion-based 3D printing, focusing on hydrogels and their applications in tissue engineering, regenerative medicine, and drug delivery. Understanding rheological properties such as shear-thinning behavior, thixotropy, and viscoelasticity is essential for optimizing the printing process and achieving the desired product quality.
2. The review "3D-Printed Hydrogel for Diverse Applications: A Review" by Agrawal et al. explores the intersection of hydrogels and 3D printing, covering current research, technological advancements, and future directions in various applications, including biomedical engineering. It discusses hydrogel basics, materials, 3D printing methods, and challenges, in addition to predicting future trends.
3. The review "Biopolymers for Tissue Engineering: Crosslinking, Printing Techniques, and Applications" by Patrocinio et al. explores the use of proteins and polysaccharides

in bioprinting for tissue engineering, focusing on their biocompatibility, biodegradability, and biomimicry. It addresses challenges related to rheological behaviors and the need for modifications or crosslinking. The manuscript also discusses tissue engineering applications, crosslinking, and bioprinting techniques, aiming to achieve bioprinted structures mimicking extracellular matrix properties with good levels of printability and stability.

4. The article "Microbial Polysaccharide-Based Formulation with Silica Nanoparticles; A New Hydrogel Nanocomposite for 3D Printing" by Marin et al. develops printable hydrogel nanocomposites by adding silica nanoparticles to a microbial polysaccharide polymer network, aiming for biomedical applications. Their morpho-structural characteristics, swelling behavior, and mechanical stability were assessed, revealing excellent biocompatibility, making them suitable for regenerative medicine.

5. The article "Three-Dimensional Printing of Red Algae Biopolymers: Effect of Locust Bean Gum on Rheology and Processability" by Oliveira et al. explores the extraction of gelling biopolymers from red seaweeds to produce sustainable food gels and assess their potential as bioinks for 3D printing, focusing on improving gel matrix definition through the addition of locust bean gum, as well as adjusting printing temperature.

6. The article "A New Method for the Production of High-Concentration Collagen Bioinks with Semiautonomic Preparation" by Matejkova et al. introduces a novel method for preparing highly concentrated collagen bioinks, utilizing a two-step neutralization process based on bicarbonate buffering mechanisms and pH adjustment. The automated bioink preparation process ensures consistent quality, demonstrating sustained cell proliferation and viability, in addition to offering potential for advancing tissue engineering applications.

7. The article "Alginate–Gelatin Hydrogel Scaffolds; An Optimization of Post-Printing Treatment for Enhanced Degradation and Swelling Behavior" by Kaliampakou et al. addresses the optimization of the post-printing treatment phase in 3D structure generation, focusing on enhancing scaffold degradation while maintaining targeted swelling behavior.

8. The article "Characterization of Bioinks Prepared via Gelifying Extracellular Matrix from Decellularized Porcine Myocardia" by Sanz-Fraile et al. introduces a novel cardiac bioink derived exclusively from decellularized porcine myocardium and loaded with human-bone-marrow-derived mesenchymal stromal cells, eliminating the need for additional biomaterials or crosslinkers.

9. The article "Role of pH and Crosslinking Ions on Cell Viability and Metabolic Activity in Alginate–Gelatin 3D Prints" by Souza et al. investigates the influence of pH and crosslinking ions on the stability, printability, and cell behavior of alginate–gelatin hydrogels commonly used in bioengineering. The results reveal that buffer pH and crosslinking ions affect the swelling and degradation rates of prints, as well as hydrogel printability.

10. The article "Polydopamine Blending Increases Human Cell Proliferation in Gelatin–Xanthan Gum 3D-Printed Hydrogel" by Yerra et al. enhances the printability of gelatin–xanthan gum hydrogel by incorporating polydopamine (PDA), a mussel-inspired biopolymer, to create cell-laden 3D scaffolds demonstrating improved fibroblast and keratinocyte growth without affecting hydrogel characteristics, thereby suggesting its potential for developing innovative 3D-printed wound dressings.

The continued exploration of hydrogels for 3D printing holds promise for groundbreaking advancements in areas such as tissue engineering, drug delivery, and regenerative medicine. We extend our gratitude to the authors for their valuable contributions to this Special Issue and look forward to the transformative impact of their work.

Author Contributions: E.A., conceptualisation; writing—original draft preparation; writing—review and editing. H.H.-M., conceptualisation; writing—original draft preparation; writing—review and editing. All authors have read and agreed to the published version of the manuscript.

Acknowledgments: We would like to thank the Ministerio de Ciencia, Innovación y Universidades of Spain (Agencia Estatal de Investigación: PID2022-140635NB-I00 funded by MICIU/AEI/10.13039/501100011033/ERDF/EU) and the Centro para el Desarrollo Tecnológico y la Innovación (CDTI) for the program "Ayudas Cervera a Centros Tecnológicos" (CER-20231013) of the State Plan for Scientific and Technical Research and Innovation 2021-2023, within the framework of the Recovery, Transformation and Resilience Plan—NextGenerationEU, for funding.

Conflicts of Interest: The authors declare no conflicts of interest.

List of Contributions

1. Herrada-Manchón, H.; Fernández, M.A.; Aguilar, E. Essential Guide to Hydrogel Rheology in Extrusion 3D Printing: How to Measure It and Why It Matters? *Gels* **2023**, *9*, 517. https://doi.org/10.3390/gels9070517.
2. Agrawal, A.; Hussain, C.M. 3D-Printed Hydrogel for Diverse Applications: A Review. *Gels* **2023**, *9*, 960. https://doi.org/10.3390/gels9120960.
3. Patrocinio, D.; Galván-Chacón, V.; Gómez-Blanco, J.C.; Miguel, S.P.; Loureiro, J.; Ribeiro, M.P.; Coutinho, P.; Pagador, J.B.; Sanchez-Margallo, F.M. Biopolymers for Tissue Engineering: Crosslinking, Printing Techniques, and Applications. *Gels* **2023**, *9*, 890. https://doi.org/10.3390/gels9110890.
4. Marin, M.M.; Gifu, I.C.; Pircalabioru, G.G.; Albu Kaya, M.; Constantinescu, R.R.; Alexa, R.L.; Trica, B.; Alexandrescu, E.; Nistor, C.L.; Petcu, C.; et al. Microbial Polysaccharide-Based Formulation with Silica Nanoparticles; A New Hydrogel Nanocomposite for 3D Printing. *Gels* **2023**, *9*, 425. https://doi.org/10.3390/gels9050425.
5. Oliveira, S.; Sousa, I.; Raymundo, A.; Bengoechea, C. Three-Dimensional Printing of Red Algae Biopolymers: Effect of Locust Bean Gum on Rheology and Processability. *Gels* **2024**, *10*, 166. https://doi.org/10.3390/gels10030166.
6. Matejkova, J.; Kanokova, D.; Supova, M.; Matejka, R. A New Method for the Production of High-Concentration Collagen Bioinks with Semiautonomic Preparation. *Gels* **2024**, *10*, 66. https://doi.org/10.3390/gels10010066.
7. Kaliampakou, C.; Lagopati, N.; Pavlatou, E.A.; Charitidis, C.A. Alginate–Gelatin Hydrogel Scaffolds; An Optimization of Post-Printing Treatment for Enhanced Degradation and Swelling Behavior. *Gels* **2023**, *9*, 857. https://doi.org/10.3390/gels9110857.
8. Sanz-Fraile, H.; Herranz-Diez, C.; Ulldemolins, A.; Falcones, B.; Almendros, I.; Gavara, N.; Sunyer, R.; Farré, R.; Otero, J. Characterization of Bioinks Prepared via Gelifying Extracellular Matrix from Decellularized Porcine Myocardia. *Gels* **2023**, *9*, 745. https://doi.org/10.3390/gels9090745.
9. Souza, A.; Parnell, M.; Rodriguez, B.J.; Reynaud, E.G. Role of pH and Crosslinking Ions on Cell Viability and Metabolic Activity in Alginate–Gelatin 3D Prints. *Gels* **2023**, *9*, 853. https://doi.org/10.3390/gels9110853.
10. Yerra, P.; Migliario, M.; Gino, S.; Sabbatini, M.; Bignotto, M.; Invernizzi, M.; Renò, F. Polydopamine Blending Increases Human Cell Proliferation in Gelatin–Xanthan Gum 3D-Printed Hydrogel. *Gels* **2024**, *10*, 145. https://doi.org/10.3390/gels10020145.

Disclaimer/Publisher's Note: The statements, opinions and data contained in all publications are solely those of the individual author(s) and contributor(s) and not of MDPI and/or the editor(s). MDPI and/or the editor(s) disclaim responsibility for any injury to people or property resulting from any ideas, methods, instructions or products referred to in the content.

Review

Essential Guide to Hydrogel Rheology in Extrusion 3D Printing: How to Measure It and Why It Matters?

Helena Herrada-Manchón [1,*], Manuel Alejandro Fernández [1] and Enrique Aguilar [2]

1. Fundación Idonial, Parque Científico y Tecnológico de Gijón, Avda, Jardín Botánico 1345, 33203 Gijón, Spain; alejandro.fernandez@idonial.com
2. Centro de Innovación en Química Avanzada (ORFEO-CINQA), Instituto Universitario de Química Organometálica "Enrique Moles", Departamento de Química Orgánica e Inorgánica, Universidad de Oviedo, C/Julián Clavería 8, 33006 Oviedo, Spain; eah@uniovi.es
* Correspondence: helena.herrada@idonial.com

Abstract: Rheology plays a crucial role in the field of extrusion-based three-dimensional (3D) printing, particularly in the context of hydrogels. Hydrogels have gained popularity in 3D printing due to their potential applications in tissue engineering, regenerative medicine, and drug delivery. The rheological properties of the printing material have a significant impact on its behaviour throughout the 3D printing process, including its extrudability, shape retention, and response to stress and strain. Thus, understanding the rheological characteristics of hydrogels, such as shear thinning behaviour, thixotropy, viscoelasticity, and gelling mechanisms, is essential for optimising the printing process and achieving desired product quality and accuracy. This review discusses the theoretical foundations of rheology, explores different types of fluid and their properties, and discusses the essential rheological tests necessary for characterising hydrogels. The paper emphasises the importance of terminology, concepts, and the correct interpretation of results in evaluating hydrogel formulations. By presenting a detailed understanding of rheology in the context of 3D printing, this review paper aims to assist researchers, engineers, and practitioners in the field of hydrogel-based 3D printing in optimizing their printing processes and achieving desired product outcomes.

Keywords: 3D printing; rheology; hydrogel; extrusion

Citation: Herrada-Manchón, H.; Fernández, M.A.; Aguilar, E. Essential Guide to Hydrogel Rheology in Extrusion 3D Printing: How to Measure It and Why It Matters? Gels 2023, 9, 517. https://doi.org/10.3390/gels9070517

Academic Editor: Bae Hoon Lee

Received: 6 June 2023
Revised: 23 June 2023
Accepted: 25 June 2023
Published: 26 June 2023

Copyright: © 2023 by the authors. Licensee MDPI, Basel, Switzerland. This article is an open access article distributed under the terms and conditions of the Creative Commons Attribution (CC BY) license (https://creativecommons.org/licenses/by/4.0/).

1. Introduction

Rheology is defined as the study of the physical principles that regulate the flow and deformation of matter subjected to forces. The practical use of rheology is found in areas such as quality control and production, chemical and mechanical engineering, industrial research and development, or materials science in numerous industries (pharmaceuticals, cosmetics, agriculture, food, ceramics, paints, etc.) where plastics, synthetic fibres, pastes, lubricants, creams, suspensions or emulsions constitute the raw material and are the object of study [1]. For this reason, it is currently accepted that rheology is an interdisciplinary science whose development is carried out not only by physicists, but also by engineers of various specialties, mathematicians, chemists, biologists, pharmacists, etc., giving rise to a wide range of practical application possibilities [2].

Among all the different 3D printing technologies, extrusion-based three-dimensional (3D) printing is one of the most common methods for freely digital fabrication due to its ease of use, low initial investment, and high variety of materials available. In this technique, 3D structures are built in a layer-by-layer mode, using sequential extrusion of materials from a nozzle, following a predesigned computer model [3,4]. In particular, extrusion-based 3D printing of hydrogels, namely cross-linked networks of flexible polymer chains containing a large amount of water as fill solvent, is a technology that is gaining popularity due to its potential applications in tissue engineering, drug delivery, and regenerative medicine [5–8]. Hydrogels are highly biocompatible materials that can mimic the properties of natural tissues, and the fact that 3D printing allows for the creation of high-resolution customised structures with precise control over their size, shape, and mechanical properties is one reason why this technology is so widely used at present.

Predictably, the study of rheology is of great importance within the extrusion-based 3D printing of hydrogels since, throughout the printing process, the material is exposed to various external forces and recovery or rest stages, which have a considerable impact on its properties and behaviour. Therefore, the rheological properties of the printing material determine how well it can be extruded through the printing nozzle and how it will behave during the printing process. The viscosity of the printing material and its shear thinning properties, for example, will determine how easily it can flow through the nozzle and will affect its ability to maintain its shape after being extruded. The elasticity of the material will also determine how it will respond to stress and strain during printing. Subsequently, different hydrogel parameters, such as shear thinning behaviour, thixotropy, viscoelasticity, and the gelling mechanism or temperature, are fundamental, and a good understanding of the rheological properties is necessary to optimise the printing process and achieve the desired quality and accuracy of the final product [3,9].

To understand the different types of existing fluid, their properties, and the tests necessary for their characterisation, familiarisation with the theoretical foundations of rheology, the different analyses to be carried out, and the correct interpretation of their results is required. Thus, the purpose of this work is to provide a complete review of terminology, concepts, and rheological tests and their applicability to 3D printing in the evaluation of the printability of different types of hydrogels.

2. Rheology Theoretical Basics

The most common classification scheme for fluids is based on their response to an externally imposed shear stress, known as rheological classification (Figure 1).

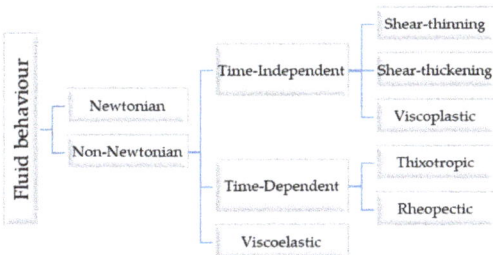

Figure 1. Rheological classification of fluids.

Understanding this classification requires reference to Newton's law (1)), where τ represents shear stress (in Pa), μ represents viscosity (in Pa·s), and $\dot{\gamma}$ represents shear rate (s^{-1}). This shear stress or external shear force is described mathematically as the force (F) applied over a unit area (A).

$$\frac{F}{A} = \tau = \mu \cdot \dot{\gamma} \tag{1}$$

This law indicates that for an incompressible Newtonian fluid in laminar flow, the resulting shear stress equals the product of the shear rate and the fluid's viscosity. Thus, μ is a constant of proportionality characteristic of the material, its temperature, and pressure, and does not depend on $\dot{\gamma}$ or τ [10,11]. Viscosity reflects resistance to flow, and decreases as temperature increases and pressure decreases [12]. For a Newtonian fluid, the rheogram or flow curve, which is the graphical representation of shear stress with respect to shear rate, is a straight line with slope μ (or Newtonian viscosity) that passes through the origin. On the contrary, a non-Newtonian fluid is characterized by either a non-linear flow curve or a flow curve that does not intersect the origin. This means that the viscosity of the fluid is not constant at a given temperature and pressure and depends on flow conditions, such as flow geometry or shear rate. It can even depend on the kinematic history of the fluid element under consideration [10,13,14]. Thus, non-Newtonian fluids can be classified into three general groups:

1. Time-independent fluids, for which the shear stress at any point is determined solely by the value of the shear rate at that point.
2. Time-dependent fluids, which refer to more intricate fluid systems, where the correlation between shear stress and shear rate depends on the duration of the applied shear stress and the fluid's kinematic history.
3. Viscoelastic fluids, which show partial elastic recovery after deformation.

It should be noted that this classification scheme is slightly arbitrary, since most real materials often exhibit a combination of two, or even three, distinct non-Newtonian behaviours. However, it is usually possible to identify the dominant non-Newtonian feature and take this as the basis for further calculations and discussions.

2.1. Time-Independent Fluids

In general, the behaviour of materials of this type can be described by the formula:

$$\dot{\gamma} = f(\tau) \qquad (2)$$

This formula suggests that the value of τ at any given point within the sheared fluid is solely determined by the value of the shear rate at that point, or vice versa. Depending on the shape of the function f, these fluids can be classified into one of three categories: pseudoplastic or shear-thinning, dilatant or shear-thickening, and viscoplastic. The literature contains numerous mathematical expressions that offer empirical relationships between shear stress and shear rate or apparent viscosity, ranging in complexity and form. Some of these expressions involve direct attempts at curve fitting, while others are based on statistical mechanics [15]. While these theoretical models can be useful in silico tools for evaluating material capabilities, it is important to note that they often involve simplifications and approximations. Therefore, experimental validation is necessary to confirm the accuracy of the models in practice [9].

Time-independent fluids and the applicable mathematical models (Figure 2) are classified and detailed in the following subsections.

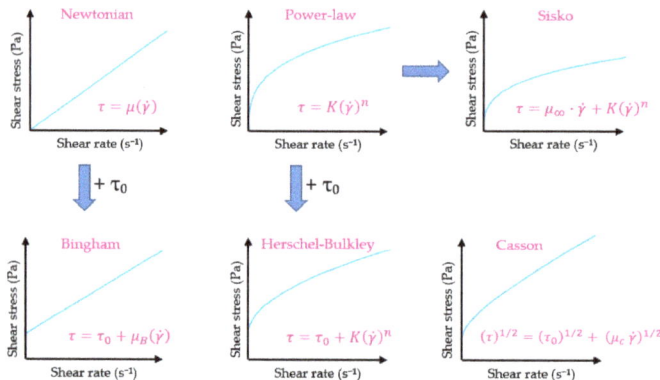

Figure 2. Overview of mathematical models for flow behaviour.

2.1.1. Shear-Thinning or Pseudoplastic Fluids

The most common types of non-Newtonian and time-independent fluids are shear-thinning fluids. In these materials, the apparent viscosity decreases as the shear rate increases.

The Ostwald-de Waele, or Power-law model, is the most commonly used mathematical model for describing the behaviour of pseudoplastic fluids, followed by the Sisko model. The Power-law model is based on the fact that the relationship between shear stress and shear rate (represented in double logarithmic coordinates) can be approximated by a straight line over a limited range of values (Figure 3). The reason behind its widespread use is that in practice, a single device cannot measure the entire rheological spectrum due to a lack of sensitivity (at very low values of $\dot{\gamma}$) or robustness (at high values of $\dot{\gamma}$). In this way, the range of shear rate usually used is from 0.1 a 10^4 s^{-1}, which corresponds to the straight part of the rheogram to which the said model can be applied [16].

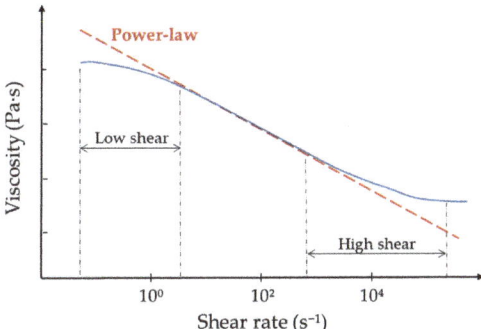

Figure 3. Example of application of the Power-law model (red dashed line) in an idealized flow curve (blue line).

Thus, for this part of the flow curve, the Power-law expression is applied:

$$\tau = K(\dot{\gamma})^n \tag{3}$$

Being the apparent viscosity:

$$\mu = \frac{\tau}{\dot{\gamma}} = K(\dot{\gamma})^{n-1} \tag{4}$$

These equations involve two empirical parameters, K and n, which are known as the consistency index (measured in Pa·s^n) and the flow index, respectively. If the value of n is less than 1, the fluid exhibits pseudoplastic properties. If n is equal to 1, the fluid is Newtonian, and if n is greater than 1, the fluid is dilatant. The lower the value of n, the more the fluid displays shear-thinning properties [10,12,13].

On the other hand, the Sisko model is an extension of the Power-law model that considers the viscosity value when stress rates approach infinity. As a result, it can be used in high-velocity gradient operations such as pumping liquid foods and mixing processes [17–21]. The mathematical expression of three parameters developed by Sisko is shown in (5):

$$\tau = \mu_\infty \cdot \dot{\gamma} + K(\dot{\gamma})^n \tag{5}$$

In this equation, the new parameter μ_∞ represents the viscosity at an infinite shear rate. Therefore, the apparent viscosity is calculated using the following expression:

$$\mu = \mu_\infty + K(\dot{\gamma})^{n-1} \tag{6}$$

Similar to the Power-law model, the values of μ_∞, K, and n are determined experimentally and are specific to each fluid.

2.1.2. Shear-Thickening or Dilatant Fluids

Dilatant fluids, in contrast to pseudoplastic fluids, demonstrate an increase in apparent viscosity as the shear rate rises. This behaviour can be explained easily: at rest, the liquid in the fluid fills the space between particles and lubricates the movement of each particle over the others at low shear levels. However, at high shear speeds, the particles pack together, blocking the passage of liquid between them, resulting in increased solid-solid friction and generating greater resistance (Figure 4) [10,12]. Dilatant fluids are commonly found in materials such as cornmeal mixed with water or fine beach sand when wet. Similar to shear-thinning fluids, dilatant fluids can be described using the Power-law or Sisko model, which have been previously mentioned.

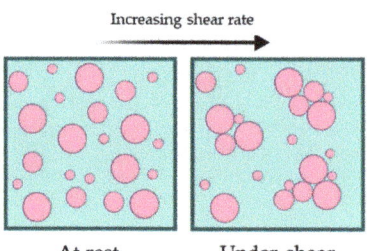

Figure 4. Representative diagram of the behaviour of a dilatant fluid under shear.

2.1.3. Viscoplastic Fluids

Viscoplastic fluids are characterized by the presence of a yield point τ_0 that must be surpassed for the fluid to deform or flow. At rest, the material has an internal three-dimensional structure that is rigid enough to withstand external stresses that are below τ_0, and it deforms elastically [10,22,23]. However, for forces that exceed τ_0, the structure breaks down, and the material behaves similar to a fluid. If the relationship between shear stress and shear rate is linear, it is classified as a Bingham plastic. On the other hand, if there is no such linear relationship, it is classified as a Bingham pseudoplastic or yield-pseudoplastic [15]. In most cases, the transition from a solid-like material to a liquid-like fluid in viscoplastic materials is reversible. The extent of recovery of the structure varies for each material. Toothpaste, jam, and egg white are some examples of fluids that exhibit such behaviour.

There are also different mathematical models used to describe viscoplastic fluids, the most common being the Bingham model and the Herschel–Bulkley model, which are equivalent to the Newtonian and Power Law model, respectively, but add the yield stress [24,25]. Finally, the Casson model is a third model used for viscoplastic materials, which provides the most precise mathematical representation for studying the behaviour of biological fluids such as blood [26].

The Bingham model is the simplest equation to describe a fluid with yield stress, representing a linear relationship between shear stress and shear rate, with a displacement from the origin equal to the value of τ_0. [10,14,24,27]. Hence, this model is characterized by two parameters: the yield point τ_0 and the plastic viscosity μ_B.

$$\begin{aligned} \tau &= \tau_0 + \mu_B(\dot{\gamma}) \quad & if\,|\tau| > |\tau_0| \\ \dot{\gamma} &= 0 \quad & if\,|\tau| < |\tau_0| \end{aligned} \tag{7}$$

However, this model is not suitable for most complex fluids that do not have a linear relationship between $\dot{\gamma}$ and τ, which can result in extremely high values of τ_0 [19]. In such cases, the Herschel–Bulkley model is used, which is an adaptation of the Bingham model that includes a third parameter and provides a better fit to some experimental data.

$$\begin{array}{ll} \tau = \tau_0 + K(\dot{\gamma})^n & if |\tau| > |\tau_0| \\ \dot{\gamma} = 0 & if |\tau| < |\tau_0| \end{array} \qquad (8)$$

Thus, the Herschel–Bulkley model is characterized by the yield point τ_0, the flow index n and the consistency index K [24,27,28]. This model is commonly used to characterize foods such as melted chocolate, yogurt, or purées [27,29–31].

Lastly, the Casson model is another two-parameter model that is commonly used to describe the rheological behaviours of different types of materials. Compared to the Bingham model, the Casson model can more precisely predict the behaviour of a fluid at low shear rates [10,19,32].

$$\begin{array}{ll} (\tau)^{1/2} = (\tau_0)^{1/2} + (\mu_c \dot{\gamma})^{1/2} & if |\tau| > |\tau_0| \\ \dot{\gamma} = 0 & if |\tau| < |\tau_0| \end{array} \qquad (9)$$

2.2. Time-Dependent Fluids

Time-dependent fluids are those for which shear stress is determined by both the magnitude and the duration of the shear rate, as well as the time lapse between consecutive applications. This phenomenon is reversible, as the apparent viscosity of the fluid recovers once the shear stress is no longer applied. There are two primary categories of time-dependent fluids: thixotropic fluids and antithixotropic or rheopectic fluids. Thixotropic fluids experience a decrease in viscosity over time when subjected to a constant shear rate at a fixed temperature. However, rheopectic fluids show an increase in viscosity over time under similar conditions (Figure 5a) [33,34].

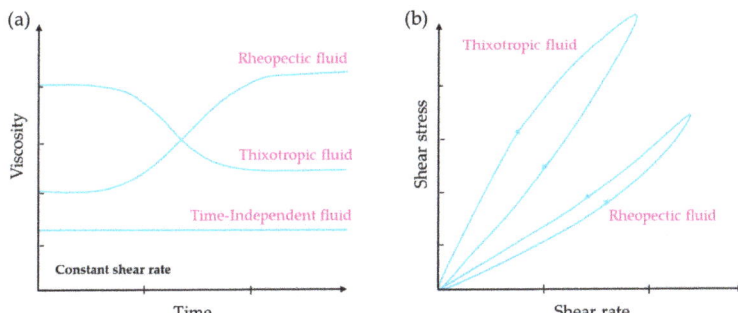

Figure 5. (a) Viscosity profiles over time at a constant shear rate and temperature. (b) Thixotropic loop test, showing shear stress plotted against shear rate.

Even though the theory on time-dependent fluids is not extensively developed, a widespread agreement exists that the underlying cause of this behaviour is a reversible alteration in the fluid's structure that takes place during the flow process [35]. In contrast to Newtonian fluids or time-independent fluids, it is not feasible to employ straightforward mathematical equations to depict the behaviour of thixotropic or rheopectic fluids. Instead, it becomes essential to conduct measurements within the desired range of conditions in order to accurately characterize their properties [10,36]. In the case of time-dependent fluids, when a flow curve develops by progressively increasing the shear rate from zero to a maximum value and subsequently decreasing it back to zero at the same rate, a hysteresis loop is observed (Figure 5b). On the other hand, for time-independent fluids, which are not influenced by time, such a loop is not observed.

2.2.1. Thixotropic Fluids

As mentioned previously, thixotropic materials exhibit greater fluidity when subjected to constant shear stress over time. However, the viscosity behaviour of thixotropic materials during the process of structure decomposition and recovery differs. During shear rate acceleration, the breakdown of the structure lags behind the shear rate, resulting in transient viscosities that are higher than what would be expected at a steady state. On the descending limb of the shear rate curve, the structure continuously rebuilds as the shear rate decreases; however, the viscosities obtained are lower than those observed at steady state [33,37]. Graphically, the surface area enclosed by the hysteresis loop has been suggested as a quantitative measure of thixotropy. However, this method has notable limitations, since the height, shape, and closed area of the loop are influenced by factors such as the duration at which the shear occurs, the rate of shear increases or decreases, and the previous kinematic history of the sample. Therefore, relying solely on the loop surface area may not provide a comprehensive and accurate characterisation of the thixotropic behaviour.

Thixotropy should not be conflated with pseudoplasticity or dilatancy, which are time-independent phenomena mentioned earlier, where viscosity is directly influenced by the magnitude of the shear rate. However, it is not uncommon for a thixotropic fluid to exhibit one of these two phenomena simultaneously [36,38].

2.2.2. Rheopectic Fluids

While a thixotropic fluid's viscosity decreases over time when subjected to a constant shear rate, a rheopectic fluid's viscosity increases when sheared. Again, hysteresis effects are observed in the flow curve (Figure 5b). However, in this case, the loop is in reverse compared to a thixotropic material. In a rheopectic fluid, shearing promotes internal structure formation; however, when the fluid is left undisturbed, it will return to its original lower viscosity [33].

2.3. Viscoelastic Fluids

Viscoelastic fluids possess properties that lie between those of elastic solids and purely viscous fluids. In rheology, dynamic oscillatory shear tests are a common tool used to study a broad range of complex fluids and soft matter, including but not limited to polymer solutions and melts, biological macromolecules, polyelectrolytes, surfactants, suspensions, and emulsions.

Performing dynamic oscillatory shear tests involves subjecting a material to a sinusoidal deformation and monitoring its mechanical response over time. While there are two distinct categories of oscillatory shear tests, namely large amplitude oscillatory shear (LAOS) and small amplitude oscillatory shear (SAOS), the latter has emerged as the preferred method for investigating the linear viscoelastic characteristics of these intricate fluids due to its solid theoretical foundation and the straightforward implementation of appropriate test procedures [39–41]. Consequently, this work will only cover these SAOS measurements. For those seeking a more comprehensive understanding of the utilization of LAOS assays, Hyun et al. published a detailed review on the subject in 2011 [40].

The sinusoidal curves exhibit distinct behaviour depending on the mechanical characteristics of the sample, as illustrated in Figure 6. In the case of a pure elastic material, the maximum stress coincides with the maximum strain, and both stress and strain are in phase, with a phase angle (δ) of $0°$. On the contrary, in a pure viscous material, the maximum stress occurs when the strain rate is at its peak, resulting in a phase shift of $90°$ between stress and strain. However, most samples demonstrate viscoelastic behaviour, where the phase difference between stress and strain falls between these two extremes, i.e., $0° < \delta < 90°$ [1,42].

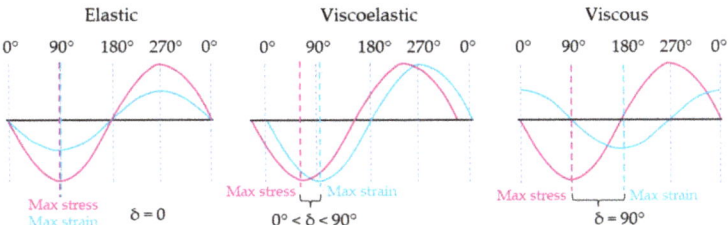

Figure 6. Stress and strain wave relationships for a purely elastic, viscoelastic, and purely viscous material.

The ratio of the applied stress to the measured strain gives the complex modulus (G*), which is a quantitative measure of material stiffness or resistance to deformation [25].

$$G^* = \text{Stress}^*/\text{Strain} \tag{10}$$

The viscoelastic behaviour of a fluid is characterised by two primary functions: the elastic or storage modulus (G′), which represents the energy accumulated by the material during each oscillation cycle, and the viscous or loss modulus (G″), which signifies the energy dissipated when the material undergoes structural changes, such as partial or complete flow of the sample. The ratio G″/G′, commonly referred to as the loss tangent or tan(δ), provides a simplified indication of whether the material exhibits solid-like or liquid-like behaviour [1,9,43].

$$G' = G * \cos\delta \tag{11}$$

$$G'' = G * \sin\delta \tag{12}$$

It is important to avoid confusing viscoplastic materials with viscoelastic ones because of the similarity of concepts. The key rheological distinction between the two lies in the presence of a yield point (τ_0). A viscoplastic fluid exhibits a yield stress below which it does not undergo deformation, while a viscoelastic fluid deforms under any applied stress.

3. Practical Assessment of the Rheological Properties of Hydrogels for 3D Printing

The investigation of a fluid's rheology plays a crucial role in assessing its printability, specifically its suitability as a hydrogel for 3D printing via semi-solid extrusion. The resolution, shape accuracy, and reproducibility of the printing process are significantly influenced by the fluid's behaviour. Consequently, several studies have identified parameters that have the potential to control and evaluate the printability of various materials, with a particular emphasis on rheological properties that are well suited for this type of process [9,44–48]. As an illustration, Yang et al. assessed the rheology of lemon juice gels with different starches by experimental studies and simulated the effect of different material properties and process parameters, such as material viscosity, inlet volume flow rate, and nozzle diameter, on the velocity and shear rate during printing. The results showed that inlet volume flow rate determined the velocity and shear velocity fields, and changes in process parameters affected the pressure field in the flow channel, with the nozzle diameter having the greatest influence [49]. Also in the field of food 3D printing, Liu et al. discovered that strong viscoelastic properties in the printed material could lead to a phenomenon known as "die swell". This effect causes viscoelastic fluids to expand beyond the nozzle diameter when they exit the tip due to the reduction of the constraining force exerted by the tip wall. As a result, the printed samples may slightly deviate from the target geometries [50]. In a completely different domain, Blaeser et al. employed a direct fluid-dynamics model to precisely control the shear stress at the nozzle site. They conducted an extensive study to examine the impact of varying shear stress levels on cell viability and proliferation potential, aiming to achieve the printing of high-resolution and cell-viable structures. The results demonstrated that cell viability and membrane integrity were minimally affected by low rates of shear stress (<5 kPa) during the printing process, with 96% cell viability. However, higher shear stress levels (5–10 kPa and >10 kPa) resulted in a significant decrease in average cell viability, with 91% and 76% cell viability, respectively. These findings support the hypothesis that cell damage caused by the printing process is directly linked to the applied nozzle shear stress [51]. In a similar vein, He et al. performed a series of experiments and determined that pressure, feed rate and printing distance are important factors that can also influence the printing quality and cell viability [52].

In that way, in the context of semi-solid extrusion, several authors reported that the printability of the ink is determined by rheological properties at three critical stages, each playing a crucial role in the process (Figure 7) [9,34,44,46].

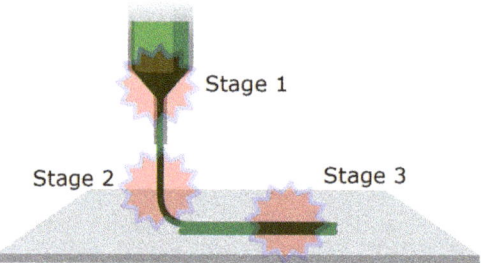

Figure 7. Critical stages of the extrusion process affected by the rheology of the ink.

The first critical stage is located at the nozzle, just before extrusion. At this point, the rheological properties of the hydrogel are crucial. When an external force is applied to the initially at-rest fluid, its viscosity should decrease rapidly, allowing for easy extrusion through a small-diameter hole. However, once the force is removed, the viscosity must be sufficiently recovered to prevent material deposition during head movement. Therefore, it is of utmost necessity for inks to exhibit pseudoplastic behaviour, characterised by a decrease in viscosity under shear stress, and to possess a yield stress that must be surpassed for the ink to initiate flow. Highly pseudoplastic fluids require less effort to flow, enabling better control over deposition through extrusion speed (mechanical or pneumatic) and other parameters, such as nozzle diameter (smaller diameters result in higher shear rates on the fluid). The presence of a yield stress in the inks reduces the likelihood of accidental drips or unwanted deposits on the printed object. Furthermore, some researchers associate the yield point with the self-supporting capability and the formation of continuous filaments with minimal deformation [46,48,50,53,54].

At the second critical point, the hydrogel needs to exhibit a certain level of mechanical resistance once it exits the nozzle. This resistance is necessary to minimise deformation, recover a higher degree of viscosity, and ensure uniform flow. The viscoelastic nature of the fluid plays a significant role at this stage, determining whether the ink demonstrates predominantly elastic behaviour ($G' \gg G''$) or viscous behaviour ($G'' \gg G'$) once it leaves the nozzle. This feature greatly influences the shape of the filament formed and consequently impacts the fidelity of the final print's shape. Additionally, the thixotropic nature of the hydrogel is of great importance in the recovery of viscosity at this stage. Rapid thixotropic behaviour is crucial in maintaining the shape and integrity of the printed filament. It helps to control the flow of the ink, ensuring that it retains its desired form once deposited and adheres to the intended pattern during the printing process.

In a similar way, the recovery time of viscosity and the internal structure of the gel after stress also impact the third critical stage: the layer-by-layer construction. During this phase of the printing process, it is crucial for the hydrogel to rapidly develop a self-supporting capacity to withstand the weight of the subsequent layers of material being deposited on top. This self-supporting capacity ensures that the printed structure remains intact and free from deformation caused by gravity or the merging of consecutive layers [45,46]. Inks with thermosensitive behaviour are beneficial for 3D printing due to their fast-gelling speed, facilitated by the temperature difference between the extruder and the printing bed. For inks that undergo a sol-gel transition, the parameter tan(δ) is also important. Strong gels, characterised by covalent or ionic bonds, exhibit tan(δ) values below 0.1. These gels provide structural integrity to the printed object but may result in inconsistent extrusion because of gel fracturing during the nozzle passage. Weak gels have a tan(δ) greater than 0.1 and offer a more uniform extrusion but are prone to collapse after deposition unless further cross-linking is made at that point. Therefore, achieving adequate printability requires a balance between various factors [55].

To thoroughly evaluate the characteristics of a hydrogel and determine its suitability for extrusion-based 3D printing, it is essential to conduct a series of tests using a rheometer. By examining the hydrogel's overall appearance and properties, such as its firmness, flowability, and presence of internal structures, researchers can determine the appropriate rheological testing equipment to employ. In hydrogel rheology, researchers can distinguish between a soft solid, a structured fluid, and a low-viscosity hydrogel based on the mechanical behaviour, internal structure, and flow properties. These differences can be identified through a preliminary visual assessment, which helps guide the selection of suitable testing geometries and gaps. The following sections provide an overview of the key considerations involved in this process.

3.1. Choosing the Right Measuring Geometry and Gap

The choice of the geometry of the measuring system as well as its dimensions should be carefully considered and be based on factors such as the viscosity of the hydrogel, the desired shear rate range, and the sample volume available. The standard geometries used include parallel plate, cone and plate, or cup and bob (Figure 8), but many others that come in different sizes and surface finishes can be used.

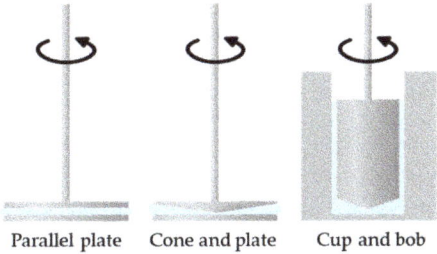

Figure 8. Common geometries for hydrogel rheology assessment.

Parallel plates are simple sets of flat upper and lower plates, with sizes ranging from 4 to 60 mm in diameter, allowing for accommodation for different viscosities. The size of the gap in the rheological measurements also plays a role in determining the shear stress and the shear rate experienced by the hydrogel. For parallel plate geometries, the gap height can be adjusted to the needs of the user. Smaller gaps result in higher shear rates (at the same angular velocity), while larger gaps lead to lower shear rates. Additionally, as a general guideline, if particles are present in the hydrogel, it is recommended to select a measuring gap that is at least 10 times larger than the largest particles [25,56]. This prevents particle jamming during the measurement, which can introduce artefacts into the results.

The cone and plate combinations consist of a flat lower plate with an upper cone-shaped geometry, typically with a geometry angle of 0.5° to 4°. The smaller the cone angle, the higher the achievable shear rate. However, cone and plate geometry should be used cautiously when dealing with hydrogels that contain particles in the micrometre size range. The cone and plates have a fixed (nominal) measurement gap; for a 1° cone, the gap is 30 microns; 70 microns for 2° cones; and 150 microns for 4° [56]. This limited distance can cause particles to jam at the apex of the geometry, potentially affecting the accuracy and reliability of the measurement results [10,57]. Therefore, careful consideration should be given to the choice of measurement geometry when working with such structured hydrogels to ensure the integrity and validity of the rheological data obtained.

Finally, cup and bob geometries consist of a lower cup to hold the sample and an upper bob to measure its rheological properties. In certain cases, other accessories such as vane bobs or helical bobs can be used instead of a cylindrical bob to accommodate different fluid behaviours. Due to the large surface area, cup and bob geometries are particularly useful for low-viscosity materials and also for samples that are prone to sedimentation. Since settling occurs parallel to the geometry, the concentration of particles near the surface of the geometry remains relatively constant. This is advantageous for accurate measurements. Furthermore, the relatively large gap between the upper bob and the lower wall of the cup is beneficial when working with samples containing larger particles, as it helps prevent jamming [56,58].

In summary, there is no simple rule for selecting a geometry, as a number of factors can come into play. However, when a new sample and geometry selection are considered, the distinction between soft solids, structured fluids, and low-viscosity hydrogels may help in the selection of proper geometries and gaps.

3.1.1. Low-Viscosity Hydrogels

Low-viscosity hydrogels are characterised by relatively low resistance to flow and deformation. These hydrogels have a lower overall viscosity, allowing them to flow more easily compared to soft solids or structured fluids, and also have a lower elastic modulus and a higher viscous modulus, indicating a prevalence of fluid-like behaviour. They are often used in applications where rapid and efficient flow is desired, such as injectable drug delivery systems or coatings.

When conducting rheological measurements, large-diameter cone plate or parallel plate geometries (e.g., 60 mm, 40 mm) are commonly used. In cases where the sample has a very low viscosity and is to be measured at low shear rates, it is ideal to use a measuring system with a large surface area. This allows for maximizing the torque response from the applied shear rate. However, for higher shear rate measurements, there is less need for a large surface area, since the stress and torque levels are naturally higher. In such cases, the focus shifts to avoiding turbulence, which requires the use of a geometry with a narrower gap. Narrow-angled cones or parallel plates (e.g., cone $1°/60$ mm, plate 60 mm) are well-suited for this purpose, as they offer both a large surface area and a narrow gap [58]. Another option is the aforementioned cup and bob system because of the extra surface area that makes them more sensitive. However, when sample volume is limited, plate systems are generally preferred over concentric cylinders.

3.1.2. Structured Fluids

Structured fluids, specifically in the context of hydrogels, refer to substances that exhibit fluid-like behaviour while still maintaining a certain level of internal structure. Structured fluids possess the ability to flow and deform under applied stress, similar to a fluid, but they also retain a degree of structural organization. They have a lower elastic modulus compared to soft solids and a higher viscous modulus, indicating a greater influence of fluid-like behaviour. Structured fluids are commonly used in applications where controlled flow, shear thinning, or shape recovery are desired, such as drug delivery systems or printable inks for 3D bioprinting.

These fluids can also contain particles, droplets, or air bubbles, which further contribute to their complex behaviour. When it comes to rheological testing, common accessories used for structured fluids include parallel plates of various sizes (40 mm, 25 mm, or 20 mm), or a cup combined with a vane rotor or a helical rotor to avoid slippage.

3.1.3. Soft Solids

A soft solid refers to a hydrogel that exhibits solid-like characteristics under certain conditions. It possesses a defined structure that enables it to resist deformation and maintain its shape, similar to a solid material. Soft solids have a significant elastic component in their rheological response, meaning they can store and recover energy upon deformation. They typically exhibit a high elastic modulus (G′) compared to their viscous modulus (G″), indicating a predominance of solid-like behaviour. Soft solids are often used in applications where mechanical strength and structural integrity are crucial, such as load-bearing tissues or scaffold materials for tissue engineering. The use of a parallel plate geometry in its smaller size (<25 mm) is recommended for testing soft solids [25,56]. Specifically, it is beneficial to employ a roughened surface on the parallel plate, such as crosshatched or sandblasted plates, to prevent slippage and ensure accurate measurements. Again, cup and vane/helical rotor systems can be used for the same reason.

3.2. Selecting and Setting the Correct Test Methods

Hydrogels are versatile materials that can also undergo transitions between these categories because of their physical properties and chemical compositions. For instance, when the gelling of a hydrogel is temperature dependent, it can exist in a low-viscosity liquid phase and be transformed into a structured fluid (or even a soft solid) with a temperature change. Similarly, crosslinking processes (such as chemical, light, pH-dependent, or enzymatic) can convert a low viscosity hydrogel or a structured fluid into a final soft solid with mechanical properties that are better suited for its intended application. These transitions highlight the dynamic nature of hydrogels and their ability to adapt their physical properties based on external stimuli or internal changes, making them highly versatile materials in various fields, including biomedicine, drug delivery, or tissue engineering. However, those transitions also reflect a need to define the appropriate testing protocols as an essential tool to accurately characterize the hydrogel for its intended use and printing conditions.

Different tests, including flow curves, oscillatory tests, and creep/recovery tests, can provide valuable insights into the viscoelastic properties, flow behaviour, or time-dependent response of the hydrogel, among others. The selection of tests should be based on the specific rheological characteristics that play a crucial role in achieving a successful 3D printing and final product, such as the ink's shear thinning behaviour, elastic modulus, and recovery time. The range and resolution of these parameters should be adjusted to match the expected rheological properties of the hydrogel during printing and during its final use.

3.2.1. Fluid Behaviour and Yield Stress Determination

The fundamental test in rheology involves measuring viscosity and shear stress by varying the shear rate within a specific range while maintaining a constant temperature, typically between 10^{-2} and 10^3 s^{-1} for hydrogels. The measured shear rate can be compared with the typical ranges of shear rates for common industrial processes. For example, low shear rates (10^{-2}–10^{-1} s^{-1}) provide information on fluid behaviour in processes such as storage, levelling, or sedimentation, which are useful for assessing the stability of a hydrogel at rest. The range between 10^{-1} and 10^3 s^{-1} represents physically active processes such as mixing, stirring, pumping, or extruding, which helps determine the suitability of the hydrogel for extrusion-based 3D printing. Shear rates greater than 10^4 s^{-1} are associated with spraying, brushing, or high-speed coatings, which are not commonly applicable to hydrogels suitable for this 3D printing technology [59].

This test generates data that can be used to plot a flow curve, providing insight into the material's behaviour (whether it exhibits pseudoplastic, dilatant, or viscoplastic properties). By analysing the data obtained, the most appropriate mathematical model can be selected to extract and interpret the model parameters. For instance, Li et al. employed the Power-law model and experimental data to assess the shear-thinning characteristics of various compositions and determine the shear rate experienced by the hydrogel during the printing process [60]. Other studies have utilized curve fitting with mathematical models to evaluate specific parameters such as the yield stress [61,62]. While fitting a flow curve to a mathematical model is one of the simplest methods to evaluate τ_0, as explained later, it is not the only approach.

3.2.2. Amplitude Sweeps and Determination of LVR, γ_c and τ_0

To measure viscoelasticity, two commonly performed small amplitude oscillatory tests (SAOS) are used to evaluate the two main parameters: the storage modulus (G′) and the loss modulus (G″). These tests are amplitude sweeps and frequency sweeps.

Amplitude sweeps are obtained by subjecting the sample to a range of deformations (from 10^{-1}–10^3%) at a fixed frequency (typically between 0.1 and 1 Hz for hydrogels). It is very rare to require values above or below this range for measuring hydrogels suitable for 3D extrusion printing. However, the values can be adjusted on the basis of the initial characteristics of the sample (viscosity, presence of particles, level of crosslinking, etc.) if necessary. The measurement results are typically presented in a diagram where strain (or shear stress) is plotted on the x-axis, and storage modulus (G′) and loss modulus (G″) are plotted on the y-axis, with both axes displayed on a logarithmic scale. This representation helps to estimate the linear viscoelastic region (LVR).

The LVR indicates the range in which an oscillatory rheology test can be performed without destroying the structure of the sample. The limit of the LVR, also known critical strain (γ_c), is the amplitude value at which the storage modulus values cease to be approximately constant. γ_c is typically determined as the first point at which G′ reaches a value equal to or lower than 95–90% of its initial value [63,64]. The extent of LVR is inversely proportional to the solid character of the sample and is sensitive to frequency and temperature [43,57,65,66]. Furthermore, it is common to evaluate the values of G′ and G″ in the linear viscoelastic region because this analysis provides insight into the viscoelastic nature of the sample. When G′ > G″, the sample exhibits a gel-like or solid structure, indicating a viscoelastic solid material (Figure 9a). Conversely, when G″ > G′, the sample demonstrates a fluid-like structure, indicating a viscoelastic liquid (Figure 9b). It is important to note that these classifications are specific to the applied measuring conditions, particularly to the preset (angular) frequency.

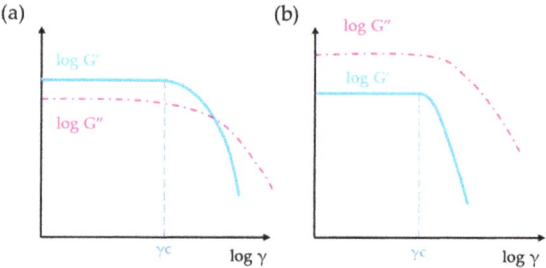

Figure 9. Diagram of two amplitude sweeps showing the LVR: (**a**) gel-like sample (G′ > G″); (**b**) fluid-like sample (G″ > G′).

Stress sweeps can also be useful in determining the yield stress when it is not possible to do so through flow curves, such as in the case of materials with high viscosity at room temperature. In these types of fluid, an increase in shear stress results in sample slippage between plates, and the measured yield stress is lower than the actual yield stress of the sample [65,67]. In an amplitude sweep, τ_0 can be estimated using two different methods. The first method involves determining the intersection of the tangents of the storage modulus at high and low oscillatory stresses (Figure 10a). The first line is fitted to the values within the LVR, and the second line is derived from the storage modulus measured for stresses exceeding the LVR [68,69]. However, there is some ambiguity in choosing the number of points required to obtain the second line, which can reduce the reliability of the method.

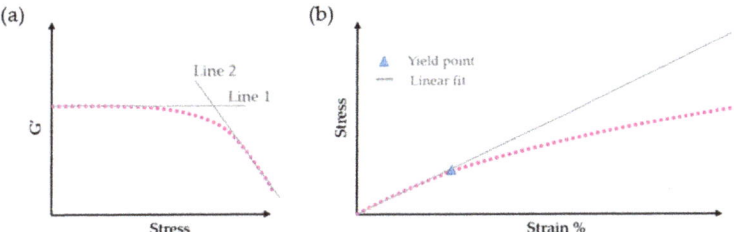

Figure 10. Methods for yield stress detection using amplitude sweeps: (**a**) intersection of the tangents of the storage modulus (G′); (**b**) point of linearity deviation in the stress—strain% curve. The pink dotted line represents the experimental results.

In the second method, a plot of stress (Pa) versus oscillatory strain (%) is used (Figure 10b). The stress-strain curve is linear at low strain because the fluid response is predominantly elastic. The point at which the curve deviates from linearity is determined as the yield stress [68,69]. It is important to note that the value of τ_0 obtained differs depending on the method used, sometimes called the apparent yield stress [67].

3.2.3. Frequency Sweeps and Determination of tan(δ)

As mentioned above, the second SAOS test most commonly performed is the frequency sweep. In this case, the strain amplitude is kept constant and within the LVR, while the oscillation frequency is increased. This test helps to better understand the internal structure of the material and its time-dependent behaviour in the non-destructive deformation range. For example, high frequencies represent short-term behaviours, such as those occurring in a mixing or extrusion process, while low frequencies represent long-term behaviours, such as settling or resting [46,68]. Figure 11 illustrates the frequency sweep measurement of different materials, including viscoelastic solids, viscoelastic liquids, and gel-like materials. In the case of viscoelastic solids, the storage modulus G′ remains constant and dominates at low frequencies. However, as the frequency increases, the loss modulus G″ becomes more prominent. These materials exhibit a firm consistency when at rest but can be easily deformed under sufficient force. Viscoelastic liquid materials, on the other hand, demonstrate a different behaviour. At low frequencies, the loss modulus G″ surpasses the storage modulus G′, indicating liquid-like behaviour. As the frequency increases, the storage modulus G′ becomes higher than G″, suggesting a solid-like behaviour. For gel-like materials, both G′ and G″ remain parallel across the entire frequency range. This indicates that the material maintains consistent behaviour regardless of the frequency applied. Gels often exhibit a solid-like structure while retaining some fluid-like properties.

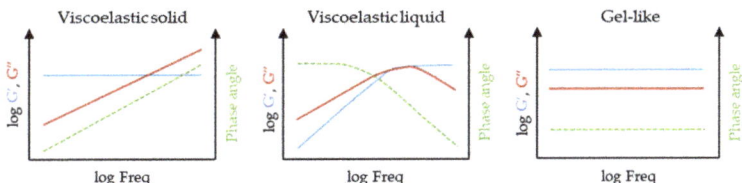

Figure 11. Frequency sweeps at constant temperature that can be found in hydrogels. Blue lines represent the storage modulus (G′), red lines represent the loss modulus (G″) and green lines represent the phase angle.

Frequency sweeps can be used to evaluate the value of tan(δ) for each fluid and its variability with respect to angular frequency. Practically, the value of tan(δ) serves as a significant criterion for analysing hydrogel formation, hardening, and curing processes. In the fluid or liquid state ("sol state"), tan(δ) is greater than 1 (as G″ > G′), indicating a more liquid-like behaviour. On the contrary, in the gel-like or solid state, tan(δ) is less than 1 (as G′ > G″), reflecting solid-like characteristics. At the point of maintaining the sol-gel transition, tan(δ) has a value equal to 1 [1,70]. In this context, tan(δ) has been employed as a parameter to assess the balance between viscous flow and elasticity, which is crucial for extrusion-based 3D printing. For instance, Petta et al. established a correlation between tan(δ) values and the extrudability of tyramine hyaluronan derivative hydrogels, indicating that formulations with tan(δ) values ranging from 0.5 to 0.6 exhibited the optimal printability for their printing conditions [71]. Similarly, Cheng et al. highlighted that for the specific 3D printing platform used by their team, the printing material should ideally have a tan(δ) ranging from 0.2 to 0.7 [72].

3.2.4. Thixotropy and Viscosity Recovery

As stated above, thixotropy is a time-dependent property that characterizes the reversibility of changes occurring in the internal structure of a gel. In thixotropic fluids, the apparent viscosity decreases when a constant shear rate is applied over time and gradually recovers when the force on the fluid decreases or ceases. This phenomenon is measured through recovery tests at different shear rates and a constant temperature, known as the Stepped Flow Method (SFM) or the Stepped Dynamic Method (SDM), depending on whether the test is performed using linear or oscillatory rheology [57,73].

These tests involve three distinct stages that examine the behaviour of the fluid over time by applying different shear rates or deformations (Figure 12a). In the SFM test, the initial stage entails the application of a low shear rate (around 0.4 s^{-1}) for a relatively extended period (60 to 180 s). Subsequently, a high shear rate (around 100 s^{-1}) is implemented, resulting in disruption of the internal gel structure and a significant reduction in viscosity for a brief duration (20–40 s). Finally, the sample undergoes a recovery stage with a low shear rate lasting an additional 120–180 s [46,48]. Similarly, in the SDM test, the stages remain the same, but the amplitude of the oscillation varies. The amplitude is set at a low level during the first and third stages (approximately 0.1% strain or an appropriate value within the linear viscoelastic region, LVR). In contrast, during the second stage, the amplitude is adjusted to a sufficiently high level, exceeding the LVR, to induce the destruction of the colloidal gel network [57,61]. By manipulating the applied stresses on the material in this manner, it becomes possible to evaluate the extent and nature of the internal structure destruction, as well as determine the percentage of regeneration at various time intervals. From a practical standpoint, the percentages of viscosity recovery can be calculated using numerical methods. Viscosity recovery refers to the percentage of viscosity achieved during the initial 30 s and/or the final 60 s of the third step (after undergoing high deformation), based on the average viscosity obtained during the last 30 s of the initial step.

Additionally, thixotropic fluids can be categorized based on their viscosity recovery rate after the shear stress is removed. A rapid thixotropic fluid quickly regains its viscosity, making it ideal for applications that require precise control over material flow and shape, such as 3D printing. In contrast, a slow thixotropic fluid takes more time to restore its original viscosity, resulting in prolonged spreading or deformation. In this way, Paxton et al. conducted a comparative analysis of viscosity recovery in hydrogels, specifically Poloxamer 407 and alginate, emphasizing the significance of rapid viscosity recovery for 3D printing. The printable concentrations of 25 wt% and 30 wt% Poloxamer 407 exhibited fast recovery, ensuring shape fidelity. Conversely, the unprintable 15 wt% Poloxamer 407 took longer to stabilize, resulting in reduced shape retention. All Poloxamer 407 samples fully regained their initial viscosity within 200 s, indicating no permanent changes in polymer structure or properties due to the high shear conditions of bioprinting. Similarly, the 8% w/v alginate sample demonstrated satisfactory recovery over the entire 200 s period, albeit with a slight delay in viscosity recovery observed by the slope in the curve after transitioning from high to low shear rate. In contrast, the pre-crosslinked alginate sample exhibited more rapid recovery, further supporting its suitability as a printable ink compared to the uncrosslinked sample [48]. Nevertheless, other studies demonstrate cases where fluids with slower recovery times can still be successfully printed by incorporating temperature-induced gelation or utilizing in situ rapid crosslinking reactions during the printing process [45,74]. Thus, thixotropic analysis allows researchers to determine if their material may require additional crosslinking processes to increase rapidly viscosity after extrusion. In that sense, thixotropy loops can also provide valuable information (Figure 12b). If the sample exhibits rapid thixotropic recovery, the downsweep curve will closely overlay the upsweep curve, indicating a minimal time-dependent delay in recovery. However, if the thixotropic recovery is slow or the sample has potentially been damaged, the downsweep curve will deviate from the expected path. The extent of deviation indicates the duration of recovery or the severity of potential damage, with larger deviations indicating longer recovery times or more significant damage [73]. This delayed viscosity recovery presents challenges to maintaining precise control over flow and shape.

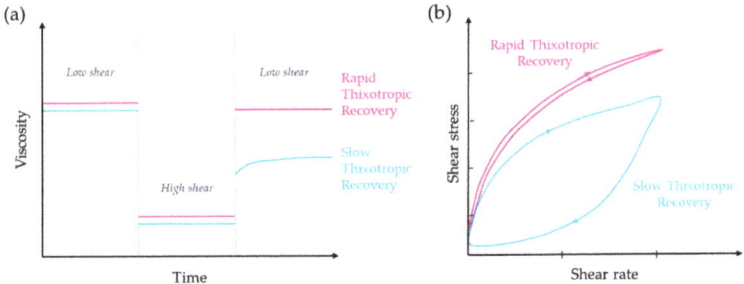

Figure 12. Thixotropy evaluation diagrams: (**a**) viscosity recovery evaluation by the SFM/SDM method, (**b**) thixotropy loop.

3.2.5. Gelation Kinetics and Gel Point Determination

Hydrogels can be formed through either physical polymer chain association or different cross-linking reactions. The gelation mechanism employed significantly impacts the mechanical and thermal properties of the hydrogel. To ensure consistent properties of the hydrogel for a particular application, it is crucial to comprehend the chemistry of gelation and quantitatively analyse the physical properties of the gel. Rheological techniques can be employed to determine the gel point, enabling the use of a rheometer to measure the kinetics of gelation.

In a curing system, the gel point can be utilized to monitor the progress of the curing reaction and determine the kinetics of gelation. The commonly employed method for determining the gelation point involves conducting isothermal dynamic time sweep experiments at a fixed temperature. Oscillatory shear is continuously applied at fixed strain deformation in the LVR to monitor gelation at a sampling interval of 20–30 s. Throughout the measurement, the changes in modulus (G', G'', and G^*) and complex viscosity are continuously monitored over time, recording the sol-gel transition time, which is the point at which G' crosses over G'' [75–77]. This transition time can vary depending on factors such as the concentration of the crosslinking agent and the reaction temperature. Once gel formation occurred, the G' and G'' curves reached plateaus, indicating the completion of the crosslinking reaction (Figure 13a). Additionally, in their work, Ravanbakhsh et al. mentioned an alternative method for determining the gelation point that involves identifying the point at which the ratio between G' and G'' becomes independent of frequency [78].

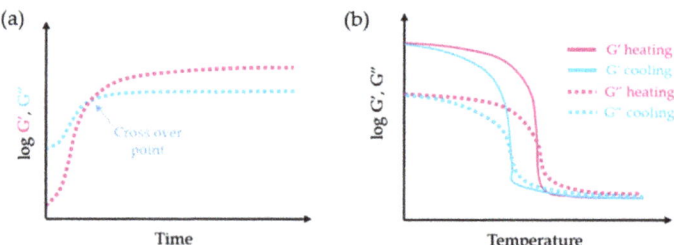

Figure 13. Gelation kinetics diagrams: (**a**) isothermal hydrogel cross-linking reaction, (**b**) temperature ramp.

For thermal reversible hydrogels, the gelation and gel-melting process can be assessed using dynamic temperature ramp tests. These tests involve gradually decreasing the temperature (ramp down) or increasing it (ramp up) at rates typically around 1 °C/min and 2 °C/min, respectively. During the cooling process, the hydrogel solution undergoes a transition from a liquid to a solid state, indicated by the crossover of the elastic modulus (G′) and viscous modulus (G″) at small range of temperatures (Figure 13b). It is worth noting that the mechanical strength of the sample, represented by the elastic modulus (G′), may increase by more than five orders of magnitude during the gelation process. On the other hand, after heating, the hydrogel starts to melt at values where G″ > G′. Thus, this temperature-raising procedure is an effective means to monitor the gelation process of a thermally reversible hydrogel [57,79].

To conduct these tests and analyse the viscoelastic properties of hydrogels, a rotational rheometer equipped with a temperature control system is required. Additionally, to prevent slippage during measurements, the use of a roughened plate, such as a sandblasted or crosshatched plate, is recommended. The measurements should be performed within the linear viscoelastic region of the material, typically at a constant frequency of 1 Hz [57,79].

3.2.6. Gel Strength

The mechanical properties in terms of stiffness of 3D structures can be accurately examined using oscillatory rheology. The modulus of elasticity (G′) and, therefore, the stiffness of a sample can be determined through two types of oscillatory tests: amplitude sweeps, where the hydrogel is subjected to a range of deformations (strain %) at a fixed frequency as mentioned earlier, or dynamic frequency sweeps, where the sample is subjected to a specific frequency range at a fixed strain percentage. Both tests are carried out at a constant temperature [57,80].

This test is particularly relevant in the case of hydrogels used in bioprinting, as the biomechanical properties of tissues vary significantly between organs and tissues and are inherently related to their function. Mechanically static tissues, such as the brain, and adaptable tissues, such as the lung, exhibit low rigidity. On the other hand, tissues exposed to high mechanical loads, such as bone or skeletal muscle, have elastic moduli that are several orders of magnitude higher [81]. In this regard, it has been demonstrated that the biochemical and biomechanical properties of inks influence cellular and tissue compatibility [74,82,83]. For example, studies conducted with chondrocytes have revealed that the rigidity of the hydrogel that encapsulates them is important for the distribution, organization, and secretion of type II collagen, which is essential for repairing articular lesions [84]. Other reports have shown that cells proliferate faster when grown on the surface of high-strength hydrogels. However, when cells are cultured in a 3D microenvironment, their survival and proliferation decrease as the matrix strength increases [85]. Therefore, it is advisable to perform this test on inks involved in bioprinting processes whose firmness can be modulated through cross-linking reactions and come into contact with cellular loads: (i) directly, by containing cells embedded within them or serving as a support or scaffold for cell seeding, or (ii) indirectly, by being designed for implantation in the body.

One of the most common protocols for the practical assessment of the stiffness of formulated and/or printed systems is described by Cox and Madsen [80]. In this protocol, samples are subjected to stress sweeps with a small range of deformation (approximately 0.2% to 2%), allowing the extraction of the storage modulus (G') at 1%. This enables the comparison of multiple measurements within the same ink, multiple inks, multiple structures, or any combination of interest. Such tests can also be employed to evaluate the crosslinking of hydrogels. For example, they can be conducted at different concentrations of cross-linking agents or exposure times, regardless of the type of cross-linking that occurs [86].

4. Conclusions

In conclusion, the field of rheology plays a vital role in the extrusion-based 3D printing of hydrogels, where the material's flow and deformation properties significantly impact its behaviour during the printing process. The rheological properties, including viscosity, shear thinning behaviour, and viscoelasticity, are crucial factors for determining the extrudability, shape retention, and overall quality of the printed structures. Understanding these properties is essential to optimising the printing process and achieving the desired accuracy and quality of the final product. A thorough knowledge of the rheological concepts, terminology, and tests applicable to 3D printing facilitates the evaluation of the printability of different hydrogels and has the potential to drive progress in the fields of tissue engineering, drug delivery, and regenerative medicine. By bridging the gap between rheology and hydrogel 3D printing, researchers and scientists from diverse disciplines can unlock new opportunities for innovation and application within this interdisciplinary field.

Author Contributions: Conceptualization, H.H.-M. and E.A.; methodology, H.H.-M.; investigation, H.H.-M.; resources, H.H.-M. and E.A.; writing—original draft preparation, H.H.-M.; writing—review and editing, H.H.-M., M.A.F. and E.A.; visualization, H.H.-M.; supervision, M.A.F. and E.A.; project administration, H.H.-M., M.A.F. and E.A.; funding acquisition, H.H.-M., M.A.F. and E.A. All authors have read and agreed to the published version of the manuscript.

Funding: This research was funded by the Ministerio de Ciencia e Innovación of Spain (Agencia Estatal de Investigación: PID2019-107580GB-I00/AEI/10.13039/501100011033), and the FICYT—Consejería de Ciencia, Innovación y Universidad—Gobierno del Principado de Asturias through the 2018–2022 Science, Technology and Innovation Plan, with the European Union through the European Regional Development Fund (ERDF) (AYUD/2021/57608).

Institutional Review Board Statement: Not applicable.

Informed Consent Statement: Not applicable.

Data Availability Statement: Data sharing is not applicable to this article.

Acknowledgments: The authors express their gratitude to E. Diez for his assistance in preparing the figures.

Conflicts of Interest: The authors declare no conflict of interest.

References

1. Mezger, T.G. *The Rheology Handbook: For Users of Rotational and Oscillatory Rheometers*, 2nd ed.; Vincentz Network GmbH & Co.: Hanover, Germany, 2006; ISBN 3-87870-174-8.
2. Murata, H. Rheology—Theory and Application to Biomaterials. In *Polymerization*; InTech: London, UK, 2012.
3. Tajik, S.; Garcia, C.N.; Gillooley, S.; Tayebi, L. 3D Printing of Hybrid-Hydrogel Materials for Tissue Engineering: A Critical Review. *Regen. Eng. Transl. Med.* **2023**, *9*, 29–41. [CrossRef] [PubMed]
4. Hu, F.; Mikolajczyk, T.; Pimenov, D.Y.; Gupta, M.K. Extrusion-Based 3D Printing of Ceramic Pastes: Mathematical Modeling and In Situ Shaping Retention Approach. *Materials* **2021**, *14*, 1137. [CrossRef] [PubMed]
5. Kaliaraj, G.; Shanmugam, D.; Dasan, A.; Mosas, K. Hydrogels—A Promising Materials for 3D Printing Technology. *Gels* **2023**, *9*, 260. [CrossRef] [PubMed]
6. Amorim, S.; Soares da Costa, D.; Pashkuleva, I.; Reis, C.A.; Reis, R.L.; Pires, R.A. 3D Hydrogel Mimics of the Tumor Microenvironment: The Interplay among Hyaluronic Acid, Stem Cells and Cancer Cells. *Biomater. Sci.* **2021**, *9*, 252–260. [CrossRef]
7. Kadajji, V.G.; Betageri, G.V. Water Soluble Polymers for Pharmaceutical Applications. *Polymers* **2011**, *3*, 1972–2009. [CrossRef]

8. Díaz, A.; Herrada-Manchón, H.; Nunes, J.; Lopez, A.; Díaz, N.; Grande, H.; Loinaz, I.; Fernández, M.A.; Dupin, D. 3D Printable Dynamic Hydrogel: As Simple as It Gets! *Macromol. Rapid Commun.* **2022**, *43*, 2200449. [CrossRef]
9. Schwab, A.; Levato, R.; D'Este, M.; Piluso, S.; Eglin, D.; Malda, J. Printability and Shape Fidelity of Bioinks in 3D Bioprinting. *Chem. Rev.* **2020**, *120*, 11028–11055. [CrossRef]
10. Chhabra, R.P.; Richardson, J.F. *Non-Newtonian Flow in the Process Industries—Fundamentals and Engineering Applications*; Butterworth-Heinemann: Oxford, UK, 1999; ISBN 0750637706.
11. Walters, K.; Hutton, J.F.; Barnes, H.A. *An Introduction to Rheology*; Walters, K., Ed.; Elsevier Science Publishers B.V.: Amsterdam, The Netherlands, 1993; ISBN 9780080933696.
12. Ramírez-Navas, J.S. Introducción a La Reología de Alimentos. *Rev. ReCiTeIA* **2006**, *1*, 1–46.
13. Masoliver i Marcos, G.; Pérez-Sánchez, M.; López-Jiménez, P.A. Modelo Experimental Para Estimar La Viscosidad de Fluidos No Newtonianos: Ajuste a Expresiones Matemáticas Convencionales. *Model. Sci. Educ. Learn.* **2017**, *10*, 5. [CrossRef]
14. Zengeni, B.T. Bingham Yield Stress and Bingham Plastic Viscosity of Homogeneous Non-Newtonian Slurries. Ph.D. Thesis, Cape Peninsula University of Technology, Cape Town, South Africa, 2016.
15. Chhabra, R.P.; Richardson, J.F. Non-Newtonian Fluid Behaviour. In *Non-Newtonian Flow and Applied Rheology: Engineering Applications*; Elsevier Ltd.: Amsterdam, The Netherlands, 2008; ISBN 9780750685320.
16. Lakkanna, M.; Mohan Kumar, G.C.; Kadoli, R. Computational Design of Mould Sprue for Injection Moulding Thermoplastics. *J. Comput. Des. Eng.* **2016**, *3*, 37–52. [CrossRef]
17. Nindo, C.I.; Tang, J.; Powers, J.R.; Takhar, P.S. Rheological Properties of Blueberry Puree for Processing Applications. *LWT-Food Sci. Technol.* **2007**, *40*, 292–299. [CrossRef]
18. Quintáns Riveiro, L.C. Reología de Productos Alimentarios. Ph.D. Thesis, Universidad de Santiago de Compostela, Santiago de Compostela, Spain, 2008.
19. Li, M.-C.; Wu, Q.; Song, K.; De Hoop, C.F.; Lee, S.; Qing, Y.; Wu, Y. Cellulose Nanocrystals and Polyanionic Cellulose as Additives in Bentonite Water-Based Drilling Fluids: Rheological Modeling and Filtration Mechanisms. *Ind. Eng. Chem. Res.* **2016**, *55*, 133–143. [CrossRef]
20. TA Instruments. Rheology Aplications Note: Rheology Software Models (Flow). Available online: http://www.tainstruments.com/pdf/literature/RN9.pdf (accessed on 1 March 2023).
21. Sisko, A.W. The Flow of Lubricating Greases. *Ind. Eng. Chem.* **1958**, *50*, 1789–1792. [CrossRef]
22. Balmforth, N.J.; Frigaard, I.A.; Ovarlez, G. Yielding to Stress: Recent Developments in Viscoplastic Fluid Mechanics. *Annu. Rev. Fluid Mech.* **2014**, *46*, 121–146. [CrossRef]
23. El-Borhamy, M. Numerical Study of the Stationary Generalized Viscoplastic Fluid Flows. *Alexandria Eng. J.* **2018**, *57*, 2007–2018. [CrossRef]
24. Bird, R.B.; Dai, G.C.; Yarusso, B.J. The Rheology and Flow of Viscoplastic Materials. *Rev. Chem. Eng.* **1983**, *1*, 1–70. [CrossRef]
25. Chen, T. Rheology Basic Theory & Applications Training Section #1. Available online: https://www.tainstruments.com/wp-content/uploads/2020-Rheology-Online-Training-1.pdf (accessed on 1 March 2023).
26. Oke, A.S.; Mutuku, W.N.; Kimathi, M.; Animasaun, I.L. Insight into the Dynamics of Non-Newtonian Casson Fluid over a Rotating Non-Uniform Surface Subject to Coriolis Force. *Nonlinear Eng.* **2020**, *9*, 398–411. [CrossRef]
27. Adewale, F.J.; Lucky, A.P.; Oluwabunmi, A.P.; Boluwaji, E.F. Selecting the Most Appropriate Model for Rheological Characterization of Synthetic Based Drilling Mud. *Int. J. Appl. Eng. Res.* **2017**, *12*, 7614–7629.
28. Talens Oliag, P. Caracterización Del Comportamiento Reológico de Un Alimento Fluido Plástico. *RiuNet Repos. UPV.* **2016**, 1–6.
29. Fernández, C.; Alvarez, M.D.; Canet, W. Rheological Behaviour of Fresh and Frozen Potato Puree in Steady and Dynamic Shear at Different Temperatures. *Eur. Food Res. Technol.* **2004**, *218*, 544–553. [CrossRef]
30. Fernández, C.; Canet, W.; Alvarez, M.D. Quality of Mashed Potatoes: Effect of Adding Blends of Kappa-Carrageenan and Xanthan Gum. *Eur. Food Res. Technol.* **2009**, *229*, 205–222. [CrossRef]
31. Sokmen, A.; Gunes, G. Influence of Some Bulk Sweeteners on Rheological Properties of Chocolate. *LWT-Food Sci. Technol.* **2006**, *39*, 1053–1058. [CrossRef]
32. HadjSadok, A.; Moulai-Mostefa, N.; Rebiha, M. Rheological Properties and Phase Separation of Xanthan-Sodium Caseinate Mixtures Analyzed by a Response Surface Method. *Int. J. Food Prop.* **2010**, *13*, 369–380. [CrossRef]
33. TA Instruments. Understanding Rheology of Structured Fluids. Available online: http://www.tainstruments.com/pdf/literature/AAN016_V1_U_StructFluids.pdf (accessed on 1 January 2023).
34. Cooke, M.E.; Rosenzweig, D.H. The Rheology of Direct and Suspended Extrusion Bioprinting. *APL Bioeng.* **2021**, *5*, 011502. [CrossRef] [PubMed]
35. Sochi, T. Non-Newtonian Flow in Porous Media. *Polymer* **2010**, *51*, 5007–5023. [CrossRef]
36. Mewis, J.; Wagner, N.J. *Colloidal Suspension Rheology*; Cambridge University Press: Cambridge, UK, 2011; ISBN 9780521515993.
37. Armelin, E.; Martí, M.; Rudé, E.; Labanda, J.; Llorens, J.; Alemán, C. A Simple Model to Describe the Thixotropic Behavior of Paints. *Prog. Org. Coat.* **2006**, *57*, 229–235. [CrossRef]
38. Sha, J.; Zhang, F.; Zhang, H. Thixotropic Flow Behaviour in Chemical Pulp Fibre Suspensions. *BioResources* **2016**, *11*, 3481–3493. [CrossRef]
39. Malvern Instruments. A Basic Introduction to Rheology. Available online: https://cdn.technologynetworks.com/TN/Resources/PDF/WP160620BasicIntroRheology.pdf (accessed on 1 February 2023).

40. Hyun, K.; Wilhelm, M.; Klein, C.O.; Cho, K.S.; Nam, J.G.; Ahn, K.H.; Lee, S.J.; Ewoldt, R.H.; McKinley, G.H. A Review of Nonlinear Oscillatory Shear Tests: Analysis and Application of Large Amplitude Oscillatory Shear (LAOS). *Prog. Polym. Sci.* **2011**, *36*, 1697–1753. [CrossRef]
41. Tschoegl, N.W. *The Phenomenological Theory of Linear Viscoelastic Behavior*; Springer: Berlin/Heidelberg, Germany, 1989; ISBN 978-3-642-73604-9.
42. Ramli, H.; Zainal, N.F.A.; Hess, M.; Chan, C.H. Basic Principle and Good Practices of Rheology for Polymers for Teachers and Beginners. *Chem. Teach. Int.* **2022**, *4*, 307–326. [CrossRef]
43. Trujillo, L.A.; Santos, J.; Calero, N.; Alfaro, M.C.; Muñoz, J. Caracterización Reológica de Una Suspoemulsión Comercial Para Uso Agroquimico. *Afinidad* **2013**, *561*, 54–59.
44. Kyle, S.; Jessop, Z.M.; Al-Sabah, A.; Whitaker, I.S. 'Printability' of Candidate Biomaterials for Extrusion Based 3D Printing: State-of-the-Art. *Adv. Healthc. Mater.* **2017**, *6*, 1700264. [CrossRef] [PubMed]
45. Diañez, I.; Gallegos, C.; Brito-de la Fuente, E.; Martínez, I.; Valencia, C.; Sánchez, M.C.; Diaz, M.J.; Franco, J.M. 3D Printing in Situ Gelification of κ-Carrageenan Solutions: Effect of Printing Variables on the Rheological Response. *Food Hydrocoll.* **2019**, *87*, 321–330. [CrossRef]
46. Liu, Z.; Bhandari, B.; Prakash, S.; Mantihal, S.; Zhang, M. Linking Rheology and Printability of a Multicomponent Gel System of Carrageenan-Xanthan-Starch in Extrusion Based Additive Manufacturing. *Food Hydrocoll.* **2019**, *87*, 413–424. [CrossRef]
47. Moore, C.A.; Shah, N.N.; Smith, C.P.; Rameshwar, P. 3D Bioprinting and Stem Cells. In *Methods in Molecular Biology*; Humana Press: New York, NY, USA, 2018; Volume 1842, pp. 93–103, ISBN 9781489980656.
48. Paxton, N.; Smolan, W.; Böck, T.; Melchels, F.; Groll, J.; Jungst, T. Proposal to Assess Printability of Bioinks for Extrusion-Based Bioprinting and Evaluation of Rheological Properties Governing Bioprintability. *Biofabrication* **2017**, *9*, 044107. [CrossRef] [PubMed]
49. Yang, F.; Guo, C.; Zhang, M.; Bhandari, B.; Liu, Y. Improving 3D Printing Process of Lemon Juice Gel Based on Fluid Flow Numerical Simulation. *LWT* **2019**, *102*, 89–99. [CrossRef]
50. Liu, Z.; Bhandari, B.; Prakash, S.; Zhang, M. Creation of Internal Structure of Mashed Potato Construct by 3D Printing and Its Textural Properties. *Food Res. Int.* **2018**, *111*, 534–543. [CrossRef]
51. Blaeser, A.; Duarte Campos, D.F.; Puster, U.; Richtering, W.; Stevens, M.M.; Fischer, H. Controlling Shear Stress in 3D Bioprinting Is a Key Factor to Balance Printing Resolution and Stem Cell Integrity. *Adv. Healthc. Mater.* **2016**, *5*, 326–333. [CrossRef]
52. He, Y.; Yang, F.; Zhao, H.; Gao, Q.; Xia, B.; Fu, J. Research on the Printability of Hydrogels in 3D Bioprinting. *Sci. Rep.* **2016**, *6*, 29977. [CrossRef]
53. Smith, P.T.; Basu, A.; Saha, A.; Nelson, A. Chemical Modification and Printability of Shear-Thinning Hydrogel Inks for Direct-Write 3D Printing. *Polymer* **2018**, *152*, 42–50. [CrossRef]
54. Sun, A.; Gunasekaran, S. Yield Stress in Foods: Measurements and Applications. *Int. J. Food Prop.* **2009**, *12*, 70–101. [CrossRef]
55. Saha, D.; Bhattacharya, S. Hydrocolloids as Thickening and Gelling Agents in Food: A Critical Review. *J. Food Sci. Technol.* **2010**, *47*, 587–597. [CrossRef]
56. NETZSCH-Gerätebau GmbH Rheology—How to Select the Appropriate Measuring Geometry. Available online: https://analyzing-testing.netzsch.com/en/training-know-how/tips-tricks/rheology/rheology-how-to-select-the-appropriate-measuring-geometry (accessed on 15 May 2023).
57. Chen, T. Rheology: Basic Theory and Applications Training Section #2. Available online: https://www.tainstruments.com/wp-content/uploads/2020-Rheology-Online-Training-2.pdf (accessed on 1 March 2023).
58. Rolfe, P. Viscosity Flow Curve. Part 2. How to Best Measure a Viscosity Flow Curve? Available online: https://www.materials-talks.com/blog/2017/06/08/viscosity-flow-curve-part-2/ (accessed on 27 March 2019).
59. Carnicer, V.; Alcázar, C.; Orts, M.J.; Sánchez, E.; Moreno, R. Microfluidic Rheology: A New Approach to Measure Viscosity of Ceramic Suspensions at Extremely High Shear Rates. *Open Ceram.* **2021**, *5*, 100052. [CrossRef]
60. Li, H.; Liu, S.; Lin, L. Rheological Study on 3D Printability of Alginate Hydrogel and Effect of Graphene Oxide. *Int. J. Bioprint.* **2016**, *2*, 10–12. [CrossRef]
61. Herrada-Manchón, H.; Rodríguez-González, D.; Fernández, M.A.; Kucko, N.W.; Barrère-de Groot, F.; Aguilar, E. Effect on Rheological Properties and 3D Printability of Biphasic Calcium Phosphate Microporous Particles in Hydrocolloid-Based Hydrogels. *Gels* **2022**, *8*, 28. [CrossRef] [PubMed]
62. Haring, A.P.; Thompson, E.G.; Tong, Y.; Laheri, S.; Cesewski, E.; Sontheimer, H.; Johnson, B.N. Process- and Bio-Inspired Hydrogels for 3D Bioprinting of Soft Free-Standing Neural and Glial Tissues. *Biofabrication* **2019**, *11*, 025009. [CrossRef]
63. Kreimendahl, F.; Köpf, M.; Thiebes, A.L.; Duarte Campos, D.F.; Blaeser, A.; Schmitz-Rode, T.; Apel, C.; Jockenhoevel, S.; Fischer, H. Three-Dimensional Printing and Angiogenesis: Tailored Agarose-Type I Collagen Blends Comprise Three-Dimensional Printability and Angiogenesis Potential for Tissue-Engineered Substitutes. *Tissue Eng. Part C Methods* **2017**, *23*, 604–615. [CrossRef]
64. TA Instruments. Determining the Linear Viscoelastic Region in Oscillatory Measurements. Available online: https://www.tainstruments.com/pdf/literature/RH107.pdf (accessed on 1 February 2023).
65. Anton Paar GmbH Amplitude Sweeps. Available online: https://wiki.anton-paar.com/en/amplitude-sweeps/ (accessed on 1 November 2021).

66. Talens Oliag, P. Cómo Caracterizar El Comportamiento Viscoelástico de Un Alimento. 2018. Available online: https://riunet.upv.es/bitstream/handle/10251/103393/Talens%20-%20C%C3%B3mo%20caracterizar%20el%20comportamiento%20viscoel%C3%A1stico%20de%20un%20alimento.pdf?sequence=1 (accessed on 5 June 2023).
67. TA Instruments. Rheological Techniques for Yield Stress Analysis. Available online: https://www.tainstruments.com/pdf/literature/RH025.pdf (accessed on 1 February 2023).
68. Huang, C.Y. Extrusion-Based 3D Printing and Characterization of Edible Materials. Master's Thesis, University of Waterloo, Waterloo, ON, Canada, 2018.
69. Cyriac, F.; Lugt, P.M.; Bosman, R. On a New Method to Determine the Yield Stress in Lubricating Grease. *Tribol. Trans.* **2015**, *58*, 1021–1030. [CrossRef]
70. Duty, C.; Ajinjeru, C.; Kishore, V.; Compton, B.; Hmeidat, N.; Chen, X.; Liu, P.; Hassen, A.A.; Lindahl, J.; Kunc, V. What Makes a Material Printable? A Viscoelastic Model for Extrusion-Based 3D Printing of Polymers. *J. Manuf. Process.* **2018**, *35*, 526–537. [CrossRef]
71. Petta, D.; Grijpma, D.W.; Alini, M.; Eglin, D.; D'Este, M. Three-Dimensional Printing of a Tyramine Hyaluronan Derivative with Double Gelation Mechanism for Independent Tuning of Shear Thinning and Postprinting Curing. *ACS Biomater. Sci. Eng.* **2018**, *4*, 3088–3098. [CrossRef]
72. Cheng, Y.; Qin, H.; Acevedo, N.C.; Jiang, X.; Shi, X. 3D Printing of Extended-Release Tablets of Theophylline Using Hydroxypropyl Methylcellulose (HPMC) Hydrogels. *Int. J. Pharm.* **2020**, *591*, 119983. [CrossRef]
73. TA Instruments. Introduction to Thixotropy Analysis Using a Rotational Rheometer. Available online: https://www.tainstruments.com/pdf/literature/RH106.pdf (accessed on 1 February 2023).
74. Herrada-Manchón, H.; Celada, L.; Rodríguez-González, D.; Alejandro Fernández, M.; Aguilar, E.; Chiara, M.-D. Three-Dimensional Bioprinted Cancer Models: A Powerful Platform for Investigating Tunneling Nanotube-like Cell Structures in Complex Microenvironments. *Mater. Sci. Eng. C* **2021**, *128*, 112357. [CrossRef]
75. Sun Han Chang, R.; Lee, J.C.W.; Pedron, S.; Harley, B.A.C.; Rogers, S.A. Rheological Analysis of the Gelation Kinetics of an Enzyme Cross-Linked PEG Hydrogel. *Biomacromolecules* **2019**, *20*, 2198–2206. [CrossRef] [PubMed]
76. Cordobés, F.; Partal, P.; Guerrero, A. Rheology and Microstructure of Heat-Induced Egg Yolk Gels. *Rheol. Acta* **2004**, *43*, 184–195. [CrossRef]
77. TA Instruments. Gelation Kinetics from Rheological Experiments. Available online: https://www.tainstruments.com/pdf/literature/RH103.pdf (accessed on 1 February 2023).
78. Ravanbakhsh, H.; Bao, G.; Latifi, N.; Mongeau, L.G. Carbon Nanotube Composite Hydrogels for Vocal Fold Tissue Engineering: Biocompatibility, Rheology, and Porosity. *Mater. Sci. Eng. C* **2019**, *103*, 109861. [CrossRef]
79. Avallone, P.R.; Raccone, E.; Costanzo, S.; Delmonte, M.; Sarrica, A.; Pasquino, R.; Grizzuti, N. Gelation Kinetics of Aqueous Gelatin Solutions in Isothermal Conditions via Rheological Tools. *Food Hydrocoll.* **2021**, *111*, 106248. [CrossRef]
80. Cox, T.; Madsen, C. Relative Stiffness Measurements of Cell-Embedded Hydrogels by Shear Rheology in Vitro. *Bio-Protocol* **2017**, *7*, e2101. [CrossRef]
81. Cox, T.R.; Erler, J.T. Remodeling and Homeostasis of the Extracellular Matrix: Implications for Fibrotic Diseases and Cancer. *Dis. Model. Mech.* **2011**, *4*, 165–178. [CrossRef]
82. Dell, A.C.; Wagner, G.; Own, J.; Geibel, J.P. 3D Bioprinting Using Hydrogels: Cell Inks and Tissue Engineering Applications. *Pharmaceutics* **2022**, *14*, 2596. [CrossRef]
83. Marques, C.F.; Diogo, G.S.; Pina, S.; Oliveira, J.M.; Silva, T.H.; Reis, R.L. Collagen-Based Bioinks for Hard Tissue Engineering Applications: A Comprehensive Review. *J. Mater. Sci. Mater. Med.* **2019**, *30*, 32. [CrossRef]
84. Bachmann, B.; Spitz, S.; Schädl, B.; Teuschl, A.H.; Redl, H.; Nürnberger, S.; Ertl, P. Stiffness Matters: Fine-Tuned Hydrogel Elasticity Alters Chondrogenic Redifferentiation. *Front. Bioeng. Biotechnol.* **2020**, *8*, 373. [CrossRef]
85. Ren, Y.; Zhang, H.; Wang, Y.; Du, B.; Yang, J.; Liu, L.; Zhang, Q. Hyaluronic Acid Hydrogel with Adjustable Stiffness for Mesenchymal Stem Cell 3D Culture via Related Molecular Mechanisms to Maintain Stemness and Induce Cartilage Differentiation. *ACS Appl. Bio Mater.* **2021**, *4*, 2601–2613. [CrossRef] [PubMed]
86. Herrada-Manchón, H. Formulación y Caracterización de Tintas Orgánicas Para (Bio)Impresión 3D. Ph.D. Thesis, Universidad de Oviedo, Oviedo, Spain, 2022.

Disclaimer/Publisher's Note: The statements, opinions and data contained in all publications are solely those of the individual author(s) and contributor(s) and not of MDPI and/or the editor(s). MDPI and/or the editor(s) disclaim responsibility for any injury to people or property resulting from any ideas, methods, instructions or products referred to in the content.

Review

3D-Printed Hydrogel for Diverse Applications: A Review

Arpana Agrawal [1] and Chaudhery Mustansar Hussain [2,*]

1. Department of Physics, Shri Neelkantheshwar Government Post-Graduate College, Khandwa 450001, India; agrawal.arpana01@gmail.com
2. Department of Chemistry and Environmental Science, New Jersey Institute of Technology, Newark, NJ 07102, USA
* Correspondence: chaudhery.m.hussain@njit.edu

Abstract: Hydrogels have emerged as a versatile and promising class of materials in the field of 3D printing, offering unique properties suitable for various applications. This review delves into the intersection of hydrogels and 3D printing, exploring current research, technological advancements, and future directions. It starts with an overview of hydrogel basics, including composition and properties, and details various hydrogel materials used in 3D printing. The review explores diverse 3D printing methods for hydrogels, discussing their advantages and limitations. It emphasizes the integration of 3D-printed hydrogels in biomedical engineering, showcasing its role in tissue engineering, regenerative medicine, and drug delivery. Beyond healthcare, it also examines their applications in the food, cosmetics, and electronics industries. Challenges like resolution limitations and scalability are addressed. The review predicts future trends in material development, printing techniques, and novel applications.

Keywords: hydrogels; 3D printing; biomedical applications; natural hydrogels; synthetic hydrogels

Citation: Agrawal, A.; Hussain, C.M. 3D-Printed Hydrogel for Diverse Applications: A Review. Gels 2023, 9, 960. https://doi.org/10.3390/gels9120960

Academic Editors: Enrique Aguilar and Helena Herrada-Manchón

Received: 5 November 2023
Revised: 25 November 2023
Accepted: 28 November 2023
Published: 7 December 2023

Copyright: © 2023 by the authors. Licensee MDPI, Basel, Switzerland. This article is an open access article distributed under the terms and conditions of the Creative Commons Attribution (CC BY) license (https://creativecommons.org/licenses/by/4.0/).

1. Introduction

The rapid advancement of 3D printing technology has revolutionized how we conceive, design, and produce objects across various industries. From aerospace to healthcare, 3D printing has ushered in an era of unparalleled customization, reduced lead times, and intricate geometries that were previously unattainable using traditional manufacturing methods [1]. By layering materials to construct 3D objects directly from digital designs, 3D printing has transcended conventional limitations, allowing for unprecedented levels of innovation. Applications of 3D printing are extensive and span a myriad of sectors and materials/composite materials [2]. In aerospace, the technology has enabled the production of complex lightweight components with enhanced performance characteristics. In the automotive industry, it has facilitated the rapid prototyping of parts and the creation of specialized tools on-demand. Similarly, healthcare has experienced transformative changes, with 3D printing employed to craft patient-specific implants, anatomical models for surgical planning, and even bioengineered tissues [3,4].

Amidst this technological evolution, one class of materials has emerged as particularly intriguing for 3D printing: hydrogels. Hydrogels are three-dimensional networks of hydrophilic polymers that possess the remarkable ability to absorb and retain large amounts of water or biological fluids [5]. This unique property imbues hydrogels with an inherent compatibility with living systems, making them an ideal candidate for a range of applications that intersect with the life sciences. Hydrogels can be of several types, including natural, synthetic, hybrid, and physically and chemically crosslinked hydrogels. Notably, hydrogels' high water content and tunable mechanical properties render them an exceptional material for 3D printing. Various 3D printing technologies can be successfully employed for the printing of hydrogels, including extrusion-based 3D printing [6], inkjet-based printing [7], stereolithography [8], laser-assisted bioprinting [9], digital light processing [10], etc. Naghieh et al. [11] explored the printability of 3D hydrogel scaffolds, focusing on the influence of hydrogel composition and printing parameters, and contributedto optimizing 3D printing processes for creating intricate structures used in tissue engineering and other applications. Kalyan and Kumar [12] provided insights into the broad spectrum of 3D printing applications, including tissue engineering, medical devices, and drug delivery, and served as a comprehensive overview of the current state of 3D printing technology in healthcare and biomedicine.

The capability to engineer the stiffness, porosity, and degradation rate of hydrogels precisely allows for tailoring constructs to match the specific requirements of the target application. This adaptability opens doors to advancements in tissue engineering, drug delivery, wound healing, and beyond. Moreover, the compatibility of hydrogels with cells and tissues offers the potential to create biomimetic structures that closely mimic natural physiological environments, thereby enhancing their utility in regenerative medicine and disease modeling. Figure 1 schematically shows the various types of hydrogels and their printing techniques that can be employed for several applications in various sectors. In a recent review, Kapusta and colleagues [13] delved into utilizing antimicrobial natural hydrogels in the realm of biomedicine, examining attributes, potential applications, and the challenges presented by hydrogels. This work sheds light on the manifold opportunities these hydrogels hold within various medical fields. Kaliaraj et al. [14] directed their attention to the auspicious uses of hydrogels in the realm of 3D printing technology, where the emphasis lies in highlighting the versatile nature of hydrogels as viable materials for 3D printing, underscoring their pivotal role in this innovative manufacturing technique. Zhang et al. [15] presented an exhaustive review that encapsulates the latest advancements in the domain of 3D printing concerning sturdy hydrogels. Their in-depth exploration provides valuable insights into the evolution of resilient hydrogel materials and their applications, thereby contributing significantly to the burgeoning field of 3D printing.

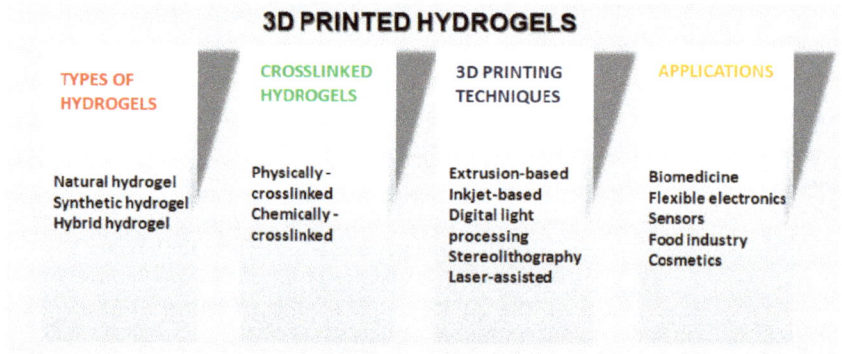

Figure 1. Schematic representation of the various types of hydrogels and their printing techniques that can be employed in several applications in various sectors, including biomedicine, flexible electronics, sensors, food industries, and cosmetics.

As we embark on this journey through the realm of hydrogels in 3D printing, this review article aims to explore their multifaceted potential, discussing the diverse range of hydrogel materials, printing techniques, and applications within which they play a pivotal role. By delving into the synergy between hydrogels and 3D printing, we seek to unravel the novel avenues for innovation that this convergence has unlocked, ultimately shaping the landscape of industries ranging from healthcare to consumer products.

2. Hydrogels: An Overview

Hydrogels are a class of materials with a three-dimensional network structure composed of hydrophilic polymers that can absorb and retain significant amounts of water or biological fluids. This unique property gives hydrogels a resemblance to natural soft tissues, making them valuable for a wide range of applications across various fields. This section will discuss the classification of hydrogels, their properties, and their applications.

2.1. Classification of Hydrogels

The choice of hydrogel material significantly influences the printability, biocompatibility, and mechanical properties of 3D-printed constructs. Researchers need to carefully consider these factors when selecting a hydrogel material for their specific application to ensure optimal performance and desired outcomes. It should be noted here that the hydrogels can be classified based on their composition and crosslinking methods. On the basis of composition, they can be natural hydrogels or synthetic hydrogels.

2.1.1. Natural Hydrogels

Natural hydrogels can be derived from natural polymers such as alginate, collagen, chitosan, and hyaluronic acid and often possess inherent bioactivity and biocompatibility.These hydrogels can be challenging to 3D print due to their complex rheological properties, including viscosity and shear-thinning behavior. Modifications, such as optimizing gelation kinetics or mixing with other materials, might be necessary to enhance printability. Also, they tend to be biocompatible due to their similarity to the extracellular matrix of tissues and hence promote cell adhesion, proliferation, and differentiation, making them valuable for applications involving direct interaction with living cells. Başyiğit et al. [16] explored soy-protein-based hydrogels enhanced with locust bean gum, investigating their mechanical properties and release characteristics. This research advances our comprehension of the potential utility of these hydrogels in applications such as controlled drug delivery systems. In other work, Xin and colleagues [17] developed a specialized okra-based hydrogel for chronic diabetic wounds, addressing a significant healthcare challenge and underscoring the promise of natural hydrogels in wound care. In the same year, Haghbin et al. [18] embarked on creating a Persian gum-based hydrogel loaded with gentamicin-loaded natural zeolite, conducting a comprehensive study of its properties both in vitro and in silico. This investigation carries implications for drug delivery and infection management, underscoring the adaptability of natural hydrogels. The biodeterioration of stone monuments and its correlation with cyanobacterial biofilm growth was also investigated [19]. They also explored the use of essential oils in natural hydrogels. This study has practical applications in preserving cultural heritage and the environmentally friendly management of biofilms. Huang et al. [20] directed their efforts toward the biofabrication of a natural Au/bacterial cellulose hydrogel for bone tissue regeneration via insitu fermentation. This research addresses the pressing need for biocompatible materials in regenerative medicine, underscoring the potential of natural hydrogels in the field of tissue engineering.

The construction and properties of physically cross-linked hydrogels based on natural polymers werecarried out by Yang et al. [21]. Their work bears wide-ranging implications in the field of biomedicine, including applications in drug delivery, wound healing, and tissue engineering. Xu et al. [22] delved into examining chitosan-based high-strength supramolecular hydrogels intended for 3D bioprinting, which advances the bioprinting field and highlights the promise of natural hydrogels as bioink materials. Cai and colleagues [23] reported on the potential of Laponite®-incorporated oxidized alginate–gelatin composite hydrogels for use in extrusion-based 3D printing, which holds great significance within the realm of 3D printing technology, illustrating how natural hydrogels can be seamlessly integrated into advanced manufacturing processes. These collective studies contribute to the growing domain of natural hydrogels, showcasing their adaptability and potential across diverse applications, from healthcare to cultural preservation. It should be noted that the mechanical strength and stability of these hydrogels may be relatively lower, which can limit their use in load-bearing applications. However, they excel in creating biomimetic environments for cell growth.

2.1.2. Synthetic Hydrogels

Contrary to natural hydrogels, synthetic hydrogels can be obtained from synthetic polymers like polyethylene glycol (PEG), polyacrylamide, and polyvinyl alcohol (PVA), and hence, the properties of synthetic hydrogels can be precisely controlled through chemical synthesis. These hydrogels generally exhibit more consistent and predictable printing behavior, and their properties can be finely tuned to match the desired viscosity and flow characteristics for different printing techniques. The biocompatibility of such hydrogels depends on the specific polymer and crosslinking chemistry used. Modifications can be made to enhance biocompatibility, and biofunctionalization can be applied to improve cell interactions. Apart from this, they also offer superior control over mechanical properties, which can be engineered to mimic a wide range of tissue stiffness, making them suitable for applications where mechanical support is crucial. Van Velthoven et al. [24] conducted a study on the entrapment of growth factors within synthetic hydrogels, specifically examining the bioactive bFGF-functionalized polyisocyanide hydrogels. Their goal was to enhance the controlled release and biological activity of growth factors, potentially benefiting biomedical applications.

Saccone et al. [25] pioneered a novel additive manufacturing (AM) technique based on visible light photopolymerization, called Hydrogel Infusion Additive Manufacturing (HIAM), which allows for the creation of a wide range of micro-architected metals and alloys using a single photoresin composition. In this process, 3D-architected hydrogel scaffolds are employed as platforms for in situ material synthesis reactions, as depicted schematically in Figure 2a. The first step in producing metal microlattices involves using DLP to print architected organogels based on N,N-dimethylformamide (DMF), and polyethylene glycol diacrylate (PEGda). The DLP printing phase determines the final shape of the part, and following printing, a solvent exchange replaces DMF with water, transforming the organogels into hydrogels. These hydrogel structures are then soaked in a solution containing metal salt precursors, allowing the metal ions to infiltrate and expand the hydrogel scaffold. Calcination in an air environment converts the metal–salt-swollen hydrogels into metal oxides, and a subsequent reduction in a forming gas mixture (95% N_2, 5% H_2) yields metal or alloy replicas that match the initially designed architecture. Throughout this process, the shape of the part, defined during DLP printing, is retained, with each dimension experiencing approximately 60–70% linear shrinkage, accompanied by an approximate mass loss of 65–90% during calcination. To illustrate the versatility of HIAM when compared to previous visible light photopolymerization AM techniques, they employed HIAM to create octet lattice structures using various materials such as copper (as shown in process steps in Figure 2b–e), nickel, silver, their alloys, and more complex materials like the high-entropy alloy CuNiCoFe and the refractory alloy W-Ni (Figure 2f). Additionally, they demonstrated the fabrication of multi-material structures, such as Cu/Co (Figure 2g–h). HIAM stands out for its capacity for parallelization, allowing multiple organogels to be printed simultaneously, swelled in distinct solutions, and subsequently calcined/reduced collectively. In Figure 2i, one can observe the simultaneous calcination of eight hydrogel lattices (precursors for Cu, CuNi, CuNiCoFe, and CuNiCoFeCr), resulting in the formation of oxides. The scale bars shown in Figure 2b,c are 5 mm; Figure 2d–f 1 mm; Figure 2g 1 cm; Figure 2h 2 mm; and Figure 2i 2 cm. Figure 2j–m represents the SEM image of Cu and CuNi samples from an overhead view (Figure 2j,l) and a single junction node (Figure 2k,m), respectively, and revealed that the Cu and CuNi samples retained their octet lattice shape throughout the thermal treatment, with beam diameters approximately around 40 μm.

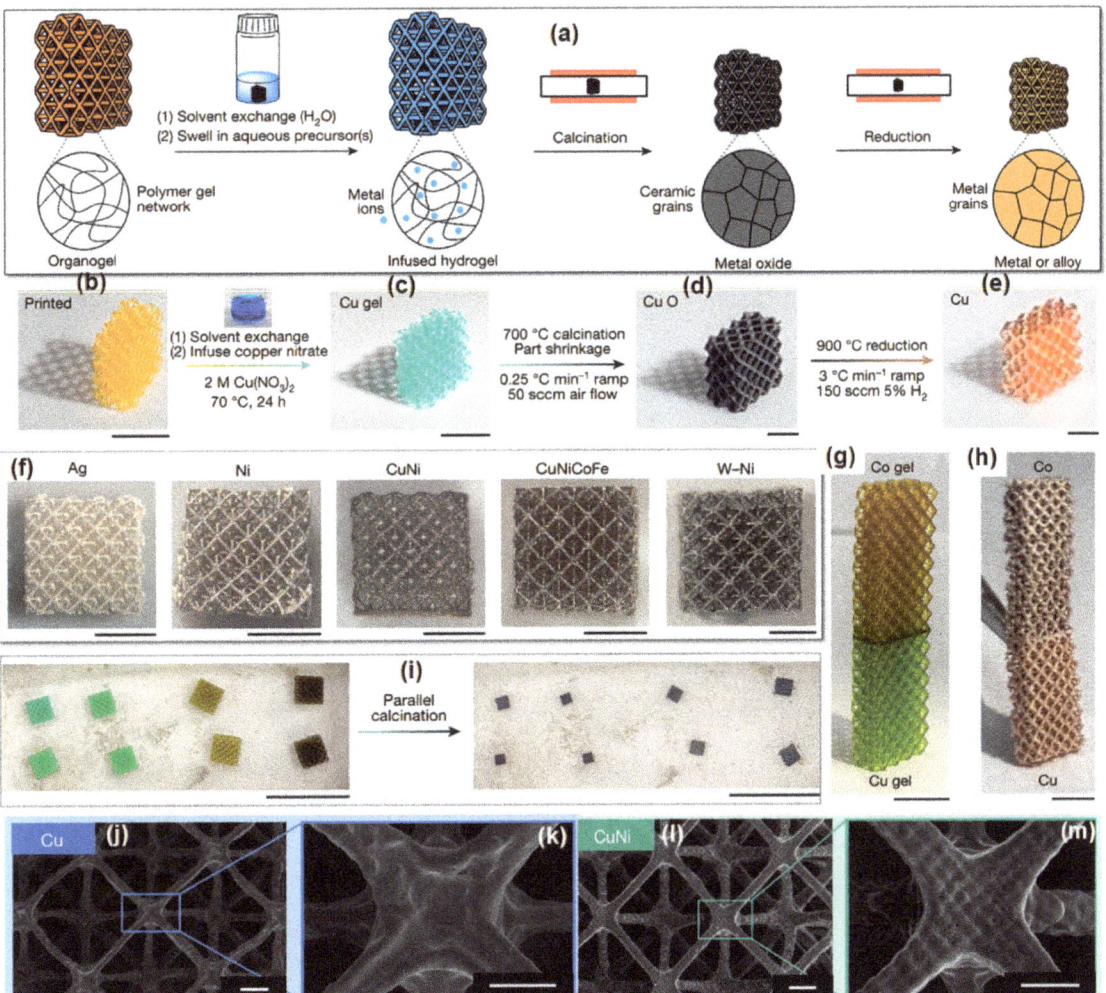

Figure 2. Hydrogel-infused additive manufacturing process and materials. (**a**) Illustration of the HIAM process. A 3D-printed organogel structure, composed of DMF/PEGda, undergoes a transformation into an infused hydrogel replica through the sequential steps of photoactive compound removal, solvent exchange, and infusion with a suitable aqueous precursor. Subsequent calcination in an oxygen environment leads to the formation of metal oxide structures, which are subsequently reduced to metals in forming gas. (**b**–**e**) Visual representations of the HIAM process for Cu metal: the printed organogel (**b**), the infused hydrogel (**c**), the calcined metal oxide (**d**), and the reduced metal (**e**). (**f**) Diverse metals and alloys fabricated using HIAM, including Ag, Ni, a binary CuNi alloy, a high-entropy CuNiCoFe alloy, and a refractory W–Ni alloy. (**g**) An octet lattice infused with $Cu(NO_3)_2$ on one end and $Co(NO_3)_2$ on the other. (**h**) After calcination and reduction, the Cu/Co gel transforms into a Cu/Co multi-material. (**i**) Simultaneous calcination of various infused gels. (**j**,**k**) and (**l**,**m**) SEM micrographs of Cu and CuNi octet lattices, depicting several unit cells from an overhead view (**j**,**l**) and a single junction node (**k**,**m**), respectively. (Scale bars: (**j**,**l**), 100 μm; (**k**,**m**), 50 μm). Reprinted from [25] under a Creative Commons Attribution License 4.0 (CC BY) "https://creativecommons.org/licenses/by/4.0/ (accessed on 1 November 2023)".

Adjuik et al. [26] presented an extensive review that delves into the degradability of both bio-based and synthetic hydrogels, with a particular emphasis on their potential as sustainable soil amendments. This comprehensive review explores the environmental implications and ecological consequences of using hydrogels in agriculture, shedding light on the sustainability and compatibility of these materials when applied to soil-enhancement practices. The 3D-printable synthetic hydrogel designed as an immobilization matrix for continuous synthesis involving fungal peroxygenases has also been reported [27]. This innovative approach holds promise in the field of biocatalysis, enabling efficient and uninterrupted production of valuable compounds. The synthetic hydrogel matrix offers a stable and controlled environment for enzyme activity, facilitating the development of sustainable and efficient biochemical processes. Yuk et al. [28] created a high-performance 3D-printable ink utilizing one of the most commonly used conducting polymers, poly(3,4-ethylenedioxythiophene):polystyrene sulfonate (PEDOT:PSS), to harness the capabilities of advanced 3D printing in producing conducting polymer structures. To ensure suitable rheological properties for 3D printing, they developed a paste-like conducting polymer ink, which is derived from cryogenically freezing an aqueous PEDOT:PSS solution Figure 3a, followed by lyophilization and controlled re-dispersion in a mixture of water and dimethyl sulfoxide (DMSO) (as shown in Figure 3b). The resulting conducting polymer ink demonstrates exceptional 3D-printing capabilities, enabling high-resolution printing (with a resolution of over 30 μm), the creation of structures with a high aspect ratio (exceeding 20 layers), and consistent fabrication of conducting polymers. These structures can also be seamlessly integrated with other 3D-printable materials, such as insulating elastomers, using multi-material 3D printing. By subjecting the 3D-printed conducting polymers to dry annealing, they achieved highly conductive microstructures that remain flexible in their dry state, which can be easily transformed into a soft, highly conductive PEDOT:PSS hydrogel through subsequent swelling in a wet environment (Figure 3c). To showcase the achievement of high-resolution microscale printing, they used a 7wt% PEDOT:PSS nanofibril-based conducting polymer ink to create intricate mesh patterns through nozzles of various diameters: 200 μm, 100 μm, 50 μm, and 30 μm (referred to as Figure 3d–g). The conducting polymer ink's favorable rheological properties also facilitate the construction of multi-layered microstructures with high aspect ratios, using a 100 μm nozzle and incorporating up to 20 layers.

Figure 3h illustrates the step-by-step progression of creating a 20-layered meshed structure using the conducting polymer ink. Following the printing, Figure 3i displays the 3D-printed conducting polymer mesh after undergoing a dry-annealing process, which is crucial for enhancing its conductivity. In contrast, Figure 3j showcases the same 3D-printed conducting polymer mesh but in its hydrogel state, highlighting the adaptability of this material to different environmental conditions. Figure 3k provides a visual representation of the sequential snapshots for 3D-printing overhanging features across high aspect ratio structures using the conducting polymer ink, demonstrating the ink's ability to maintain structural integrity during complex printing. Finally, Figure 3l reveals the 3D-printed conducting polymer structure with overhanging features, which has been transformed into a hydrogel state. Collectively, these figures demonstrate the versatility, resilience, and transformative capabilities of conducting polymer inks in advanced 3D printing applications.

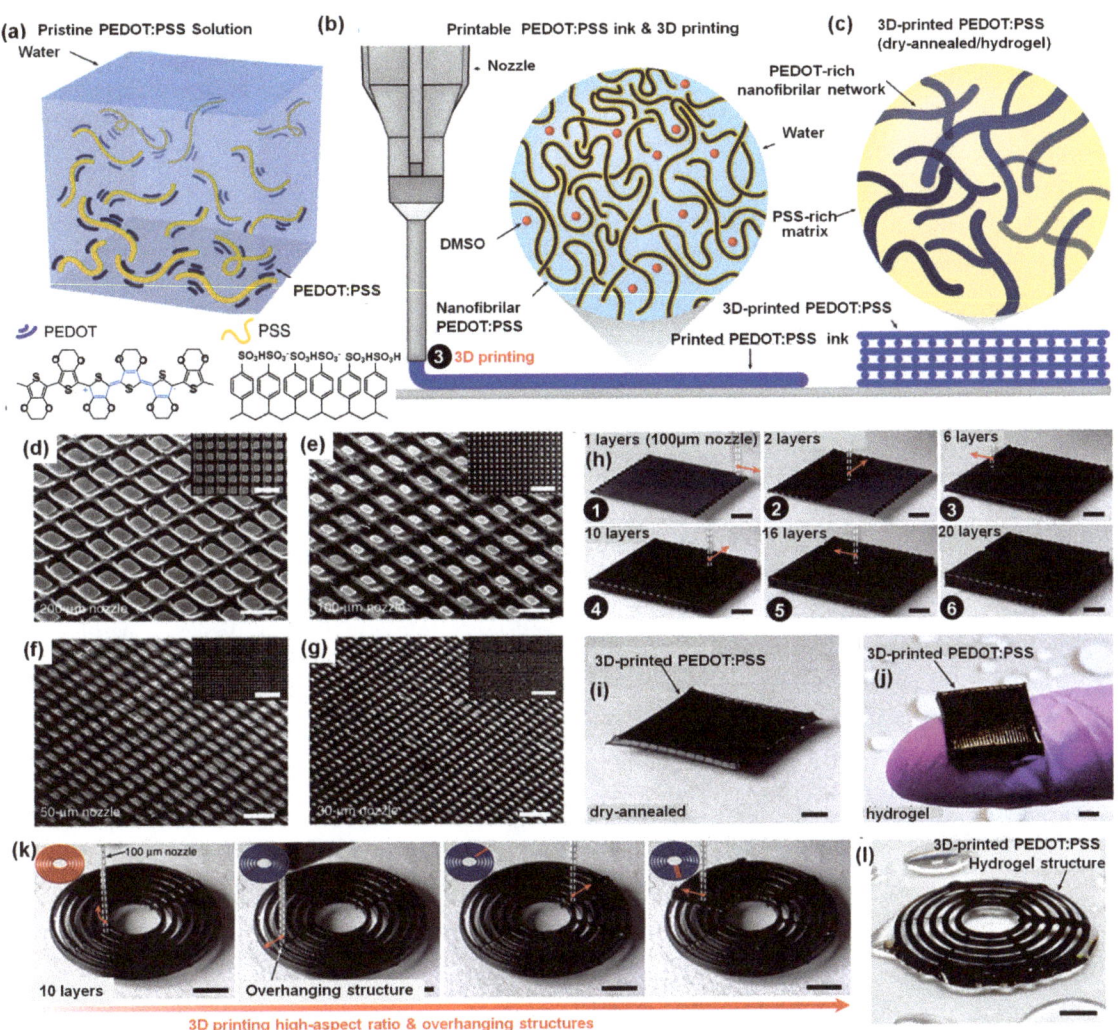

Figure 3. Development of a 3D-printable conductive polymer ink design. (**a,b**) The initial pristine PEDOT:PSS solution (**a**) can undergo a transformation into a 3D-printable conductive polymer ink (**b**) through lyophilization under cryogenic conditions followed by re-dispersion using an appropriate solvent. (**c**) 3D-printed conductive polymers can be transformed into pure PEDOT:PSS, whether in their dry state or hydrogel state, through a process involving dry annealing and subsequent expansion in a wet environment. (**d–g**) SEM micrographs of 3D-printed conducting polymer meshes. (**h**) Step-by-step progression of creating a 20-layered meshed structure using the conducting polymer ink. (**i**) 3D-printed conducting polymer mesh after undergoing a dry-annealing process, crucial for enhancing its conductivity. (**j**) 3D-printed conducting polymer mesh in its hydrogel state, highlighting its adaptability to different environmental conditions. (**k**) Sequential snapshots for 3D printing overhanging features across high aspect ratio structures using the conducting polymer ink, demonstrating its ability to maintain structural integrity during complex printing. (**l**) 3D-printed conducting polymer structure with overhanging features transformed into a hydrogel state. Reprinted from [28] under a Creative Commons Attribution License 4.0 (CC BY) "https://creativecommons.org/licenses/by/4.0/ (accessed on 1 November 2023)".

2.1.3. Hybrid Hydrogels

Apart from natural and synthetic hydrogels, hybrid hydrogels can also be prepared by combining the components from both natural and synthetic sources, leveraging the advantages of each material type. Hybrid hydrogels often strike a balance between natural and synthetic hydrogels, aiming to improve printability by combining the advantages of both material types. They can also harness the biocompatibility of natural components while introducing controlled synthetic elements to enhance mechanical properties and stability. Also, these hydrogels aim to combine the mechanical advantages of synthetic materials with the biocompatibility of natural materials, offering a balanced approach for applications requiring both properties. Tang et al. [29] delved into the domain of hybrid hydrogels for three-dimensional cell culture and employed stereolithography to craft nanocellulose/PEGDA aerogel scaffolds, thereby introducing tunability to the modulus of these structures. This investigation is of notable significance within the realms of tissue engineering and regenerative medicine as it provides a platform for 3D cell cultivation, effectively emulating natural tissue environments. Cao and their collaborators [30] have also explored the potential of antibacterial hybrid hydrogels as a robust tool for combatting microbial infections. This research represents a vital area of study with broad-reaching implications in the field of medicine, where the development of infection-resistant materials can significantly elevate patient care and mitigate the transmission of infectious diseases.

Palmese et al. [31] provided a comprehensive investigation and multifaceted applications of hybrid hydrogels within the biomedical field, which underscores their versatility and potential, showcasing the diverse ways they can be applied to address various medical challenges, from developing drug delivery systems to advancing tissue engineering. Vasile and his research team also examined the incorporation of natural polymers into hybrid hydrogels for medical applications [32]. Their study sheds light on the diverse utility of these materials in healthcare, indicating their roles in drug delivery, wound healing, and various other medical interventions. This research emphasizes the increasing interest in harnessing the properties of natural polymers within medical contexts. Liao et al. [33] introduced a unique hybrid hydrogel with the specific aim of preventing bone tumor recurrence and promoting bone regeneration. This specialized hydrogel incorporates gold nanorods and nanohydroxyapatite and utilizes photothermal therapy. This innovative approach holds great promise for addressing bone-related health concerns and managing bone tumors, offering a novel and effective treatment strategy. Anisotropic hybrid hydrogels with mechanical properties reminiscent of tendons or ligaments werealso explored [34] for tissue engineering, providing the potential to mimic the mechanical characteristics of specific tissues. This development offers a valuable resource for creating artificial ligaments and tendons. Yang et al. [35] introduced 3D macroporous oxidation-resistant Ti_3C_2Tx MXene hybrid hydrogels, which demonstrate remarkable supercapacitive performance with an extraordinarily long cycle life. This research holds significant implications for energy storage applications, particularly in supercapacitors, where long-term stability and performance are imperative. Collectively, these studies contribute to the expanding field of hybrid hydrogels, underscoring their potential and versatility in a wide range of biomedical and materials science applications. These applications span from healthcare to advanced materials used in sustainable technologies. Table 1 comparatively summarizes the various properties of natural, synthetic, and hybrid hydrogels in terms of preparation, printability, biocompatibility, and mechanical properties.

Table 1. Comparative summary of the properties of natural, synthetic, and hybrid hydrogels in terms of preparation, printability, biocompatibility, and mechanical properties.

Property	Natural Hydrogels	Synthetic Hydrogels	Hybrid Hydrogels
Preparation	- Derived from biological sources like alginate, collagen, or agarose. - Typically requires minimal chemical modification.	- Chemically synthesized with precise control over composition. - Tailored to specific requirements through chemical synthesis.	- Combination of natural and synthetic components. - Utilizes the advantages of both natural and synthetic components.
Printability	- Varies depending on source material. - May require additional modifications for 3D printing.	- Generally good printability due to controlled composition. - Compatible with various 3D printing techniques.	- Printability depends on the combination and ratio of components. - Printability can be customized for specific applications.
Biocompatibility	- Generally biocompatible due to natural origin. - Often suitable for cell encapsulation and tissue engineering.	- Biocompatibility can vary based on the polymer used. - Careful selection of synthetic components can enhance biocompatibility.	- Biocompatibility influenced by natural components. - Potential for improved biocompatibility through hybridization.
Mechanical Properties	- Mechanical properties can vary widely depending on the source. - May have lower mechanical strength compared to synthetics.	- Can be precisely tuned for specific applications. - Offers a wide range of mechanical properties (soft to stiff).	- Mechanical properties can be tailored to desired levels. - Balances natural properties with synthetic enhancements.

2.1.4. Crosslinked Hydrogels

Depending upon crosslinking, hydrogels can be classified as physically crosslinked hydrogels and chemically crosslinked hydrogels. Physically crosslinked hydrogels are formed through non-covalent interactions, such as hydrogen bonding or physical entanglements, and are reversible and responsive to environmental changes, while chemically crosslinked hydrogels are formed by covalent bonds between polymer chains, resulting in higher stability and mechanical strength. Examples include photopolymerization and chemical crosslinking agents. Iwanaga et al. [36] focused their efforts on designing and producing fully developed engineered pre-cardiac tissue using 3D bioprinting techniques coupled with enzymatic crosslinking hydrogels. This pioneering investigation shows great potential in the fields of regenerative medicine and cardiac tissue engineering, addressing the significant hurdle of generating functional heart tissues. Recently, Ianchis et al. [37] also delved into creating innovative green crosslinked hydrogels derived from salecan and initiated their preliminary exploration of these hydrogels in the context of 3D printing. The prospects of utilizing sustainable and environmentally friendly hydrogel materials in additive manufacturing processes have been highlighted in this research, with implications spanning drug delivery and tissue engineering. Another groundbreaking work by Farsheed and his collaborators has been witnessed in the realm of 3D printing of self-assembling nanofibrous multidomain peptide hydrogels [38]. This research has the potential to revolutionize the fabrication of intricate and adaptable nanofibrous structures through 3D printing, introducing novel applications in regenerative medicine, drug delivery, and tissue engineering.

2.2. Properties of Hydrogels

Hydrogels showcase a myriad of captivating features, notably their substantial water content, customizable mechanical attributes, swelling dynamics, biocompatibility, permeability, and more. With water constituting up to 90% of their composition, hydrogels offer a conducive environment for interactions with biological systems. Their mechanical properties, encompassing factors like stiffness, elasticity, and resilience, are finely adjustable through parameters such as polymer type, crosslinking density, and composition. The responsive nature of hydrogels, exhibiting swelling and deswelling reactions to environmental shifts like pH and temperature, positions them as ideal candidates for controlled drug delivery systems. Many hydrogels exhibit biologically inert and non-toxic traits, ensuring their safe utilization in medical applications without adverse effects on cells and tissues. Moreover, their ability to selectively allow the diffusion of small molecules while impeding larger ones holds promise for applications in separation and filtration processes.

The unique characteristics, diverse compositions, and tunable crosslinking methods of hydrogels make them invaluable materials with wide-ranging applications across biomedical, pharmaceutical, and various other industries, fostering innovation and influencing multiple aspects of contemporary life. Table 2 provides a summary of the various properties of hydrogels along with their corresponding applications. While biomedical and pharmaceutical applications of hydrogel materials are notably prominent, their extensive promise in tissue engineering stands out. Hydrogels, owing to their biocompatibility and resemblance to natural tissues, serve as scaffolds for tissue regeneration, facilitating the repair of damaged tissues and organs. In drug delivery systems, hydrogels play a pivotal role by enabling the controlled release of therapeutic agents over time. Their capacity to encapsulate drugs and proteins ensures targeted delivery, minimizing side effects and enhancing treatment outcomes.

Table 2. Description of the various properties of hydrogels along with their corresponding applications.

Property	Description	Applications
Water Absorption and Retention	Ability to absorb and retain a significant amount of water or biological fluids.	Wound dressings, contact lenses, diapers, and drug delivery systems.
Biocompatibility	Compatibility with living tissues, making hydrogels suitable for medical and biological applications.	Tissue engineering, drug delivery, wound healing, and surgical implants.
Tunable Mechanical Properties	Adjustability of mechanical characteristics like elasticity and stiffness for specific applications.	Cartilage replacements, soft tissue engineering, and drug delivery matrices.
Swelling Behavior	Controllable ability to swell in response to factors such as pH, temperature, or ionic strength.	Controlled drug release, biosensors, and wound dressings.
Permeability	Ability to allow the passage of certain substances while restricting others.	Controlled drug delivery, filtration membranes, and biosensors.
Responsive to Stimuli	Capability to undergo changes in volume or structure in response to external stimuli (e.g., temperature, pH, and light).	Smart drug delivery, biosensing, and controlled release systems.
Adhesive Properties	Ability to adhere to biological tissues, useful for applications like wound dressings and tissue adhesives.	Surgical adhesives, wound closures, and tissue engineering.
Versatility	Wide range of options in terms of composition and structure, allowing customization for various applications.	Tissue engineering scaffolds, wound dressings, and drug delivery systems.
Electrical Conductivity	Capability to conduct electricity, often achieved by incorporating conductive polymers or nanoparticles.	Biosensors, flexible electronics, and neural interfaces.

2.3. Designing Hydrogel Formulations for 3D Printing

Designing hydrogel formulations for 3D printing involves optimizing various parameters to ensure proper printability, structural integrity, and compatibility with the chosen printing method. A delicate balance between rheological properties, gelation kinetics, and print fidelity is highly imperative for designing hydrogel formulations for 3D printing [39]. Rheological characteristics determine the hydrogel's flow behavior during printing, while gelation kinetics influence the temporal aspects of the printing process. Achieving optimal print fidelity requires a careful balance of these properties to ensure accurate layer deposition and the overall success of 3D-printed hydrogel constructs, particularly in biomedical applications. Additives also play a critical role in tailoring hydrogel properties to meet specific application requirements, enabling the creation of structures with optimized mechanical, biological, and functional characteristics.

It should be noted here that rheology is the study of the flow and deformation of materials. In the context of hydrogels for 3D printing, rheological properties such as viscosity and shear-thinning behavior are crucial. A suitable viscosity ensures proper extrusion through the printer nozzle, while shear-thinning behavior allows the hydrogel to reduce viscosity under shear stress during printing, facilitating smooth flow and layer deposition. Several researchers have shed light on the correlation between rheological properties and printing behavior. Bom et al. [39] explored this correlation, emphasizing the importance of understanding rheological properties in achieving successful 3D printing outcomes. Amorim et al. [40] provided insights into the shear rheology of inks for extrusion-based 3D bioprinting, contributing to the fundamental understanding of the behavior of printing materials under shear stress. In a related vein, Kokol et al. [41] investigated how flow- and horizontally-induced cooling rates during 3D cryo-printing affect the rheological properties of gelatine hydrogels. Kim et al. [42] focused on enhancing the rheological behaviors of alginate hydrogels with carrageenan, specifically for extrusion-based bioprinting applications. Townsend et al. [43] emphasized the significance of precursor rheology for hydrogel placement in medical applications and 3D bioprinting, underscoring the importance of flow behavior before crosslinking. Zhou et al. [44] introduced microbial transglutaminase-induced controlled crosslinking of gelatin methacryloyl, showcasing a method to tailor rheological properties for optimized 3D printing.

On the other hand, gelation refers to the process by which a liquid hydrogel transforms into a gel, solidifying the material. Gelation kinetics in 3D printing involves the study of the time taken for the hydrogel to transition from a liquid to a solid state. The speed of gelation is a critical factor influencing the printing speed and the overall printing time. Ideally, the gelation kinetics should match the printing speed to ensure proper layer-by-layer deposition. Apart from these, print fidelity encompasses the accuracy and precision of the printed structure compared to the digital design. Several factors contribute to print fidelity, including the rheological properties of the hydrogel, the gelation kinetics, and the overall printing parameters. Achieving high print fidelity is crucial for accurately replicating complex structures, especially in applications like tissue engineering, where the precise arrangement of cells and biomaterials is essential for functionality. Schwab and colleagues [45] delved into the printability and shape fidelity of bioinks, offering comprehensive insights crucial for developing precise 3D bioprinted structures. On the other hand, Mora-Boza et al. [46] explored the potential of glycerylphytate as an ionic crosslinker for 3D printing, demonstrating its effectiveness in producing multi-layered scaffolds with enhanced shape fidelity and desirable biological features. In a different avenue, Huang et al. [47] investigated the incorporation of bacterial cellulose nanofibers to improve the stress and fidelity of 3D-printed silk-based hydrogel scaffolds, introducing a novel approach to scaffold enhancement. Sheikhi et al. [48] focused on the 3D printing of jammed self-supporting microgels, presenting an alternative mechanism that addresses shape fidelity, crosslinking, and conductivity. Table 3 summarizes various key parameters and properties under rheology, gelation kinetics, print fidelity, and their effects on 3D printing processes.

Table 3. Influence and definition of various key parameters and properties under rheology, gelation kinetics, and print fidelity on 3D printing processes.

Parameter/Properties	Definition	Effect on 3D Printing
RHEOLOGICAL PROPERTIES		
Viscosity	Viscosity measures a fluid's resistance to flow.	Affects the ease of handling and deposition.
Shear-Thinning Behavior	Shear-thinning is the property where viscosity decreases under shear stress.	Facilitates smooth flow during printing; hydrogel becomes less viscous when subjected to shear stress, allowing for easy extrusion.

Table 3. Cont.

Parameter/Properties	Definition	Effect on 3D Printing
Thixotropy	Property where a material becomes less viscous over time under constant shear stress and recovers its viscosity when the stress is removed.	Allows the hydrogel to recover its original viscosity between printing layers, preventing spreading and maintaining structural integrity.
Viscoelasticity	Viscoelastic materials exhibit both viscous (flow) and elastic (deformation recovery) properties.	Affects the material's response to stress and strain during printing and ensures the printed structure retains its shape after deposition.
Gelling Mechanisms	Refers to the process by which a liquid transforms into a gel or solid.	Determines speed and control of gelation, influencing the overall printing process and final construct.
Extrudability	The ease with which a material can be extruded or forced through a nozzle.	Affects the precision and control of material deposition during 3D printing.
Shape Retention	The ability of the material to maintain its intended shape after deposition.	Critical for achieving accurate and consistent layer-by-layer printing, ensuring the final structure matches the design.
GELATION KINETICS		
Gelation Time	The time taken for a liquid to transition into a gel.	Affects the overall printing speed and duration of the 3D printing process.
Crosslinking Density	The concentration of crosslinks formed between polymer chains during gelation.	Affects the mechanical strength and stability of the resulting hydrogel.
Temperature	The degree of heat applied during the gelation process.	Affects the rate of chemical reactions or physical processes leading to gel formation.
Concentration of Crosslinking Agents	The amount of crosslinking agents, such as chemical initiators, present in the hydrogel formulation.	Higher concentrations typically result in faster gelation but may impact other material properties.
pH Level	The acidity or alkalinity of the hydrogel formulation.	Affects the ionization of functional groups and, hence, the gelation process.
Polymer Concentration	The concentration of polymer molecules in the hydrogel formulation.	Higher concentrations can lead to denser networks and affect the gelation time and mechanical properties.
Solvent Composition	The type and ratio of solvents used in the hydrogel formulation.	Solvent properties can impact the rate of gelation and the resulting structure of the hydrogel.
Initiator Concentration (Photochemical Gelation)	The concentration of photoinitiators in the hydrogel formulation.	Critical for photochemical gelation processes, where light triggers crosslink formation.
Stirring Rate (for Physical Gelation)	The speed at which the hydrogel components are mixed.	Affects the distribution of components and can influence the gelation time in physically gelled hydrogels.
Presence of Catalysts	Chemical substances that accelerate gelation reactions.	Help in enhancing the speed and efficiency of gelation processes.
PRINT FIDELITY		
Layer Resolution	The thickness of each layer deposited during printing.	Finer resolutions lead to smoother surfaces and improved details.
Extruder Calibration	Adjusting the extruder to ensure accurate material deposition.	Prevents under- or over-extrusion, enhancing print accuracy.
Bed Leveling	Ensuring the print bed is perfectly level.	Prevents uneven layer heights, promoting uniform adhesion.
Print Speed	The speed at which the printer deposits material.	Optimizing print speed balances accuracy with efficiency; too fast can lead to errors.
Temperature Control	Maintaining consistent temperatures for the printer and printing material.	Fluctuations can affect material flow and layer adhesion.
Material Quality	The quality and consistency of the printing material.	Inconsistent materials may lead to variations in print quality.
Print Bed Adhesion	Ensuring the first layer adheres well to the print bed.	Proper adhesion prevents warping and helps maintain accurate layer alignment.
Support Structures	Temporary structures to support overhanging features.	Well-designed supports prevent deformations and maintain accuracy.
Cooling Systems	Fans or other cooling mechanisms to solidify layers quickly.	Proper cooling prevents overheating and improves feature definition.

Table 3. *Cont.*

Parameter/Properties	Definition	Effect on 3D Printing
Printer Rigidity	The stability and rigidity of the printer frame.	A stable frame reduces vibrations and ensures precise movements.
Print Orientation	The angle and direction in which the object is printed.	Optimal orientation minimizes overhangs and supports, improving print fidelity.
Filament Diameter	The diameter of the printing filament.	Accurate filament diameter ensures consistent material flow.
Environmental Conditions	Factors like temperature and humidity in the printing environment.	Extreme conditions can affect material properties and printing outcomes.
Print Design	The complexity and geometry of the printed object.	Complex designs may require specific settings for accurate printing.

3. 3D Printing Methods for Hydrogels

Several 3D printing methods are employed to create hydrogel-based structures, each with its own set of advantages and limitations. In selecting a 3D printing method for hydrogel-based projects, the trade-offs between resolution, speed, and material compatibility should be carefully considered to align with the desired application and project requirements. The various 3D printing technologies include extrusion-based 3D printing, inkjet-based printing, stereolithography (SLA), laser-assisted bioprinting, digital light processing (DLP), etc. [6–10]. Figure 4 schematically illustrates the various 3D printing technologies that can be successfully employed for hydrogel printing. Kantaros et al. [49] explored the technologies and resources employed in 3D printing within the realm of regenerative medicine and shed light on the instrumental tools and methodologies used to propel the field forward, facilitating the generation of intricate tissue constructs for transplantation and mending. Bedell and colleagues [50] investigated human gelatin-based composite hydrogels tailored for osteochondral tissue engineering, adapting them into bioinks suitable for a variety of 3D printing techniques. Their investigation contributes to the advancement of novel materials and printing methodologies aimed at restoring damaged joints and cartilage. The emerging domain of 3D printing for immobilizing biocatalysts has also been reported [51,52]. This study underscores the immense potential of 3D printing in biotechnology, enabling the efficient immobilization of enzymes for diverse applications, spanning bioprocessing and biofuel production.

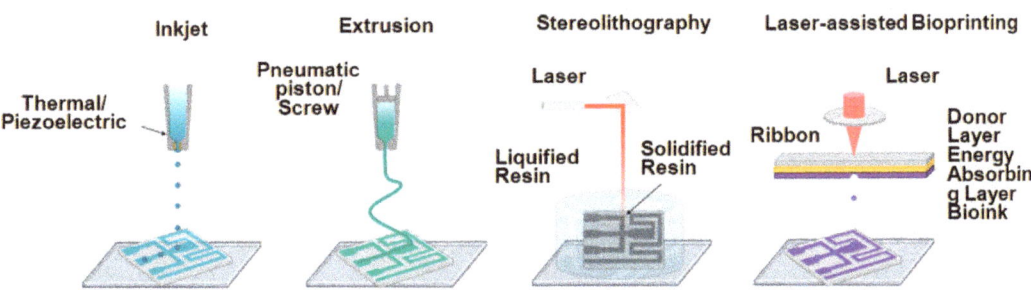

Figure 4. Illustration depicting diverse 3D printing techniques. Reprinted from [52] under a Creative Commons Attribution License 4.0 (CC BY) "https://creativecommons.org/licenses/by/4.0/ (accessed on 1 November 2023)".

3.1. Extrusion-Based Printing

Extrusion-based 3D printing for hydrogels is a sophisticated additive manufacturing process with immense potential in various fields, particularly biomedicine. This method harnesses the unique properties of hydrogel materials, which consist of a water-based gel with a polymer network, making them suitable for extrusion. A hydrogel material is meticulously prepared in this process, often as a viscous bioink, customized with specific additives like cells or therapeutic agents. A specialized extruder nozzle is employed in a 3D printer, precisely depositing the hydrogel layer by layer to construct intricate three-dimensional structures. Each layer is carefully deposited and, if needed, undergoes a crosslinking process to solidify and stabilize the structure. This technique is widely used in biomedicine to fabricate tissue scaffolds, organ models, and drug delivery systems and in other industries, such as food and cosmetics, for creating customized products. In recent years, a series of notable studies have significantly advanced the field of extrusion-based 3D printing, introducing innovative materials and technologies that broaden its application horizon. Cheng et al. [53] delved into the printability of a cellulose derivative for 3D printing, specifically focusing on its use as a biodegradable support material. Itmarkeda vital step toward more sustainable and versatile 3D printing processes, which can contribute to environmentally friendly manufacturing. Zhou et al. [54] explored the potential of gelatin-oxidized nanocellulose hydrogels in extrusion-based 3D bioprinting for tissue engineering and regenerative medicine, introducing a novel avenue in the quest for innovative and regenerative biomaterials.

Dong et al. [55] have demonstrated the extrusion 3D printing of a gelatine methacrylate/Laponite nanocomposite hydrogel, emphasizing its promise for applications in the realm of bone tissue regeneration. A unique approach by exploring advanced printable hydrogels derived from pre-crosslinked alginate was demonstrated by Falcone et al. [56]. This study highlights the suitability of these materials for semi-solid extrusion 3D printing, effectively extending the capabilities of this technology in creating intricate and complex structures for a wide array of applications, from tissue engineering to drug delivery systems. The work of Murphy et al. [57] introduced a fascinating twist by focusing on 3D extrusion printing of stable constructs composed of photoresponsive polypeptide hydrogels. This innovative approach allows for the creation of responsive and tunable 3D-printed materials, paving the way for a variety of applications where dynamic, light-responsive materials are required. Hu et al. [58] explored the extrusion 3D printing of a cellulose hydrogel skeleton, which opens new possibilities for creating eco-friendly and sustainable materials through 3D printing. Their work aligns with the growing demand for more environmentally responsible manufacturing practices. Substantial progress in advancing extrusion 3D bioprinting, with a specific focus on multicomponent hydrogel-based bioinks, was also demonstrated [59]. By pushing the boundaries of complexity and functionality in 3D-printed constructs, their research directly contributes to developing cutting-edge biomedical applications, from tissue engineering to personalized medicine.

Apart from experimental works, simulations for the extrusion 3D printing of chitosan hydrogels have also been reported [60]. Their research provides valuable insights into the virtual design and optimization of bioprinted structures, streamlining the development of precise and complex geometries for tissue engineering and other biomedical applications. Table 4 summarizes various extrusion-based 3D-printed hydrogels for diverse applications. Extrusion-based 3D printing for hydrogels offers versatility, cost-effectiveness, and the ability to work with various hydrogel formulations, making it a valuable tool for research, development, and manufacturing applications. However, moderate to lower resolution compared to other methods, potential nozzle clogging, and slower printing speeds due to layer-by-layer deposition limit its utility.

Table 4. Extrusion-based 3D-printed hydrogel varieties and their diverse applications.

Hydrogel Used	Applications/Remarks	Reference
Laponite® incorporated oxidized alginate–gelatin composite hydrogels	Integration of natural hydrogels into advanced manufacturing processes	Cai et al., (2021) [23]
Cellulose derivative	Biodegradable support material	Cheng, Y. et al., (2020) [53]
Gelatin-oxidized nanocellulose hydrogels	Bioprinting applications for tissue engineering	Zhou, S. et al., (2022) [54]

Table 4. Cont.

Hydrogel Used	Applications/Remarks	Reference
Gelatine methacrylate/Laponite nanocomposite hydrogel	High-concentration nanoclay for bone tissue regeneration	Dong, L. et al., (2021) [55]
Pre-crosslinked alginate	Advanced printable hydrogels	Falcone, G. et al., (2022) [56]
Photoresponsive polypeptide hydrogels	Stable constructs composed of photoresponsive polypeptide hydrogels	Murphy, R. D. et al., (2019) [57]
Cellulose hydrogel	Cellulose hydrogel skeleton using extrusion 3D printing of solution	Hu, X. et al., (2020) [58]
Multicomponent hydrogel-based bioinks	Bioprinting for various applications using multicomponent hydrogel-based bioinks	Cui, X. et al., (2020) [59]
Chitosan hydrogels	Simulations of extrusion 3D printing of chitosan hydrogels	Ramezani, H. et al., (2022) [60]

3.2. Inkjet-Based Printing

Inkjet-based 3D printing for hydrogels is a cutting-edge additive manufacturing technique that harnesses the precision of inkjet printing to construct three-dimensional structures using hydrogel materials. This innovative method is particularly well-suited for hydrogels, which often possess high water content and distinct rheological properties. The process begins with preparing a hydrogel ink tailored to the specific application, often containing crosslinking agents, water, and additives like cells or therapeutic compounds. Specialized inkjet 3D printers are equipped with printheads designed to accommodate these hydrogel inks. These printheads feature micro-sized nozzles that dispense minute droplets of the hydrogel ink onto a build platform. Each droplet represents a pixel; layer by layer, these droplets create the desired three-dimensional object. Following deposition, the hydrogel droplets are solidified, typically through methods such as UV light exposure or temperature adjustment. The technology is versatile, finding applications in biomedicine for creating tissue constructs, implants, and precise drug delivery systems. Additionally, it extends to various industries, enabling the customization of products with hydrogel components. In the field of 3D bioprinting, Suntornnond and colleagues [61] focused on improving the printability of hydrogel-based bioinks for thermal inkjet bioprinting applications and employed saponification and heat treatment processes to enhance the performance of these bioinks, aiming to optimize their application in creating complex biological structures.

Inkjet-based 3D printing technology also plays a vital role in fabricating microfluidic devices [62,63]. In the realm of advanced manufacturing and micro-optofluidic applications, Saitta et al. [64] employed a regression approach to model the refractive index measurements of innovative 3D-printable photocurable resins, a crucial step in micro-optofluidic applications. Marzano et al. [65] explored the potential of 3D printers and UV-cured optical adhesives in creating V-shaped plasmonic probes tailored for medical applications. Adamski et al. [66] have introduced an innovative technique for DNA analysis employing a lab-on-a-chip (LOC) system created through inkjet printing and on-chip gel electrophoresis. This novel capillary gel electrophoresis method is designed to separate genetic materials, primarily DNA and RNA, by examining the migration velocities of their fractions. The electrophoretic chip itself was produced using inkjet 3D printing. Two methods for fabricating the microfluidic electrophoretic chip are schematically depicted in Figure 5a,b. In the glass microengineering approach (Figure 5a), two glass substrates were bonded together to create the structure, including the injector and separation microchannels. Conversely, in the inkjet 3D printing method (Figure 5b), the entire structure was created in a single printing process on a chip. The LOC structure, as fabricated, encompasses an injection microchannel, two separation microchannels with lengths of 20 mm and 50 mm, and an integrated buffer and sample reservoir components. The chip's CAD design is presented in Figure 5c, which was then exported to a geometric file (Figure 5d), leading to the final structure illustrated in Figure 5e. To analyze DNA, the electrophoretic gel was injected into the chip's injection microchannel, subsequently entering the separation microchannel (Figure 5f–h). Finally, it underwent optical detection employing fluorescence modulation, accomplished with laser light and a CCD mini-camera, as shown in Figure 5i.

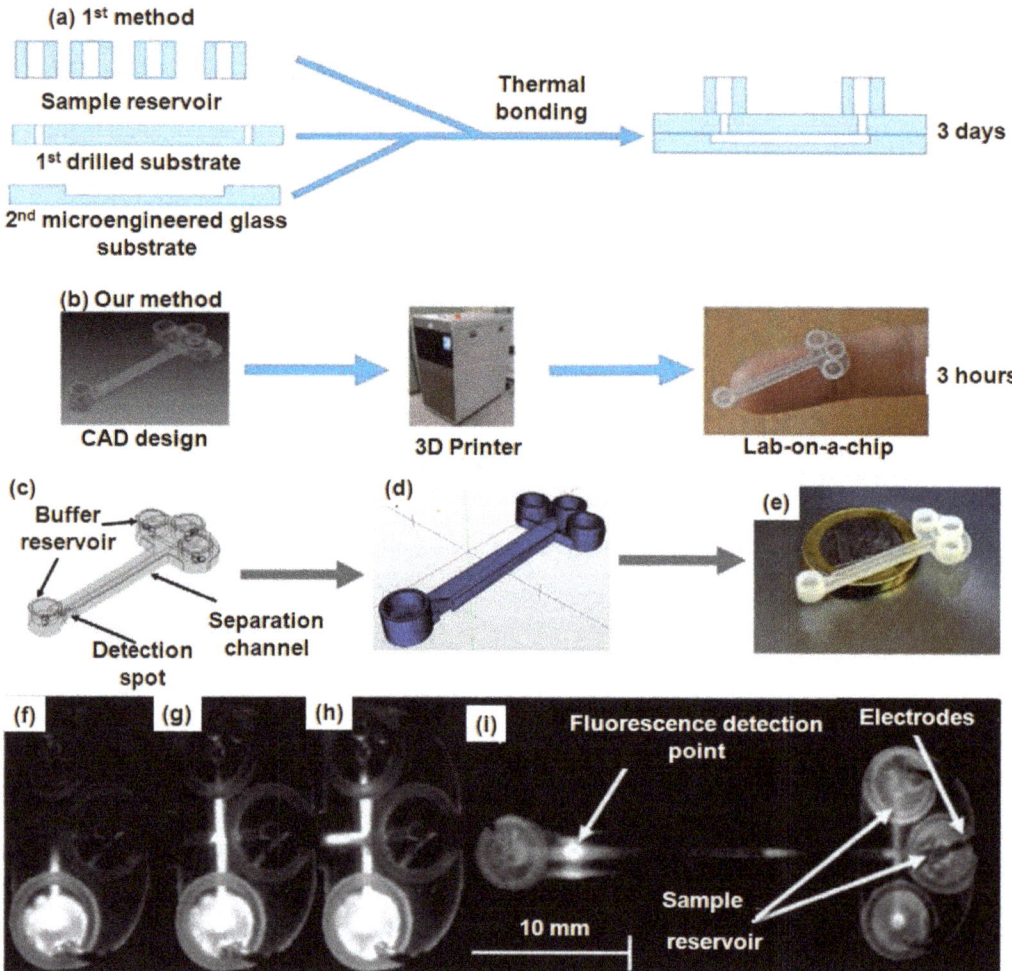

Figure 5. Sequential steps in the development and application of a microfluidic chip for gel electrophoresis. (a) Schematic presentation of the microfluidics chip manufacturing process through glass microengineering. (b) CAD design and the subsequent inkjet 3D printing of the lab-on-a-chip. (c) The original CAD design. (d) The exportation process leading to the creation of a geometric file. (e) The final structure offering a size comparison to a coin. In the investigation of DNA samples, (f–h) delineate the process of introducing an electrophoretic gel into the chip's injection microchannel and subsequently injecting it into the separation microchannel. (i) Depicts the optical detection method involving fluorescence modulation via a laser light and a CCD mini camera. Reprinted from [66] under a Creative Commons Attribution License 4.0 (CC BY) "https://creativecommons.org/licenses/by/4.0/ (accessed on 1 November 2023)".

Recent advancements in 3D inkjet printing technology have paved the way for fabricating intricate, cell-laden hydrogel structures, opening up new possibilities in tissue engineering, regenerative medicine, and drug delivery [67,68]. Teo et al. [69] contributed to this field by enabling the creation of free-standing 3D hydrogel microstructures through microreactive inkjet printing. Meanwhile, Nakagawa et al. [70] explored the use of star block copolymer hydrogels cross-linked with various metallic ions, demonstrating the versatility of inkjet printing in creating complex structures. In a study by Jiao et al. [71], inkjet printing was employed to fabricate alginate/gelatin hydrogels with tunable mechanical and biological properties, expanding the potential applications of these materials.

Yoon et al. [72] introduced an inkjet–spray hybrid printing approach for the 3D freeform fabrication of multilayered hydrogel structures, showcasing a versatile technique for complex hydrogel architectures. Peng et al. [73] demonstrated the surface patterning of hydrogels using ion inkjet printing, enabling programmable and complex shape deformations. Additionally, Duffy et al. [74] presented a 3D reactive inkjet printing method for poly-ε-lysine/gellan gum hydrogels, showing promise for potential corneal constructs. Table 5 summarizes various inkjet-based 3D-printed hydrogels for diverse applications.

Table 5. Inkjet-based 3D-printed hydrogels and their diverse applications.

Hydrogel Used	3D Printing Technology	Applications/Remarks	Reference
Itraconazole nanocrystals on hydrogel	Inkjet Printing	Ophthalmic drug delivery	Tetyczka, C. et al., (2022) [68]
Hydrogel-based bioinks	Thermal inkjet bioprinting	Printability improvement via saponification and heat treatment processes	Suntornnond, R. et al., (2022) [61]
Hyaluronic acid hydrogel	Photolithography and light-cured inkjet printing methods	Comparing different methods for creating hyaluronic acid hydrogel micropatterns	Chen, F. et al., (2022) [75]
Hydrogel	3D Inkjet Printing	Cell-laden structures Complex tissue engineering	Negro et al., (2018) [67]
Hydrogel	Microreactive Inkjet Printing	Free-standing 3D microstructures Microscale devices	Teo et al., (2019) [69]
Star block copolymer hydrogels	3D Inkjet Printing	Cross-linked with metallic ions Structural materials	Nakagawa et al., (2017) [70]
Alginate/gelatin hydrogel	Inkjet Printing	Mechanical and biological properties Tissue engineering	Jiao et al., (2021) [71]
Multilayered hydrogel	Inkjet–Spray Hybrid Printing	Multilayered hydrogel structures Multifunctional constructs	Yoon et al., (2018) [72]
Multilayered hydrogel	Ion Inkjet Printing	Surface patterning for shape deformations Programmable structures	Peng et al., (2017) [73]
Poly-ε-lysine/gellan gum hydrogels	Reactive Inkjet Printing	Corneal constructs Ophthalmic applications	Duffy et al., (2021) [74]

A comparative study involving hyaluronic acid hydrogel exquisite micropatterns using two distinct methods, photolithography and light-cured inkjet printing, to create intricate and precise microstructures has been reported by Chen et al. [75]. This research contributes to developing advanced techniques for fabricating microscale patterns with hyaluronic acid hydrogels, which are relevant in various biomedical and tissue-engineering applications. While inkjet-based 3D printing for hydrogels offers high resolution and intricate control, it also presents challenges related to material viscosity and nozzle maintenance. Nonetheless, it holds great promise for advancing research and development in fields demanding precise, biocompatible structures.

3.3. Stereolithography (SLA)

Stereolithography-based 3D printing for hydrogels represents a cutting-edge additive manufacturing technique that harnesses the power of light to create intricate three-dimensional structures using hydrogel materials. This technology builds upon the principles of traditional stereolithography but adapts them to accommodate the unique properties of hydrogels, which typically consist of water-based polymer networks. The process begins with a hydrogel resin that may contain photo-initiators, crosslinkers, and other additives. In this method, a precisely controlled light source, often in the form of a laser or projector, is used to selectively solidify specific regions of the hydrogel resin. As each layer is solidified, the build platform descends incrementally, allowing for the gradual construction of the 3D object. Stereolithography offers exceptional precision and fine detail, making it wellsuited for applications in tissue engineering, biomedical devices, and microfluidic systems. Postprinting, additional steps such as rinsing and curing may be required to enhance structural stability. In the realm of 3D printing, various studies have explored the application of stereolithography to create hydrogels with unique properties and capabilities. Kalossaka and colleagues [76] delved into the creation of 3D nanocomposite hydrogels featuring lattice vascular networks, demonstrating the potential for engineering intricate structures using this technique. Karakurt et al. [77] focused on SLA 3D printing of hydrogels loaded with ascorbic acid, investigating their controlled release properties. The study sheds light on the use of SLA in drug delivery systems.

The impact of SLA 3D printing on the properties of PEGDMA hydrogels, contributing insights into the interactions between the printing process and hydrogel characteristics, has also been discussed [78]. Magalhães et al. [79] explored the use of low-cost stereolithography technology to create 3D hydrogel structures, offering potential applications in biomedicine and biomaterials. Effects of PEGDA photopolymerization in microstereolithography on 3D-printed hydrogelstructures, addressing aspects of structural integrity and swelling behavior, havebeen reported by Alketbi et al. [80]. Sun et al. [81] introduced a new stereolithographic 3D printing strategy for hydrogels, focusing on achieving a large mechanical tunability and self-weldability, which opens doors to more versatile and functional hydrogel applications. Table 6 summarizes various stereolithography-based 3D-printed hydrogels for diverse applications.

Table 6. Stereolithography-based 3D-printed hydrogels and their diverse applications.

Hydrogel Used	Applications/Remarks	Reference
Nanocellulose/PEGDA	3D cell culture, tissue engineering	Tang, A. et al., (2019) [29]
Nanocomposite hydrogels	Creating hydrogel structures with vascular networks	Kalossaka, L.M. et al., (2021) [76]
Ascorbic acid-loaded hydrogels	Controlled drug release from 3D-printed hydrogels	Karakurt, I. et al., (2020) [77]
PEGDMA hydrogels	Investigating the impact of 3D printing on hydrogel properties	Burke, G. et al., (2020) [78]
Hydrogel structures	Low-cost 3D printing of hydrogel structures	Magalhães, L.S.S. et al., (2020) [79]
PEGDA hydrogels	Studying the impact of photopolymerization in micro-stereolithography	Alketbi, A.S. et al., (2021) [80]
Hydrogels with tunability	Developing hydrogels with mechanical tunability and self-welding properties	Sun, Z. et al., (2022) [81]

Stereolithography-based 3D printing for hydrogels is celebrated for its ability to create complex, high-resolution structures, but challenges such as limited material options and the need for specialized equipment are considerations in its implementation. Nevertheless, it remains a powerful tool in the field of hydrogel-based additive manufacturing, enabling the fabrication of advanced biomimetic structures for a wide range of applications.

3.4. Digital Light Processing (DLP)

DLP-based 3D printing for hydrogels is an innovative and precise additive manufacturing technique that harnesses digital light projection to construct intricate 3D structures using hydrogel materials. This method builds upon the principles of photopolymerization, where a liquid resin containing hydrogel components is selectively cured by a digital light source, typically a projector or UV LED. This process proceeds layer by layer, with each layer being solidified by the precise projection of light patterns, forming the desired object. DLP 3D printing for hydrogels is celebrated for its remarkable speed, high resolution, and ability to create complex geometries with fine detail. It finds particular utility in biomedical applications, including tissue engineering and the fabrication of customized implants and drug delivery systems.

Recent advancements in the DLP 3D printing of hydrogels have ushered in a new era of possibilities and applications. Several research groups have made significant contributions to this evolving field. Hosseinabadi and collaborators [82] delved into the critical aspects of ink material selection and optical design in DLP 3D printing, underlining the importance of precision and performance in hydrogel printing. A concise review summarizing the key developments and prospects of this technology, providing an overview of the current state of DLP 3D printing for hydrogels, has been presented by Ding et al. [83]. Sun et al. [84] introduced a novel approach to DLP 3D printing by presenting hydrogels with hierarchical structures enhanced through lyophilization and ionic locking, offering new possibilities for tailored hydrogel properties. In a different direction, Dong and colleagues [85] focused on creating tough supramolecular hydrogels using DLP 3D printing, emphasizing their potential application as impact-absorption elements. An effective DLP 3D printing strategy was also introduced for cellulose hydrogels with high strength and toughness, wellsuited for strain sensing applications [86]. Cafiso et al. [87] explored 3D printing of fully cellulose-based hydrogels through DLP, highlighting the sustainability and versatility of this approach. On the biomedical front, Wang et al. [88] successfully fabricated antimicrobial hydrogels using DLP 3D printing, leveraging sustainable resin and hybrid nanospheres for potential applications in healthcare.

Caproili and colleagues [89] introduced a novel 3D-printed hydrogel imbued with self-healing properties, employing readily available materials and a DLP printer commonly found in the market. The ingenious design of this system involved creating a sequential semi-interpenetrated network by introducing chemical covalent network precursors into a solution containing a linear polymer. Following polymerization, this linear polymer is effectively encapsulated within the cross-linked matrix. The photocurable ink formulation involved the blending of an aqueous solution of unmodified, non-crosslinked Poly (vinyl alcohol) (PVA) with acrylic acid (AAc), the cross-linking agent Poly (ethylene glycol) diacrylate (PEGDA), and a water-compatible photoinitiator based on diphenyl (2,4,6-trimethylbenzoyl)phosphine oxide (TPO). Figure 6a provides a visual representation of the chemical structures of the initiator, monomer, cross-linker, and mending agent within the photocurable resin. Figure 6b illustrates the creation of the semi-interpenetrated network: a physical network is blended with the precursors of the chemical network, which materializes during exposure to light. The outcome is a hydrogel composed of PVA chains that are uniformly distributed and incorporated into a cross-linked acrylic matrix. For tensile testing, two distinct specimens were produced, each featuring a different color, one printed with methyl red sodium salt dye and the other with brilliant green dye. These samples demonstrated their capacity to endure bending deformation immediately upon rejoining, as depicted in Figure 6c. Figure 6d showcases a perforated cylindrical structure printed with methyl red sodium salt dye, clearly demonstrating that the reconnected sample could withstand stretching deformation after a 2h healing process. Figure 6e reveals body-centered cubic lattice-like structures printed with both methyl red sodium salt dye and brilliant green dye, enhancing visual comprehension. The inset in this figure vividly portrays diffusion at the interface after 12 h of contact, owing to the gradient of the dye. Figure 6f,g display 3D fabricated samples with a PVA 0.8 formulation, featuring a body-centered cubic lattice-like structure printed with methyl red sodium salt dye and an axis-symmetric structure with a central pillar printed with brilliant green dye, respectively. Finally, Figure 6h offers insight into the stress–strain curves of the self-healed hydrogels, showcasing their performance over increasing healing durations.

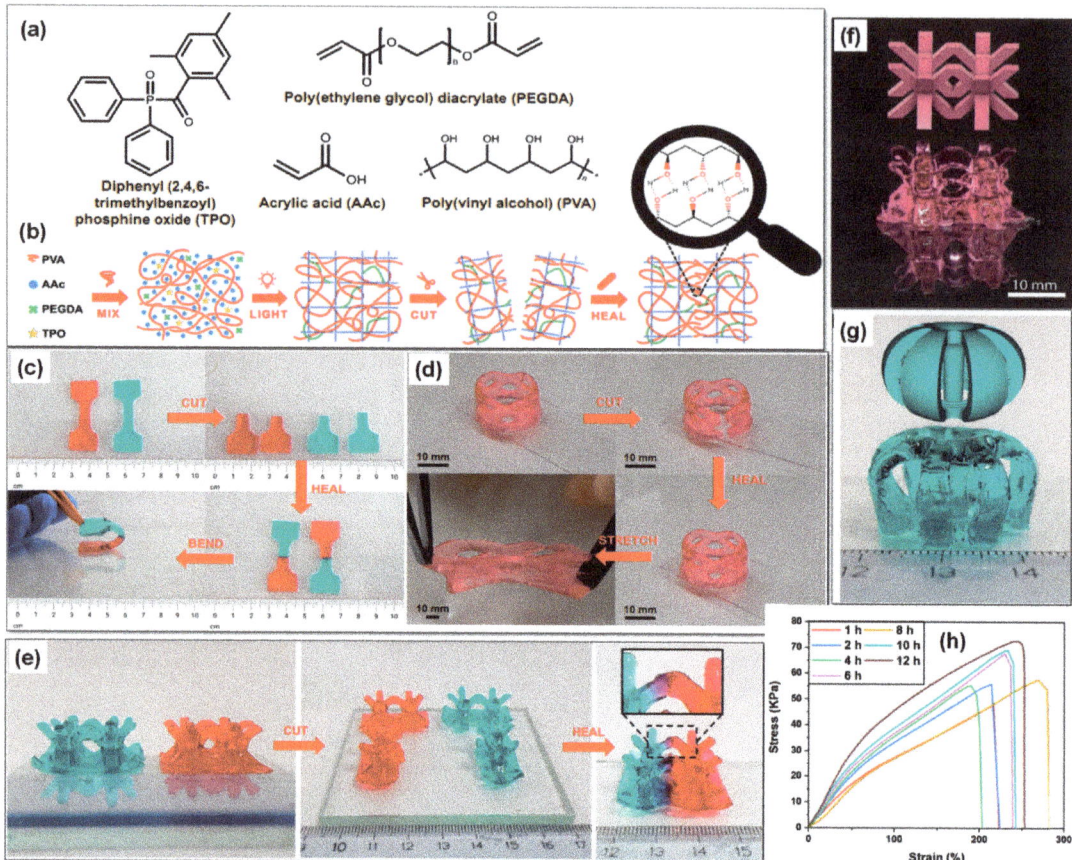

Figure 6. Composition formulation and network formation. (**a**) Molecular structure of initiator, monomer, cross-linker, and mending agent within the photocurable resin. (**b**) Schematic depiction of semi-IPN and the healing process. Illustrated recovery of cut and rejoined objects printed with an AAc/PVA ratio of 0.8. (**c**) Tensile test samples featuring methyl red sodium salt and brilliant green dye. These specimens exhibited immediate bending deformation upon rejoining. (**d**) A cylindrical structure with holes printed using methyl red sodium salt dye. The reconnected sample demonstrated stretching deformation endurance after a 2h healing process. (**e**) Lattice-like structures with a body-centered cubic pattern, printed with methyl red sodium salt and brilliant green dye. Noticeable dye diffusion at the interface is visible after 12 h (inset) due to the dye concentration gradient. (**f**,**g**) 3D-printed samples with the PVA 0.8 formulation: (**f**) body-centered cubic lattice-like structure printed with methyl red sodium salt dye. (**g**) Axisymmetric structure featuring a central pillar, printed with brilliant green dye. (**h**) Stress–strain curves of self-healed hydrogels over rising healing periods. Reprinted from [89] under a Creative Commons Attribution License 4.0 (CC BY) "https://creativecommons.org/licenses/by/4.0/ (accessed on 1 November 2023)".

Zhang et al. [90] experimented with Antheraea pernyi silk fibroin bioinks for DLP 3D printing, offering new opportunities in bioprinting and regenerative medicine. In another innovative stride, Xiang et al. [91] focused on creating 3D-printed high-toughness double network hydrogels via DLP, contributing to the development of mechanically robust hydrogel materials. Lopez-Larrea et al. [92] explored the application of PEDOT-based photopolymerizable inks for biosensing through DLP 3D printing, marking the potential for enhanced biosensor development. Table 7 summarizes various DLP-based 3D-printed hydrogels for diverse applications.

Table 7. DLP-based 3D-printed hydrogels and their diverse applications.

Hydrogel Used	Applications/Remarks	Reference
Hydrogels	Creating hydrogels with hierarchical structures	Sun, Z. et al., (2023) [84]
Supramolecular hydrogels	Tough supramolecular hydrogels with sophisticated architectures as impact-absorption elements	Dong, M. et al., (2022) [85]
Cellulose hydrogel	Cellulose hydrogel for strain sensing	Guo, Z. et al., (2023) [86]
Cellulose-based hydrogels	3D printing of fully cellulose-based hydrogels using DLP	Cafiso, D. et al., (2022) [87]
Antimicrobial hydrogel	Developing an antimicrobial hydrogel using sustainable resin and hybrid nanospheres	Wang, L. et al., (2022) [88]
Antheraea pernyi silk fibroin	Using silk fibroin bioinks for DLP 3D printing	Zhang, X. et al., (2023) [90]
Double-network hydrogels	3D printing of high-toughness double network hydrogels	Xiang, Z. et al., (2022) [91]
PEDOT-based photopolymerizable inks	DLP 3D printing of PEDOT-based photopolymerizable inks for biosensing	Lopez-Larrea, N. et al., (2022) [92]

It is noteworthy to mention here that in the case of DLP-based 3D printing technology, post-processing steps, including rinsing and additional curing, may be required after printing to enhance structural stability. However, it is important to note that material selection and compatibility with the photopolymerization process are key considerations. Despite these challenges, DLP-based 3D printing stands as a powerful tool for creating intricate hydrogel structures, pushing the boundaries of precision and complexity in various scientific and medical fields. Table 8 summarizes the principle and various properties of 3D printing technologies, including extrusion-based, inkjet-based, SLA, and DLP.

Table 8. Comparative summary of various aspects of 3D printing technologies in terms of principle, resolution, material compatibility, speed, post-processing, advantages, and limitations.

	Extrusion-Based	Inkjet-Based	SLA	DLP
Principle	Layer-by-layer extrusion	Droplet deposition	Photopolymerization	Photopolymerization
Resolution	Moderate	High	Very High	Very High
Material Compatibility	Wide range of hydrogel materials	Limited to specific hydrogel formulations	Limited choice due to compatibility	Limited choice due to compatibility
Speed	Moderate	Moderate to Fast	Moderate to Fast	Very Fast
Post-Processing	Often minimal	May require curing	Minimal	Minimal
Advantages	- Versatile for various hydrogels. - Cost-effective and simple. - Well-suited for large-scale objects. - Handles high-viscosity inks well.	- Precise control. - Multi-material printing capabilities. - Suitable for biomedicine and research. - Excellent for fine-detail structures.	- Exceptional resolution and surface finish. - Complex and intricate geometries possible. - Fast printing speed for high-detail objects.	- Fast printing speed, suitable for large-scale production. - Capable of intricate geometries with precision.
Limitations	- Viscosity can lead to nozzle clogging. - Challenges with intricate structures. - Limited in applications requiring high precision.	- Prone to nozzle clogging, especially with viscous inks. - May require crosslinking. - Limited in creating large-scale objects.	- UV light exposure may affect cell viability. - Specialized equipment with high upfront cost. - Potential resin waste if not fully used.	- UV light exposure may affect cell viability. - Requires specialized equipment and UV light source. - Potential resin waste if not fully used.

4. Applications of 3D-Printed Hydrogels

4.1. Role of a 3D-Printed Hydrogel in Biomedical Applications

Hydrogel 3D printing has emerged as a groundbreaking technology in the field of biomedical engineering, offering innovative solutions in tissue engineering, regenerative medicine, and drug delivery [93–97]. By harnessing the unique properties of hydrogels, this approach enables the creation of precise and complex structures that closely mimic the natural extracellular matrix. By creating intricate structures with tailored properties, this technology opens avenues for personalized medicine, faster wound healing, and the development of advanced organ-on-a-chip systems, ultimately reshaping the landscape of biomedical research and healthcare.

4.1.1. Tissue Engineering

Hydrogel 3D printing plays a pivotal role in fabricating scaffolds and constructs for tissue engineering, providing a supportive framework for cells to adhere, proliferate, and differentiate, ultimately aiding in regenerating damaged tissues or organs. Lan et al. [98] provided a comprehensive review of progress in 3D printing for bone tissue engineering, highlighting the potential for personalized and effective solutions in orthopedics. Varaprasad et al. [99] underscored the importance of alginate as a biomacromolecule for 3D-printing hydrogels in biomedical applications, focusing on its potential for creating bioactive scaffolds.

Xu et al. [100] presented a novel glucose-responsive antioxidant hybrid hydrogel designed to enhance diabetic wound repair, highlighting its potential in addressing critical healthcare challenges. Tsegay and colleagues [101] provided a review on smart 3D-printed hydrogelskin wound bandages, emphasizing the potential of these advanced wound dressings for improved patient care and recovery. Hydrogel 3D printing is also used to create microfluidic devices that mimic the structure and function of human organs, allowing researchers to study diseases and drug responses in a controlled environment. These systems provide insights into organ-level behavior and drug efficacy without the need for animal testing. Hydrogel-based constructs can be engineered with vascular networks, enabling the perfusion of nutrients and oxygen to cells deep within the tissue. This advancement is crucial for creating thicker tissues and organs, addressing one of the challenges in tissue engineering.

Guo et al. [102] introduced a groundbreaking 3D liver-inspired detox device. It is madeusing3D-printing hydrogels with special nanoparticles that attract, capture, and sense toxins. This innovative system efficiently traps toxins using a liver-like microstructure. Their work shows that this device effectively neutralizes toxins, offering a promising path for advanced detoxification platforms. Leveraging the benefits of a 3D biomimetic structure for enrichment, separation, and detection, a bioinspired 3D detoxification device has been devised. This innovative device incorporates polydiacetylene (PDA) nanoparticles into a precisely designed 3D matrix resembling a modified liver lobule configuration, achieved through advanced 3D printing technology known as dynamic optical projection stereolithography (DOPsL). DOPsL employs a digital mirror array device (DMD) to create dynamic photomasks, allowing for the layer-by-layer photopolymerization of biomaterials into intricate 3D structures.

Figure 7a visually illustrates the integration of PDA nanoparticles (green) within a PEGDA hydrogel matrix (grey) featuring a liver-mimicking 3D structure produced through 3D printing. This structure can attract, capture, and detect toxins (red), while the modified liver lobule-like matrix efficiently traps these harmful substances. This biomimetic 3D detoxifier holds promise for clinical applications, serving as an effective means of collecting and removing toxins. Furthermore, they have successfully immobilized toxins within 3D-printed hydrogel nanocomposites, each sporting distinct surface patterns in the form of flower-like shapes with varying diameters (large, medium, and small). When exposed to a melittin solution, as depicted in Figure 7b–d, the red fluorescence indicates interactions between PDA and melittin, offering insights into toxin localization.

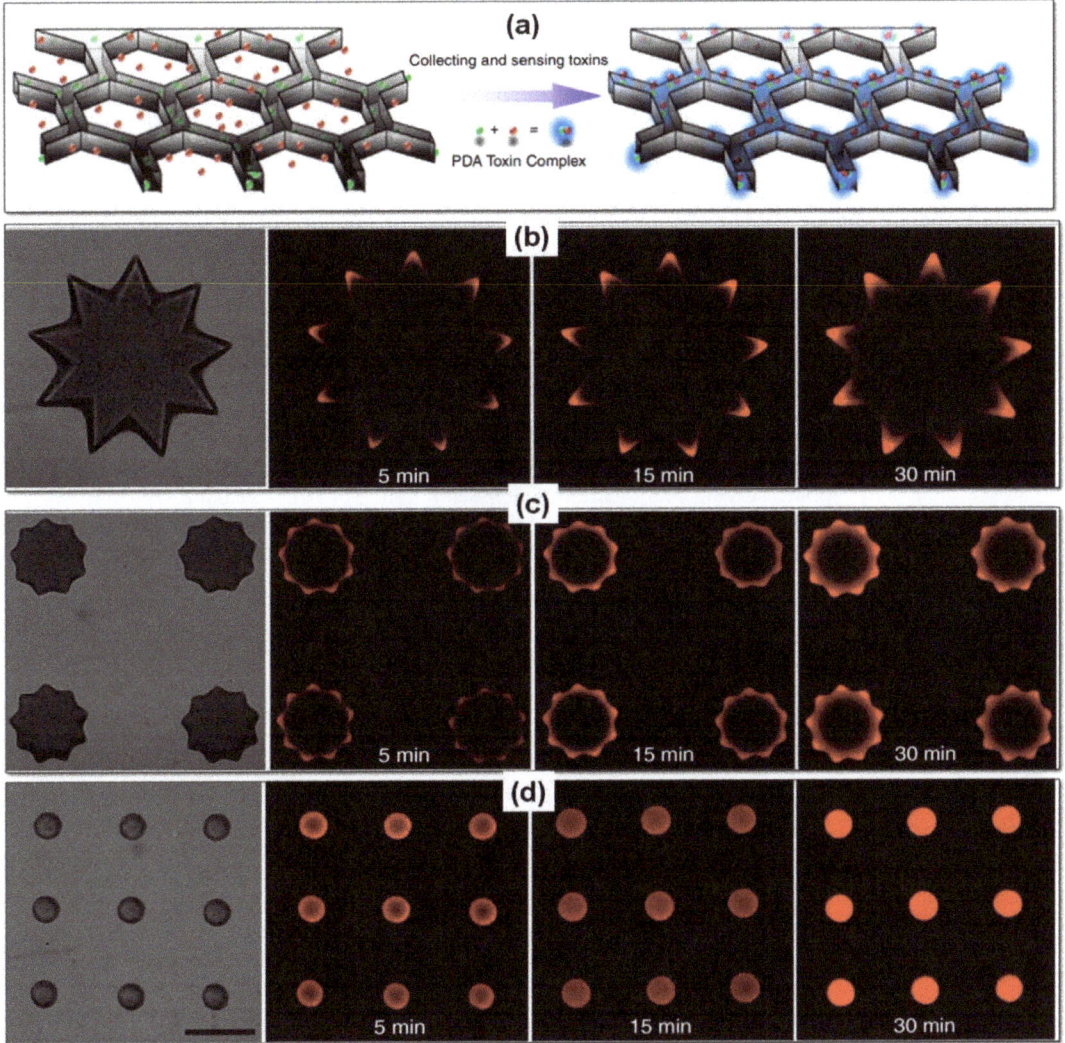

Figure 7. (**a**) PDA nanoparticles (green) embedded within a PEGDA hydrogel matrix (grey) to create a liver-mimetic 3D structure through 3D printing. Nanoparticles possess the capability to attract, capture, and sense toxins (red), while the modified liver lobule structure within the 3D matrix allows for efficient toxin entrapment. (**b–d**) Toxin capture within 3D-printed hydrogel nanocomposites featuring various surface patterns. Three distinct types of 3D structural posts, sharing a similar flower-like form and length but differing diameters, undergo incubation at 37 °C with a melittin solution (50 mg/mL). (**b**) Demonstrates a post with a large diameter. (**c**) Represents the medium-diameter post. (**d**) Features the narrow-diameter post. The red fluorescence highlights the interactions between PDA and melittin, marking the toxic region. A scale bar of 200 mm is provided for reference. Reprinted from [102] under a Creative Commons Attribution License 4.0 (CC BY) "https://creativecommons.org/licenses/by/4.0/ (accessed on 1 November 2023)".

Hydrogel-based dressings with 3D-printed structures provide an ideal environment for wound healing. These dressings can maintain a moist and protective environment, promote cell migration, and facilitate tissue regeneration. For example, 3D-printed hydrogel dressings can cover burn wounds, providing a barrier against infection, enhancing wound healing, and reducing pain and scarring. Hydrogel-based wound dressings can be infused with growth factors and antimicrobial agents, accelerating healing in chronic wounds like diabetic ulcers. Xiong et al. [103] fabricated DLP-based 3D-printed wearable sensors possessing self-adhesion and self-healing properties using polymerizable rotaxane hydrogels. In their study, acrylated β-cyclodextrin combined with bile acid is self-organized into a polymerizable pseudorotaxane through precise host–guest recognition. This pseudorotaxane is subsequently photopolymerized alongside acrylamide to create conductive polymerizable rotaxane hydrogels (PR-Gel). PR-Gel exhibited robust self-adhesion to the skin, as depicted in Figure 8a, and notably, they left no residue or caused any allergic reactions when adhered to the wrist. Lap-shear tests were conducted to assess their adhesion to various surfaces, including nitrile gloves, glass plates, silicone rubber, aluminum, and pigskin (Figure 8b). Although the adhesion strength between PR-Gel and these substrates fell within the range of 4.5 to 6.0 kPa, which is lower than that of some highly adhesive hydrogels, this moderate self-adhesion ensured a secure bond between PR-Gel and human skin, as demonstrated in Figure 8a. This feature guarantees stable signal transmission and consistent adhesion for wearable sensor applications. To showcase the self-healing capabilities, two cylindrical PR-Gel samples were cut and then reassembled at the incision point, serving as a qualitative demonstration. For better visual distinction, one of the samples was stained in a red-orange hue, as depicted in Figure 8c. These mended gels exhibited resilience to tensile deformation without fracturing at room temperature. The self-healing prowess of PR-Gel underwent further scrutiny through structural damage and recovery assessments involving continuous strain sweeps with alternating low (1%) and high (500%) oscillatory excitations. In Figure 8d, the storage modulus (G') and loss modulus (G'') of PR-Gel during continuous strain sweeps illustrate their rapid self-healing properties post-damage. This process could be repeated multiple times, underscoring the reproducible restoration of the PR-Gel network.

4.1.2. Regenerative Medicine

The 3D-printed hydrogels have emerged as a transformative technology with significant implications for regenerative medicine. By combining the precision of 3D printing with the versatility of hydrogels, researchers and clinicians can fabricate intricate structures that mimic the complexity of biological tissues. This innovative approach enables the creation of customized scaffolds tailored to match the specific geometry and mechanical properties of target tissues. The hydrogel's biocompatibility and ability to encapsulate cells make it an ideal candidate for supporting cellular growth and tissue regeneration. Moreover, the controlled release of bioactive molecules from these 3D-printed hydrogels further enhances their regenerative potential. Applications span a wide range, from engineering bone and cartilage to creating vascularized tissues. As the field advances, 3D-printed hydrogels hold promise for revolutionizing regenerative medicine, offering new avenues for personalized and effective tissue repair and regeneration. Tajik et al. [97] critically reviewed the 3D printing of hybrid hydrogel materials for tissue engineering, underlining the significance of this approach in advancing regenerative medicine.

Deptula et al. [104] reported various diverse applications of 3D-printed hydrogels in the domains of wound healing and regenerative medicine. Notably, their research highlights the creation of skin wound bandages through 3D-printed hydrogels and the integration of these hydrogels into tissue-engineering strategies. The work of Bhatnagar et al. [105] emphasized the significance of hydrogels in regenerative medicine, showcasing their potential for tissue repair and regeneration. Nanoengineered biomimetic hydrogels, a major advancement detailed by Cernencu et al. [106], have enabled the fabrication of 3D-printed constructs, offering a promising avenue for regenerative medicine applications. Natural hydrogels, as reviewed by Catoira et al. [107], contribute to the overview of materials used in regenerative medicine, emphasizing their relevance in the field. The comprehensive exploration of hydrogels in regenerative medicine, including their applications and challenges, is presented in the work by Hasirci et al. [108]. Aghamirsalim et al. [109] delved into the specific use of 3D-printed hydrogels for ocular wound healing, showcasing the versatility of this technology. Shamma et al. [110] focused on the use of triblock copolymer bioinks, specifically pluronic F127, in hydrogel 3D printing for regenerative medicine. Diverse applications of hydrogels in regenerative medicine and other biomedical fields have also been reviewed [111]. The engineering of hydrogels for personalized disease modeling and regenerative medicine wasdiscussed by Tayler and Stowers [112], highlighting the potential for tailored therapeutic approaches. Heo et al. [113] presenteda remarkable application, where 3D-printed microstructures incorporated with hybrid nano hydrogels enhance bone tissue regeneration. Foyt et al. [114] focused on exploiting advanced hydrogel technologies to address key challenges in regenerative medicine, showcasing the continuous evolution of this field.

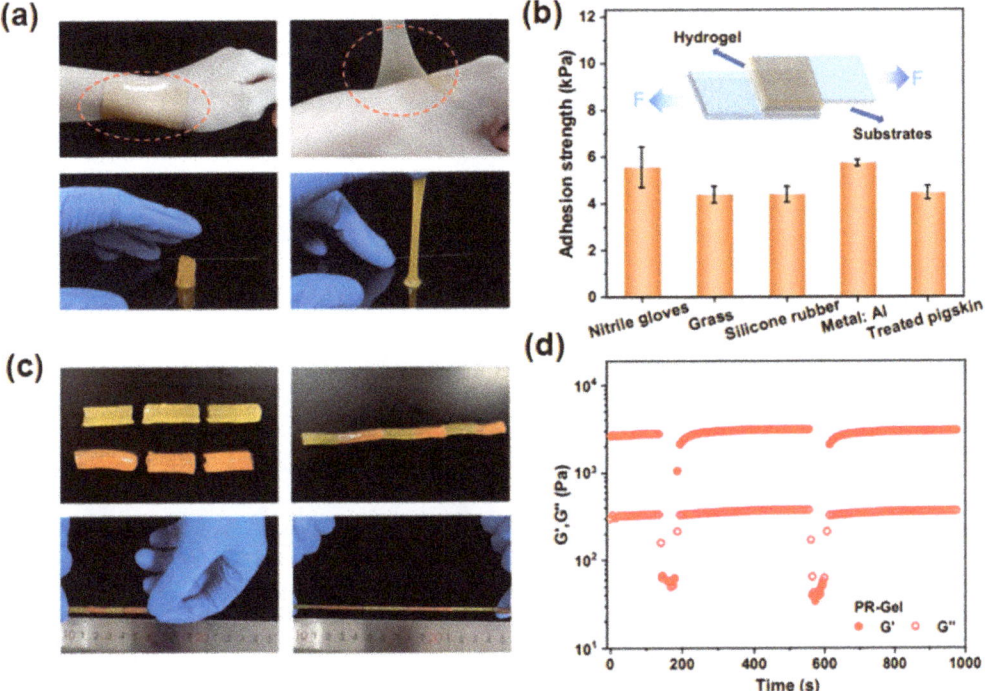

Figure 8. Shear adhesive strength and self-healing: (**a**) photograph of PR-Gel (40 w/v%) stuck onto wrist joints and a glass surface. (**b**) Represents the shear adhesive strength of PR-Gel (40 w/v%) on various substrates, including a nitrile glove, glass plate, silicone rubber, aluminum, and pigskin. The inset offers a schematic illustration of the lap-shear test conducted at 25 °C. (**c**) Self-healing behavior of the PR-Gel (40 w/v%) at 25 °C, with the visualization aided by rhodamine B. (**d**) Storage modulus (G') and loss modulus (G'') of the PR-Gel (40 w/v%) under alternating strains, oscillating between small (1%) and large strain (500%) at 25 °C. Reprinted from [103] under a Creative Commons Attribution License 4.0 (CC BY) "https://creativecommons.org/licenses/by/4.0/ (accessed on 1 November 2023)".

4.1.3. Drug Delivery

The 3D-printed hydrogelalso offers precise control over the spatial distribution of drug-loaded hydrogel matrices. This technology enables the design of drug delivery systems that release therapeutic agents in a controlled and sustained manner, improving treatment efficacy and reducing side effects. Hydrogel-based implants can be 3D printed with encapsulated drugs or growth factors, allowing for site-specific and controlled release. This approach is used in cancer therapy, wound healing, and bone regeneration. Oral drug delivery systems can also be created using a 3D-printed hydrogelthat releases drugs gradually, enhancing bioavailability and reducing the frequency of dosing. Patient-specific drug formulations can also be achieved through 3D printing, tailoring drug release profiles to individual patient needs, which is particularly relevant in cases of chronic diseases.

Tetyczka et al. [100] examined the integration of itraconazole nanocrystals on hydrogel contact lenses via inkjet printing, demonstrating the implications of this technique for ophthalmic drug delivery. Maiz-Fernández et al. [115] introduced self-healing hyaluronic acid/chitosan polycomplex hydrogels with drug release capabilities, opening avenues for customized and responsive medical interventions. Larush et al. [116] explored the potential of 3D printing responsive hydrogels for drug delivery, emphasizing the versatility of this approach. Aguilar-de-Leyva et al. [117] delved into the development of 3D-printed drug delivery systems using natural products, aligning with the growing trend toward sustainable and bio-inspired pharmaceutical solutions. Martinez et al. [118] focused on stereolithographic 3D printing to fabricate drug-loaded hydrogels, presenting a precise and controlled drug-delivery platform. NIR-triggered drug release from 3D-printed hydrogel/PCL core/shell fiber scaffolds, showcasing the potential for localized cancer therapy and wound healing, has also been reported [119,120]. Dreiss [121] provided insights into hydrogel design strategies for drug delivery, offering a comprehensive overview of the evolving landscape in this domain. Wang et al. [122] contributed to the field by 3D printing shape memory hydrogels with internal structures for drug delivery, adding a layer of sophistication to the design of drug-release systems.

4.2. Hydrogel 3D Printing across Industries

While hydrogel 3D printing has gained significant traction in biomedical applications, its versatility extends to diverse industries beyond healthcare. The unique properties of hydrogels, such as high water content and tunable mechanical properties, make them intriguing candidates for various applications in industries like food, cosmetics, and electronics.

4.2.1. Food Industry

In the food industry, 3D-printed hydrogels offer a range of applications, revolutionizing the way we produce and consume food. This technology enables the creation of customized food structures with unique textures and shapes while also allowing for precise control over the release of nutrients and flavors. The ability to tailor mechanical properties enhances the overall eating experience, mimicking desired textures in food products. Moreover, 3D printing facilitates the integration of functional ingredients, such as probiotics and antioxidants, in a spatially controlled manner. The technology's potential to reduce food waste, streamline processing steps, and contribute to personalized nutrition makes it a promising innovation in the quest for sustainable and innovative food solutions. Chefs can leverage 3D printing for artistic culinary designs, while companies benefit from rapid prototyping for quicker product development.

In the rapidly evolving landscape of food technology, hydrogels derived from natural polymers have emerged as versatile materials with transformative applications. Zhang et al. [123] extensively explored the diverse applications of natural polymer-based hydrogels in the food industry, detailing their functionalities and contributions. Meanwhile, the intersection of 3D printing and food production has been exemplified by Park et al. [124] through innovative work involving callus-based 3D printing, particularly using carrot tissues. This approach showcases the potential for revolutionizing food manufacturing processes. Another promising avenue is the application of hydrogels based on ozonated cassava starch, as investigated by Maniglia et al. [125]. Their research focused on the impact of ozone processing and gelatinization conditions, particularly in the context of enhancing 3Dprinting applications. Additionally, the exploration of pectin hydrogels in food applications has been comprehensively reviewed by Ishwarya and Nisha [126], shedding light on the recent advances and future prospects in leveraging pectin hydrogels for diverse food-related functionalities.

4.2.2. Cosmetic Industry

In the cosmetics industry, hydrogel 3D printing offers the potential to develop novel skincare products with enhanced absorption and release properties. Customizable shapes and designs enable the creation of cosmetics that adhere better to the skin. Several cosmetic industries are using hydrogel 3D printing to create intricate face masks that fit snugly on the face, enhancing the delivery of skincare ingredients and improving the overall application experience. Finny et al. [127] delved into the realm of sun protection with their work on 3D-printed hydrogel-based sensors designed for quantifying UV exposure. This innovation holds immense potential for personalized UV monitoring. Diogo et al. [128] explored the engineering of hard tissues using cell-laden biomimetically mineralized shark-skin-collagen-based 3D-printed hydrogels, offering a promising avenue for developing advanced biomedical materials. The intersection of 3D printing and dermatology is evident in the work of de Oliveira et al. [129], who focused on 3D-printed products for topical skin applications, ranging from personalized dressings to drug delivery systems. Othman et al. [130] contributed to the field of cosmetic science by developing easy-to-use and cost-effective sensors for assessing the quality and traceability of cosmetic antioxidants, addressing a critical need in the cosmetic industry.

4.2.3. Electronics

Hydrogel 3D printing is also being explored for applications in electronics due to its unique combination of properties, including high water content and the potential for conductive or responsive behavior. This technology can enable the creation of soft, flexible electronics. Hydrogel-based sensors with 3D-printed structures can be used in wearable devices to monitor physiological parameters, such as heart rate and hydration levels. Guo et al. [131] introduced the 3D printing of electrically conductive and degradable hydrogel for epidermal strain sensors, which hold promise in wearable electronics.

Liu et al. [132] introduced a cost-effective, specialized 3D printing technology, direct ink writing (DIW), for creating hydrogel sensors enhanced with two-dimensional transition metal carbides (MXenes). These sensors exhibit exceptional strain and temperature sensing capabilities for shape memory solar array hinges. In their approach, Figure 9a illustrates the MXenes preparation process, which involves a fluoride-based salt etching system to minimize the risks associated with concentrated hydrogen fluoride (HF) dilution. Notably, this process uses LiF crystals dissolved in a hydrochloric acid (HCl) aqueous solution to gradually form HF with lower concentration under low-temperature conditions. This mixture is then blended with Ti_3AlC_2 powders to create a black suspension, ensuring sufficient etching without excessive oxidation.

Subsequent steps involve washing, intercalation, sonication, and argon protection to obtain delaminated $Ti_3C_2T_x$ flakes. Subsequently, extended centrifugation was employed to separate the delaminated $Ti_3C_2T_x$ flakes from the multi-layered aggregates. Given the presence of electronegative groups on $Ti_3C_2T_x$, polymer hydrogel networks were established, as depicted in Figure 9b. The typical procedure involved the even dispersion of lyophilized $Ti_3C_2T_x$ flakes, followed by the addition of glycerol and polymers. The resulting black mixture was heated to facilitate the dissolution of PVA under an argon (Ar) atmosphere, preventing unnecessary oxidation of $Ti_3C_2T_x$ due to high temperatures. The resulting liquid precursor was then injected into a syringe and utilized in the DIW printing process. The SEM images provided in Figure 9c,d depict the $Ti_3C_2T_x$ flakes displaying a sleek 2D nanosheet morphology, with lateral sizes ranging from 0.4 to 1 μm. As shown in the atomic force microscope (AFM) in Figure 9e, these $Ti_3C_2T_x$ flakes exhibit a smooth surface with an average thickness of 1.6 nm, consistent with the thickness of a $Ti_3C_2T_x$ bilayer structure. To examine the intricate internal microstructure of the hydrogel networks, the printed hydrogel underwent a process of lyophilization following solvent replacement.

As illustrated in Figure 9f,g, the unaltered printed hydrogel showcases a porous network structure characterized by random orientation and a hierarchical distribution of pore sizes. The cross-sectional morphology of the printed hydrogel, along with its corresponding energy-dispersive X-ray spectroscopy (EDS) elemental mapping, is displayed in Figure 9h, confirming the uniform distribution of elements C, O, and N. Analyzing the EDS element distribution, it becomes evident that the relative content of N elements is lower than that of C and O. This indicates that the hard domains, comprised of nitrogen-containing groups within the polyurethane (PU) network, are less abundant than the soft domains, which in turn impacts the mechanical strength of the hydrogel. Similarly, when 0.25 wt% $Ti_3C_2T_x$ flakes are introduced into the precursor, the resulting printed hydrogel exhibits an unoriented porous framework with random pore size distribution, resembling that of the pristine hydrogel, as depicted in Figure 9i,j. Furthermore, EDS analysis of elements C, O, and Ti (Figure 9k) confirms the homogeneous distribution of Ti_3C_2Tx flakes within the hydrogel.

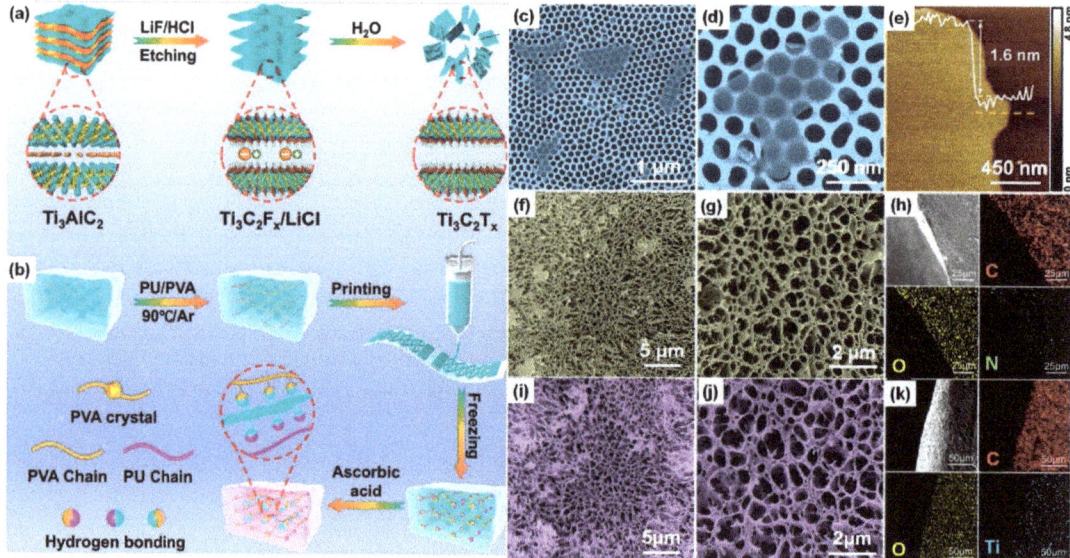

Figure 9. Synthesis of MXenes and hydrogel: (**a**) pictorial representation of the synthesis process of Ti_3C_2Tx flakes. (**b**) Demonstration of DIW printing of Mxenes-attached polyurethane/polyvinyl alcohol (PU/PVA) hydrogel. Morphology and structure of materials: (**c,d**) SEM morphology and (**e**) AFM image of Ti_3C_2Tx flakes. (**f,g**) Microstructure of pristine lyophilized gel and (**h**) corresponding elements distribution. (**i,j**) Microstructure of Ti_3C_2Tx-loaded lyophilized gel and (**k**) depicts element distribution within this structure. Reprinted from [132] under a Creative Commons Attribution License 4.0 (CC BY) "https://creativecommons.org/licenses/by/4.0/ (accessed on 1 November 2023)".

Yang et al. [133] summarized recent advances in the 3D printing of electrically conductive hydrogels for flexible electronics, demonstrating their significance in developing next-generation electronic devices. Zhang and collaborators [134] also provided a review on 3D-printing hydrogels for actuators, offering insights into the development of responsive materials for various applications. Chen et al. [135] explored the utilization of 3D-printed hydrogels in the development of soft thermo-responsive smart windows, showcasing their adaptability in architectural and environmental applications. Guo and colleagues [136] investigated the use of biomass-derived hybrid hydrogel evaporators for cost-effective solar water purification, providing sustainable solutions for clean water access.

5. Current Challenges and Future Trends in Hydrogel 3D Printing

5.1. Current Challenges

Inspite of the various versatile properties and applications of hydrogel 3D printing, several challenges associated with it persist, such as resolution limitations, post-printing processing, scalability, etc. Achieving high-resolution printing remains a challenge due to the complex rheological properties of hydrogels. The precise deposition of fine features is hindered by factors like nozzle clogging and the tendency of hydrogels to spread upon deposition. Also, many hydrogel 3D printing methods require post-processing steps, such as curing or crosslinking, to enhance mechanical stability, which can be time-consuming and may introduce variability in the final product. Scaling up hydrogel 3D printing to produce large quantities of intricate structures is also challenging. Maintaining consistent print quality, gelation kinetics, and resolution across a larger volume is a significant hurdle.

5.2. Future Trends and Advancements

Hydrogel 3D printing faces current challenges related to resolution, post-printing processing, and scalability. The future holds promising trends and advancements. As materials, techniques, and applications continue to evolve, hydrogel 3D printing is poised to reshape industries and contribute to breakthroughs in personalized medicine, tissue engineering, and various other fields. Future advancements will likely introduce innovative hydrogel formulations with improved printability, biocompatibility, and mechanical properties. Hybrid hydrogels combining natural and synthetic components might become more prevalent, allowing for tailored materials for specific applications. Researchers are likely to develop improved printing techniques that enhance resolution and speed. This might include advancements in nozzle design, improved droplet formation in inkjet-based systems, and refined light-based curing methods in SLA and DLP. Advances in multi-material 3D printing could enable the integration of hydrogels with other materials, such as polymers and ceramics, allowing for the creation of complex and functional hybrid structures.

Hydrogel 3D printing will likely explore incorporating bioactive agents directly into printed structures. This could lead to the creation of constructs with integrated therapeutic properties, reducing the need for post-printing functionalization. Research into intricate scaffold architectures will continue, focusing on enhancing tissue integration, nutrient transport, and cellular responses. This could involve the exploration of gradient structures and hierarchical designs. The field might see an expansion into personalized medicine, with the capability to tailor hydrogel formulations and structures for individual patients. Point-of-care 3D printing could become feasible for producing patient-specific implants and drug delivery systems. As hydrogel 3D printing matures, more efforts will be directed toward regulatory approval and clinical translation of hydrogel-based medical devices and implants.

As technology continues to evolve, 4D printing technologies have also been developed for hydrogel printing. Very recently, Wang and Guo [137] provided insights into recent advances in 4D-printing hydrogels designed for biological interfaces and delved into the innovative use of hydrogels that exhibit dynamic shape changes over time, showcasing the potential for applications in biological systems. Seo et al. have also introduced a hydrogel production platform that utilizes photo-cross-linkable and temperature-reversible chitosan polymer in combination with stereolithography 4D printing technology. This platform enables the creation of hydrogels with dynamic movement, opening up new possibilities in tissue engineering and regenerative medicine.

6. Conclusions

In the realm of 3D printing, the union of hydrogels and technology has fostered a dynamic landscape ripe with innovation and promise. This review has traversed the rich tapestry of hydrogel 3D printing, elucidating its profound impact on diverse domains and hinting at the limitless possibilities on the horizon. From their humble beginnings as unique, water-rich materials, hydrogels have evolved into versatile substrates capable of being sculpted into intricate structures with unparalleled precision. The convergence of hydrogels and 3D printing has unlocked doors to a myriad of applications spanning biomedical engineering, industrial production, and beyond. The biomedical sector, in particular, stands as a beacon of hope for hydrogel 3D printing's transformative power. The creation of tissue-engineered constructs, regenerative implants, and precise drug delivery systems has revolutionized healthcare and holds promise for further breakthroughs. The ability to replicate the intricacies of human biology, coupled with the advent of personalized medicine, offers a glimpse into a future where hydrogel 3D printing plays a pivotal role in enhancing the quality of life. Yet, this review acknowledges that challenges persist. Resolution limitations, post-printing processing intricacies, and the quest for scalable manufacturing methods are reminders that there is still much terrain to conquer.

Looking forward, the future appears promising. The trajectory of hydrogel 3D printing is set to be marked by advanced material development, enhanced printing techniques, and novel applications that defy current boundaries. As we journey further into this frontier, we anticipate the emergence of innovative solutions, new industries shaped by hydrogel capabilities, and a profound impact on scientific and technological advancement.

Author Contributions: Conceptualization, A.A. and C.M.H.; writing—original draft preparation, A.A.; writing—review and editing, A.A. and C.M.H.; supervision, C.M.H. All authors have read and agreed to the published version of the manuscript.

Funding: This research received no external funding.

Institutional Review Board Statement: Not applicable.

Informed Consent Statement: Not applicable.

Data Availability Statement: Data sharing not applicable.

Conflicts of Interest: The authors declare no conflict of interest.

References

1. Gajakosh, A.; Erannagari, S.; Kumar, R.S.; Thyagaraj, N.R.; Mallaradhya, H.M.; Rudresha, S. An Overview of 3D-Printed Smart Polymers and Composites. In *Development, Properties, and Industrial Applications of 3D Printed Polymer Composites*; IGI Global: Hershey, PA, USA, 2023; pp. 130–148.
2. Blanco, I. The use of composite materials in 3D printing. *J. Compos. Sci.* **2020**, *4*, 42. [CrossRef]
3. Tasneem, I.; Ariz, A.; Bharti, D.; Haleem, A.; Javaid, M.; Bahl, S. 3D printing technology and its significant applications in the context of healthcare education. *J. Ind. Integr. Manag.* **2023**, *8*, 113–130. [CrossRef]
4. Ho, T.C.; Chang, C.C.; Chan, H.P.; Chung, T.W.; Shu, C.W.; Chuang, K.P.; Duh, T.H.; Yang, M.H.; Tyan, Y.C. Hydrogels: Properties and applications in biomedicine. *Molecules* **2022**, *27*, 2902. [CrossRef] [PubMed]
5. Yang, D. Recent advances in hydrogels. *Chem. Mater.* **2022**, *34*, 1987–1989. [CrossRef]
6. Placone, J.K.; Engler, A.J. Recent advances in extrusion-based 3D printing for biomedical applications. *Adv. Healthc. Mater.* **2018**, *7*, 1701161. [CrossRef] [PubMed]
7. Agrawal, A.; Hussain, C.M. Materials and Technologies for Flexible and Wearable Sensors. In *Flexible and Wearable Sensors: Materials, Technologies, and Challenges*; Gupta, R.K., Ed.; CRC Press: Roca Raton, FL, USA, 2023.
8. Agrawal, A.; Hussain, C.M. Wearable Metal-Air Batteries. In *Metal-Air Batteries: Principles, Progress, and Perspectives*; Gupta, R.K., Ed.; CRC Press: Roca Raton, FL, USA, 2023; pp. 335–346.
9. Wang, Z.; Jin, X.; Dai, R.; Holzman, J.F.; Kim, K. An ultrafast hydrogel photocrosslinking method for direct laser bioprinting. *RSC Adv.* **2016**, *6*, 21099–21104. [CrossRef]
10. Mo, X.; Ouyang, L.; Xiong, Z.; Zhang, T. Advances in digital light processing of hydrogels. *Biomed. Mater.* **2022**, *17*, 042002. [CrossRef]
11. Naghieh, S.; Sarker, M.D.; Sharma, N.K.; Barhoumi, Z.; Chen, X. Printability of 3D printed hydrogel scaffolds: Influence of hydrogel composition and printing parameters. *Appl. Sci.* **2019**, *10*, 292. [CrossRef]
12. Pavan Kalyan, B.G.; Kumar, L. 3D printing: Applications in tissue engineering, medical devices, and drug delivery. *Aaps Pharmscitech* **2022**, *23*, 92. [CrossRef]
13. Kapusta, O.; Jarosz, A.; Stadnik, K.; Giannakoudakis, D.A.; Barczyński, B.; Barczak, M. Antimicrobial Natural Hydrogels in Biomedicine: Properties, Applications, and Challenges—A Concise Review. *Int. J. Mol. Sci.* **2023**, *24*, 2191. [CrossRef]
14. Kaliaraj, G.S.; Shanmugam, D.K.; Dasan, A.; Mosas, K.K.A. Hydrogels—A Promising Materials for 3D Printing Technology. *Gels* **2023**, *9*, 260. [CrossRef] [PubMed]

15. Zhang, X.N.; Zheng, Q.; Wu, Z.L. Recent advances in 3D printing of tough hydrogels: A review. *Compos. Part B Eng.* **2022**, *238*, 109895. [CrossRef]
16. Xin, P.; Han, S.; Huang, J.; Zhou, C.; Zhang, J.; You, X.; Wu, J. Natural okra-based hydrogel for chronic diabetic wound healing. *Chin. Chem. Lett.* **2023**, *34*, 108125. [CrossRef]
17. Haghbin, M.; Malekshah, R.E.; Sobhani, M.; Izadi, Z.; Haghshenas, B.; Ghasemi, M.; Kalani, B.S.; Samadian, H. Fabrication and characterization of Persian gum-based hydrogel loaded with gentamicin-loaded natural zeolite: An in vitro and in silico study. *Int. J. Biol. Macromol.* **2023**, *235*, 123766. [CrossRef] [PubMed]
18. Gabriele, F.; Ranaldi, R.; Bruno, L.; Casieri, C.; Rugnini, L.; Spreti, N. Biodeterioration of stone monuments: Studies on the influence of bioreceptivity on cyanobacterial biofilm growth and on the biocidal efficacy of essential oils in natural hydrogel. *Sci. Total Environ.* **2023**, *870*, 161901. [CrossRef]
19. Huang, C.; Ye, Q.; Dong, J.; Li, L.; Wang, M.; Zhang, Y.; Wang, X.; Wang, P.; Jiang, Q. Biofabrication of natural Au/bacterial cellulose hydrogel for bone tissue regeneration via in-situ fermentation. *Smart Mater. Med.* **2023**, *4*, 1–14. [CrossRef]
20. Yang, J.; Chen, Y.; Zhao, L.; Zhang, J.; Luo, H. Constructions and properties of physically cross-linked hydrogels based on natural polymers. *Polym. Rev.* **2023**, *63*, 574–612. [CrossRef]
21. Xu, J.; Zhang, M.; Du, W.; Zhao, J.; Ling, G.; Zhang, P. Chitosan-based high-strength supramolecular hydrogels for 3D bioprinting. *Int. J. Biol. Macromol.* **2022**, *219*, 545–557. [CrossRef]
22. Cai, F.F.; Heid, S.; Boccaccini, A.R. Potential of Laponite®incorporated oxidized alginate–gelatin (ADA-GEL) composite hydrogels for extrusion-based 3D printing. *J. Biomed. Mater. Res. Part B Appl. Biomater.* **2021**, *109*, 1090–1104. [CrossRef]
23. Van Velthoven, M.J.; Gudde, A.N.; Arendsen, E.; Roovers, J.P.; Guler, Z.; Oosterwijk, E.; Kouwer, P.H. Growth Factor Immobilization to Synthetic Hydrogels: Bioactive bFGF-Functionalized Polyisocyanide Hydrogels. *Adv. Healthc. Mater.* **2023**, *12*, 2301109. [CrossRef]
24. Saccone, M.A.; Gallivan, R.A.; Narita, K.; Yee, D.W.; Greer, J.R. Additive manufacturing of micro-architected metals via hydrogel infusion. *Nature* **2022**, *612*, 685–690. [CrossRef] [PubMed]
25. Adjuik, T.A.; Nokes, S.E.; Montross, M.D. Biodegradability of bio-based and synthetic hydrogels as sustainable soil amendments: A review. *J. Appl. Polym. Sci.* **2023**, *140*, e53655. [CrossRef]
26. Meyer, L.E.; Horváth, D.; Vaupel, S.; Meyer, J.; Alcalde, M.; Kara, S. A 3D printable synthetic hydrogel as an immobilization matrix for continuous synthesis with fungal peroxygenases. *React. Chem. Eng.* **2023**, *8*, 984–988. [CrossRef]
27. Yuk, H.; Lu, B.; Lin, S.; Qu, K.; Xu, J.; Luo, J.; Zhao, X. 3D printing of conducting polymers. *Nat. Commun.* **2020**, *11*, 1604. [CrossRef] [PubMed]
28. Tang, A.; Li, J.; Li, J.; Zhao, S.; Liu, W.; Liu, T.; Wang, J.; Liu, Y. Nanocellulose/PEGDA aerogel scaffolds with tunable modulus prepared by stereolithography for three-dimensional cell culture. *J. Biomater. Sci. Polym. Ed.* **2019**, *30*, 797–814. [CrossRef] [PubMed]
29. Cao, Z.; Luo, Y.; Li, Z.; Tan, L.; Liu, X.; Li, C.; Zheng, Y.; Cui, Z.; Yeung, K.W.K.; Liang, Y.; et al. Antibacterial hybrid hydrogels. *Macromol. Biosci.* **2021**, *21*, 2000252. [CrossRef]
30. Palmese, L.L.; Thapa, R.K.; Sullivan, M.O.; Kiick, K.L. Hybrid hydrogels for biomedical applications. *Curr. Opin. Chem. Eng.* **2019**, *24*, 143–157. [CrossRef]
31. Vasile, C.; Pamfil, D.; Stoleru, E.; Baican, M. New developments in medical applications of hybrid hydrogels containing natural polymers. *Molecules* **2020**, *25*, 1539. [CrossRef]
32. Liao, J.; Shi, K.; Jia, Y.; Wu, Y.; Qian, Z. Gold nanorods and nanohydroxyapatite hybrid hydrogel for preventing bone tumor recurrence via postoperative photothermal therapy and bone regeneration promotion. *Bioact. Mater.* **2021**, *6*, 2221–2230. [CrossRef]
33. Choi, S.; Choi, Y.; Kim, J. Anisotropic hybrid hydrogels with superior mechanical properties reminiscent of tendons or ligaments. *Adv. Funct. Mater.* **2019**, *29*, 1904342. [CrossRef]
34. Yang, X.; Yao, Y.; Wang, Q.; Zhu, K.; Ye, K.; Wang, G.; Cao, D.; Yan, J. 3D macroporous oxidation-resistant Ti3C2Tx MXene hybrid hydrogels for enhanced supercapacitive performances with ultralong cycle life. *Adv. Funct. Mater.* **2022**, *32*, 2109479. [CrossRef]
35. Iwanaga, S.; Hamada, Y.; Tsukamoto, Y.; Arai, K.; Kurooka, T.; Sakai, S.; Nakamura, M. Design and Fabrication of Mature Engineered Pre-Cardiac Tissue Utilizing 3D Bioprinting Technology and Enzymatically Crosslinking Hydrogel. *Materials* **2022**, *15*, 7928. [CrossRef]
36. Ianchis, R.; Alexa, R.L.; Gifu, I.C.; Marin, M.M.; Alexandrescu, E.; Constantinescu, R.; Serafim, A.; Nistor, C.L.; Petcu, C. Novel Green Crosslinked Salecan Hydrogels and Preliminary Investigation of Their Use in 3D Printing. *Pharmaceutics* **2023**, *15*, 373. [CrossRef]
37. Farsheed, A.C.; Thomas, A.J.; Pogostin, B.H.; Hartgerink, J.D. 3D Printing of Self-Assembling Nanofibrous Multidomain Peptide Hydrogels. *Adv. Mater.* **2023**, *35*, 2210378. [CrossRef]
38. Bom, S.; Ribeiro, R.; Ribeiro, H.M.; Santos, C.; Marto, J. On the progress of hydrogel-based 3D printing: Correlating rheological properties with printing behaviour. *Int. J. Pharm.* **2022**, *615*, 121506. [CrossRef]
39. Amorim, P.A.; d'Ávila, M.A.; Anand, R.; Moldenaers, P.; Van Puyvelde, P.; Bloemen, V. Insights on shear rheology of inks for extrusion-based 3D bioprinting. *Bioprinting* **2021**, *22*, e00129. [CrossRef]
40. Kokol, V.; Pottathara, Y.B.; Mihelčič, M.; Perše, L.S. Rheological properties of gelatine hydrogels affected by flow-and horizontally-induced cooling rates during 3D cryo-printing. *Colloids Surf. A Physicochem. Eng. Asp.* **2021**, *616*, 126356. [CrossRef]

41. Kim, M.H.; Lee, Y.W.; Jung, W.K.; Oh, J.; Nam, S.Y. Enhanced rheological behaviors of alginate hydrogels with carrageenan for extrusion-based bioprinting. *J. Mech. Behav. Biomed. Mater.* **2019**, *98*, 187–194. [CrossRef]
42. Townsend, J.M.; Beck, E.C.; Gehrke, S.H.; Berkland, C.J.; Detamore, M.S. Flow behavior prior to crosslinking: The need for precursor rheology for placement of hydrogels in medical applications and for 3D bioprinting. *Prog. Polym. Sci.* **2019**, *91*, 126–140. [CrossRef]
43. Zhou, M.; Lee, B.H.; Tan, Y.J.; Tan, L.P. Microbial transglutaminase induced controlled crosslinking of gelatin methacryloyl to tailor rheological properties for 3D printing. *Biofabrication* **2019**, *11*, 025011. [CrossRef]
44. Schwab, A.; Levato, R.; D'Este, M.; Piluso, S.; Eglin, D.; Malda, J. Printability and shape fidelity of bioinks in 3D bioprinting. *Chem. Rev.* **2020**, *120*, 11028–11055. [CrossRef]
45. Mora-Boza, A.; Włodarczyk-Biegun, M.K.; Del Campo, A.; Vázquez-Lasa, B.; San Román, J. Glycerylphytate as an ionic crosslinker for 3D printing of multi-layered scaffolds with improved shape fidelity and biological features. *Biomater. Sci.* **2020**, *8*, 506–516. [CrossRef]
46. Huang, L.; Du, X.; Fan, S.; Yang, G.; Shao, H.; Li, D.; Zhang, Y. Bacterial cellulose nanofibers promote stress and fidelity of 3D-printed silk based hydrogel scaffold with hierarchical pores. *Carbohydr. Polym.* **2019**, *221*, 146–156. [CrossRef]
47. Sheikhi, M.; Rafiemanzelat, F.; Ghodsi, S.; Moroni, L.; Setayeshmehr, M. 3D printing of jammed self-supporting microgels with alternative mechanism for shape fidelity, crosslinking and conductivity. *Addit. Manuf.* **2022**, *58*, 102997. [CrossRef]
48. Kantaros, A. 3D Printing in Regenerative Medicine: Technologies and Resources Utilized. *Int. J. Mol. Sci.* **2022**, *23*, 14621. [CrossRef]
49. Bedell, M.L.; Torres, A.L.; Hogan, K.J.; Wang, Z.; Wang, B.; Melchiorri, A.J.; Grande-Allen, K.J.; Mikos, A.G. Human gelatin-based composite hydrogels for osteochondral tissue engineering and their adaptation into bioinks for extrusion, inkjet, and digital light processing bioprinting. *Biofabrication* **2022**, *14*, 045012. [CrossRef]
50. Pose-Boirazian, T.; Martínez-Costas, J.; Eibes, G. 3D Printing: An Emerging Technology for Biocatalyst Immobilization. *Macromol. Biosci.* **2022**, *22*, 2200110. [CrossRef]
51. Rothbauer, M.; Eilenberger, C.; Spitz, S.; Bachmann, B.E.M.; Kratz, S.R.A.; Reihs, E.I.; Windhager, R.; Toegel, S.; Ertl, P. Recent advances in additive manufacturing and 3D bioprinting for organs-on-A-chip and microphysiological systems. *Front. Bioeng. Biotechnol.* **2022**, *10*, 837087. [CrossRef]
52. Cheng, Y.; Shi, X.; Jiang, X.; Wang, X.; Qin, H. Printability of a cellulose derivative for extrusion-based 3D printing: The application on a biodegradable support material. *Front. Mater.* **2020**, *7*, 86. [CrossRef]
53. Zhou, S.; Han, N.; Ni, Z.; Yang, C.; Ni, Y.; Lv, Y. Gelatin-oxidized nanocellulose hydrogels suitable for extrusion-based 3D bioprinting. *Processes* **2022**, *10*, 2216. [CrossRef]
54. Dong, L.; Bu, Z.; Xiong, Y.; Zhang, H.; Fang, J.; Hu, J.; Liu, Z.; Li, X. Facile extrusion 3D printing of gelatine methacrylate/Laponite nanocomposite hydrogel with high concentration nanoclay for bone tissue regeneration. *Int. J. Biol. Macromol.* **2021**, *188*, 72–81. [CrossRef]
55. Falcone, G.; Mazzei, P.; Piccolo, A.; Esposito, T.; Mencherini, T.; Aquino, R.P.; Del Gaudio, P.; Russo, P. Advanced printable hydrogels from pre-crosslinked alginate as a new tool in semi solid extrusion 3D printing process. *Carbohydr. Polym.* **2022**, *276*, 118746. [CrossRef]
56. Murphy, R.D.; Kimmins, S.; Hibbitts, A.J.; Heise, A. 3D-extrusion printing of stable constructs composed of photoresponsive polypeptide hydrogels. *Polym. Chem.* **2019**, *10*, 4675–4682. [CrossRef]
57. Hu, X.; Yang, Z.; Kang, S.; Jiang, M.; Zhou, Z.; Gou, J.; Hui, D.; He, J. Cellulose hydrogel skeleton by extrusion 3D printing of solution. *Nanotechnol. Rev.* **2020**, *9*, 345–353. [CrossRef]
58. Cui, X.; Li, J.; Hartanto, Y.; Durham, M.; Tang, J.; Zhang, H.; Hooper, G.; Lim, K.; Woodfield, T. Advances in extrusion 3D bioprinting: A focus on multicomponent hydrogel-based bioinks. *Adv. Healthc. Mater.* **2020**, *9*, 1901648. [CrossRef]
59. Ramezani, H.; Mohammad Mirjamali, S.; He, Y. Simulations of extrusion 3D printing of chitosan hydrogels. *Appl. Sci.* **2022**, *12*, 7530. [CrossRef]
60. Suntornnond, R.; Ng, W.L.; Huang, X.; Yeow CH, E.; Yeong, W.Y. Improving printability of hydrogel-based bio-inks for thermal inkjet bioprinting applications via saponification and heat treatment processes. *J. Mater. Chem. B* **2022**, *10*, 5989–6000. [CrossRef]
61. Stella, G.; Saitta, L.; Ongaro, A.; Cicala, G.; Kersaudy-Kerhoas, M.; Bucolo, M. Advanced Technologies in the Fabrication of a Micro-Optical Light Splitter. *Micro* **2023**, *3*, 338–352. [CrossRef]
62. Saitta, L.; Arcadio, F.; Celano, G.; Cennamo, N.; Zeni, L.; Tosto, C.; Cicala, G. Design and manufacturing of a surface plasmon resonance sensor based on inkjet 3D printing for simultaneous measurements of refractive index and temperature. *Int. J. Adv. Manuf. Technol.* **2023**, *124*, 2261–2278. [CrossRef]
63. Saitta, L.; Cutuli, E.; Celano, G.; Tosto, C.; Stella, G.; Cicala, G.; Bucolo, M. A Regression Approach to Model Refractive Index Measurements of Novel 3D Printable Photocurable Resins for Micro-Optofluidic Applications. *Polymers* **2023**, *15*, 26901. [CrossRef]
64. Marzano, C.; Arcadio, F.; Minardo, A.; Zeni, L.; Del Prete, D.; Cicala, G.; Saitta, L. Towards V-shaped Plasmonic probes made by exploiting 3D printers and UV-cured optical adhesives for Medical applications. In Proceedings of the IEEE International Workshop on Metrology for Industry 4.0 & IoT 2023, Brescia, Italy, 6–8 June 2023.
65. Adamski, K.; Kubicki, W.; Walczak, R. 3D printed electrophoretic lab-on-chip for DNA separation. *Procedia Eng.* **2016**, *168*, 1454–1457. [CrossRef]

66. Negro, A.; Cherbuin, T.; Lutolf, M.P. 3D inkjet printing of complex, cell-laden hydrogel structures. *Sci. Rep.* **2018**, *8*, 17099. [CrossRef]
67. Tetyczka, C.; Brisberger, K.; Reiser, M.; Zettl, M.; Jeitler, R.; Winter, C.; Kolb, D.; Leitinger, G.; Spoerk, M.; Roblegg, E. Itraconazole nanocrystals on hydrogel contact lenses via inkjet printing: Implications for ophthalmic drug delivery. *ACS Appl. Nano Mater.* **2022**, *5*, 9435–9446. [CrossRef]
68. Teo, M.Y.; Kee, S.; RaviChandran, N.; Stuart, L.; Aw, K.C.; Stringer, J. Enabling free-standing 3D hydrogel microstructures with microreactive inkjet printing. *ACS Appl. Mater. Interfaces* **2019**, *12*, 1832–1839. [CrossRef]
69. Nakagawa, Y.; Ohta, S.; Nakamura, M.; Ito, T. 3D inkjet printing of star block copolymer hydrogels cross-linked using various metallic ions. *RSC Adv.* **2017**, *7*, 55571–55576. [CrossRef]
70. Jiao, T.; Lian, Q.; Zhao, T.; Wang, H.; Li, D. Preparation, mechanical and biological properties of inkjet printed alginate/gelatin hydrogel. *J. Bionic Eng.* **2021**, *18*, 574–583. [CrossRef]
71. Yoon, S.; Park, J.A.; Lee, H.R.; Yoon, W.H.; Hwang, D.S.; Jung, S. Inkjet–spray hybrid printing for 3D freeform fabrication of multilayered hydrogel structures. *Adv. Healthc. Mater.* **2018**, *7*, 1800050. [CrossRef]
72. Peng, X.; Liu, T.; Zhang, Q.; Shang, C.; Bai, Q.W.; Wang, H. Surface patterning of hydrogels for programmable and complex shape deformations by ion inkjet printing. *Adv. Funct. Mater.* **2017**, *27*, 1701962. [CrossRef]
73. Duffy, G.L.; Liang, H.; Williams, R.L.; Wellings, D.A.; Black, K. 3D reactive inkjet printing of poly-ε-lysine/gellan gum hydrogels for potential corneal constructs. *Mater. Sci. Eng. C* **2021**, *131*, 112476. [CrossRef]
74. Chen, F.; Gu, S.; Zhang, Q.; Liu, T.; Liu, Z.; Kuang, T. A comparison study of hyaluronic acid hydrogel exquisite micropatterns with photolithography and light-cured inkjet printing methods. *e-Polymers* **2022**, *22*, 332–341. [CrossRef]
75. Kalossaka, L.M.; Mohammed, A.A.; Sena, G.; Barter, L.; Myant, C. 3D printing nanocomposite hydrogels with lattice vascular networks using stereolithography. *J. Mater. Res.* **2021**, *36*, 4249–4261. [CrossRef]
76. Karakurt, I.; Aydoğdu, A.; Çıkrıkcı, S.; Orozco, J.; Lin, L. Stereolithography (SLA) 3D printing of ascorbic acid loaded hydrogels: A controlled release study. *Int. J. Pharm.* **2020**, *584*, 119428. [CrossRef]
77. Burke, G.; Devine, D.M.; Major, I. Effect of stereolithography 3D printing on the properties of PEGDMA hydrogels. *Polymers* **2020**, *12*, 2015. [CrossRef]
78. Magalhães LS, S.; Santos FE, P.; Elias CD, M.V.; Afewerki, S.; Sousa, G.F.; Furtado, A.S.; Marciano, F.R.; Lobo, A.O. Printing 3D hydrogel structures employing low-cost stereolithography technology. *J. Funct. Biomater.* **2020**, *11*, 12. [CrossRef]
79. Alketbi, A.S.; Shi, Y.; Li, H.; Raza, A.; Zhang, T. Impact of PEGDA photopolymerization in micro-stereolithography on 3D printed hydrogel structure and swelling. *Soft Matter* **2021**, *17*, 7188–7195. [CrossRef]
80. Sun, Z.; Lu, Y.; Zhao, Q.; Wu, J. A new stereolithographic 3D printing strategy for hydrogels with a large mechanical tunability and self-weldability. *Addit. Manuf.* **2022**, *50*, 102563. [CrossRef]
81. Hosseinabadi, H.G.; Nieto, D.; Yousefinejad, A.; Fattel, H.; Ionov, L.; Miri, A.K. Ink material selection and optical design considerations in DLP 3D printing. *Appl. Mater. Today* **2023**, *30*, 101721. [CrossRef]
82. Ding, H.; Dong, M.; Zheng, Q.; Wu, Z.L. Digital light processing 3D printing of hydrogels: A minireview. *Mol. Syst. Des. Eng.* **2022**, *7*, 1017–1029. [CrossRef]
83. Sun, Z.; Zhao, Q.; Ma, S.; Wu, J. DLP 3D printed hydrogels with hierarchical structures post-programmed by lyophilization and ionic locking. *Mater. Horiz.* **2023**, *10*, 179–186. [CrossRef]
84. Dong, M.; Han, Y.; Hao, X.P.; Yu, H.C.; Yin, J.; Du, M.; Zheng, Q.; Wu, Z.L. Digital Light Processing 3D Printing of Tough Supramolecular Hydrogels with Sophisticated Architectures as Impact-Absorption Elements. *Adv. Mater.* **2022**, *34*, 2204333. [CrossRef]
85. Guo, Z.; Ma, C.; Xie, W.; Tang, A.; Liu, W. An effective DLP 3D printing strategy of high strength and toughness cellulose hydrogel towards strain sensing. *Carbohydr. Polym.* **2023**, *315*, 121006. [CrossRef] [PubMed]
86. Cafiso, D.; Septevani, A.A.; Noè, C.; Schiller, T.; Pirri, C.F.; Roppolo, I.; Chiappone, A. 3D printing of fully cellulose-based hydrogels by digital light processing. *Sustain. Mater. Technol.* **2022**, *32*, e00444. [CrossRef]
87. Wang, L.; Wang, Q.; Slita, A.; Backman, O.; Gounani, Z.; Rosqvist, E.; Peltonen, J.; Willför, S.; Xu, C.; Rosenholm, J.M.; et al. Digital light processing (DLP) 3D-fabricated antimicrobial hydrogel with a sustainable resin of methacrylated woody polysaccharides and hybrid silver-lignin nanospheres. *Green Chem.* **2022**, *24*, 2129–2145. [CrossRef]
88. Caprioli, M.; Roppolo, I.; Chiappone, A.; Larush, L.; Pirri, C.F.; Magdassi, S. 3D-printed self-healing hydrogels via Digital Light Processing. *Nat. Commun.* **2021**, *12*, 2462. [CrossRef]
89. Zhang, X.; Wu, W.; Huang, Y.; Yang, X.; Gou, M. Antheraea pernyi silk fibroin bioinks for digital light processing 3D printing. *Int. J. Bioprinting* **2023**, *9*, 760. [CrossRef]
90. Xiang, Z.; Li, N.; Rong, Y.; Zhu, L.; Huang, X. 3D-printed high-toughness double network hydrogels via digital light processing. *Colloids Surf. A Physicochem. Eng. Asp.* **2022**, *639*, 128329. [CrossRef]
91. Lopez-Larrea, N.; Criado-Gonzalez, M.; Dominguez-Alfaro, A.; Alegret, N.; Agua, I.D.; Marchiori, B.; Mecerreyes, D. Digital Light 3D Printing of PEDOT-Based Photopolymerizable Inks for Biosensing. *ACS Appl. Polym. Mater.* **2022**, *4*, 6749–6759. [CrossRef]
92. Athukorala, S.S.; Tran, T.S.; Balu, R.; Truong, V.K.; Chapman, J.; Dutta, N.K.; Roy Choudhury, N. 3D Printable Electrically Conductive Hydrogel Scaffolds for Biomedical Applications: A Review. *Polymers* **2021**, *13*, 474. [CrossRef]
93. Sánchez-Cid, P.; Jiménez-Rosado, M.; Romero, A.; Pérez-Puyana, V. Novel Trends in Hydrogel Development for Biomedical Applications: A Review. *Polymers* **2022**, *14*, 3023. [CrossRef]

94. Su, C.; Chen, Y.; Tian, S.; Lu, C.; Lv, Q. Natural Materials for 3D Printing and Their Applications. *Gels* **2022**, *8*, 748. [CrossRef]
95. Tran, T.S.; Balu, R.; Mettu, S.; Roy Choudhury, N.; Dutta, N.K. 4D Printing of Hydrogels: Innovation in Material Design and Emerging Smart Systems for Drug Delivery. *Pharmaceuticals* **2022**, *15*, 1282. [CrossRef] [PubMed]
96. Tajik, S.; Garcia, C.N.; Gillooley, S.; Tayebi, L. 3D Printing of Hybrid-Hydrogel Materials for Tissue Engineering: A Critical Review. *Regen. Eng. Transl. Med.* **2023**, *9*, 29–41. [CrossRef] [PubMed]
97. Lan, W.; Huang, X.; Huang, D.; Wei, X.; Chen, W. Progress in 3D printing for bone tissue engineering: A review. *J. Mater. Sci.* **2022**, *57*, 12685–12709. [CrossRef]
98. Varaprasad, K.; Karthikeyan, C.; Yallapu, M.M.; Sadiku, R. The significance of biomacromolecule alginate for the 3D printing of hydrogels for biomedical applications. *Int. J. Biol. Macromol.* **2022**, *212*, 561–578. [CrossRef]
99. Xu, Z.; Liu, G.; Huang, J.; Wu, J. Novel glucose-responsive antioxidant hybrid hydrogel for enhanced diabetic wound repair. *ACS Appl. Mater. Interfaces* **2022**, *14*, 7680–7689. [CrossRef] [PubMed]
100. Tsegay, F.; Elsherif, M.; Butt, H. Smart 3D Printed Hydrogel Skin Wound Bandages: A Review. *Polymers* **2022**, *14*, 1012. [CrossRef] [PubMed]
101. Gou, M.; Qu, X.; Zhu, W.; Xiang, M.; Yang, J.; Zhang, K.; Wei, Y.; Chen, S. Bio-inspired detoxification using 3D-printed hydrogel nanocomposites. *Nat. Commun.* **2014**, *5*, 3774. [CrossRef]
102. Xiong, X.; Chen, Y.; Wang, Z.; Liu, H.; Le, M.; Lin, C.; Wu, G.; Wang, L.; Shi, X.; Jia, Y.G.; et al. Polymerizable rotaxane hydrogels for three-dimensional printing fabrication of wearable sensors. *Nat. Commun.* **2023**, *14*, 1331. [CrossRef]
103. Deptuła, M.; Zawrzykraj, M.; Sawicka, J.; Banach-Kopeć, A.; Tylingo, R.; Pikuła, M. Application of 3D-printed hydrogels in wound healing and regenerative medicine. *Biomed. Pharmacother.* **2023**, *167*, 115416. [CrossRef]
104. Bhatnagar, D.; Simon, M.; Rafailovich, M.H. Hydrogels for regenerative medicine. In *Recent Advances in Biopolymers*; IntechOpen: London, UK, 2016; Volume 105.
105. Cernencu, A.I.; Dinu, A.I.; Stancu, I.C.; Lungu, A.; Iovu, H. Nanoengineered biomimetic hydrogels: A major advancement to fabricate 3D-printed constructs for regenerative medicine. *Biotechnol. Bioeng.* **2022**, *119*, 762–783. [CrossRef]
106. Catoira, M.C.; Fusaro, L.; Di Francesco, D.; Ramella, M.; Boccafoschi, F. Overview of natural hydrogels for regenerative medicine applications. *J. Mater. Sci. Mater. Med.* **2019**, *30*, 115. [CrossRef] [PubMed]
107. Hasirci, N.; Kilic, C.; Kömez, A.; Bahcecioglu, G.; Hasirci, V. Hydrogels in regenerative medicine. In *GELS Handbook: Fundamentals, Properties and Applications Volume 2: Applications of Hydrogels in Regenerative Medicine*; World Scientific: London, UK, 2016; pp. 1–52.
108. Aghamirsalim, M.; Mobaraki, M.; Soltani, M.; Kiani Shahvandi, M.; Jabbarvand, M.; Afzali, E.; Raahemifar, K. 3D printed hydrogels for ocular wound healing. *Biomedicines* **2022**, *10*, 1562. [CrossRef] [PubMed]
109. Shamma, R.N.; Sayed, R.H.; Madry, H.; El Sayed, N.S.; Cucchiarini, M. Triblock copolymer bioinks in hydrogel three-dimensional printing for regenerative medicine: A focus on pluronic F127. *Tissue Eng. Part B Rev.* **2022**, *28*, 451–463. [CrossRef]
110. Jervis, P.J. Hydrogels in Regenerative Medicine and Other Biomedical Applications. *Int. J. Mol. Sci.* **2022**, *23*, 3270. [CrossRef]
111. Tayler, I.M.; Stowers, R.S. Engineering hydrogels for personalized disease modeling and regenerative medicine. *Acta Biomater.* **2021**, *132*, 4–22. [CrossRef]
112. Heo, D.N.; Castro, N.J.; Lee, S.J.; Noh, H.; Zhu, W.; Zhang, L.G. Enhanced bone tissue regeneration using a 3D printed microstructure incorporated with a hybrid nano hydrogel. *Nanoscale* **2017**, *9*, 5055–5062. [CrossRef]
113. Foyt, D.A.; Norman, M.D.; Yu, T.T.; Gentleman, E. Exploiting advanced hydrogel technologies to address key challenges in regenerative medicine. *Adv. Healthc. Mater.* **2018**, *7*, 1700939. [CrossRef]
114. Maiz-Fernández, S.; Barroso, N.; Pérez-Álvarez, L.; Silván, U.; Vilas-Vilela, J.L.; Lanceros-Mendez, S. 3D printable self-healing hyaluronic acid/chitosan polycomplex hydrogels with drug release capability. *Int. J. Biol. Macromol.* **2021**, *188*, 820–832. [CrossRef]
115. Larush, L.; Kaner, I.; Fluksman, A.; Tamsut, A.; Pawar, A.A.; Lesnovski, P.; Magdassi, S. 3D printing of responsive hydrogels for drug-delivery systems. *J. 3D Print. Med.* **2017**, *1*, 219–229. [CrossRef]
116. Aguilar-de-Leyva, Á.; Linares, V.; Casas, M.; Caraballo, I. 3D printed drug delivery systems based on natural products. *Pharmaceutics* **2020**, *12*, 620. [CrossRef]
117. Martinez, P.R.; Goyanes, A.; Basit, A.W.; Gaisford, S. Fabrication of drug-loaded hydrogels with stereolithographic 3D printing. *Int. J. Pharm.* **2017**, *532*, 313–317. [CrossRef]
118. Liu, C.; Wang, Z.; Wei, X.; Chen, B.; Luo, Y. 3D printed hydrogel/PCL core/shell fiber scaffolds with NIR-triggered drug release for cancer therapy and wound healing. *Acta Biomater.* **2021**, *131*, 314–325. [CrossRef] [PubMed]
119. Wei, X.; Liu, C.; Wang, Z.; Luo, Y. 3D printed core-shell hydrogel fiber scaffolds with NIR-triggered drug release for localized therapy of breast cancer. *Int. J. Pharm.* **2020**, *580*, 119219. [CrossRef]
120. Dreiss, C.A. Hydrogel design strategies for drug delivery. *Curr. Opin. Colloid Interface Sci.* **2020**, *48*, 1–17. [CrossRef]
121. Wang, Y.; Miao, Y.; Zhang, J.; Wu, J.P.; Kirk, T.B.; Xu, J.; Xue, W. Three-dimensional printing of shape memory hydrogels with internal structure for drug delivery. *Mater. Sci. Eng. C* **2018**, *84*, 44–51. [CrossRef] [PubMed]
122. Zhang, H.; Zhang, F.; Yuan, R. Applications of natural polymer-based hydrogels in the food industry. In *Hydrogels Based on Natural Polymers*; Elsevier: Amsterdam, The Netherlands, 2020; pp. 357–410.
123. Park, S.M.; Kim, H.W.; Park, H.J. Callus-based 3D printing for food exemplified with carrot tissues and its potential for innovative food production. *J. Food Eng.* **2020**, *271*, 109781. [CrossRef]

124. Maniglia, B.C.; Lima, D.C.; Junior MD, M.; Le-Bail, P.; Le-Bail, A.; Augusto, P.E. Hydrogels based on ozonated cassava starch: Effect of ozone processing and gelatinization conditions on enhancing 3D-printing applications. *Int. J. Biol. Macromol.* **2019**, *138*, 1087–1097. [CrossRef] [PubMed]
125. Ishwarya, S.P.; Nisha, P. Advances and prospects in the food applications of pectin hydrogels. *Crit. Rev. Food Sci. Nutr.* **2022**, *62*, 4393–4417. [CrossRef]
126. Finny, A.S.; Jiang, C.; Andreescu, S. 3D printed hydrogel-based sensors for quantifying UV exposure. *ACS Appl. Mater. Interfaces* **2020**, *12*, 43911–43920. [CrossRef]
127. Diogo, G.S.; Marques, C.F.; Sotelo, C.G.; Pérez-Martín, R.I.; Pirraco, R.P.; Reis, R.L.; Silva, T.H. Cell-laden biomimetically mineralized shark-skin-collagen-based 3D printed hydrogels for the engineering of hard tissues. *ACS Biomater. Sci. Eng.* **2020**, *6*, 3664–3672. [CrossRef]
128. de Oliveira, R.S.; Fantaus, S.S.; Guillot, A.J.; Melero, A.; Beck, R.C.R. 3D-printed products for topical skin applications: From personalized dressings to drug delivery. *Pharmaceutics* **2021**, *13*, 1946. [CrossRef] [PubMed]
129. Othman, A.; Norton, L.; Finny, A.S.; Andreescu, S. Easy-to-use and inexpensive sensors for assessing the quality and traceability of cosmetic antioxidants. *Talanta* **2020**, *208*, 120473. [CrossRef]
130. Guo, B.; Zhong, Y.; Chen, X.; Yu, S.; Bai, J. 3D printing of electrically conductive and degradable hydrogel for epidermal strain sensor. *Compos. Commun.* **2023**, *37*, 101454. [CrossRef]
131. Liu, H.; Du, C.; Liao, L.; Zhang, H.; Zhou, H.; Zhou, W.; Ren, T.; Sun, Z.; Lu, Y.; Nie, Z.; et al. Approaching intrinsic dynamics of MXenes hybrid hydrogel for 3D printed multimodal intelligent devices with ultrahigh superelasticity and temperature sensitivity. *Nat. Commun.* **2022**, *13*, 3420. [CrossRef] [PubMed]
132. Yang, R.; Chen, X.; Zheng, Y.; Chen, K.; Zeng, W.; Wu, X. Recent advances in the 3D printing of electrically conductive hydrogels for flexible electronics. *J. Mater. Chem. C* **2022**, *10*, 5380–5399. [CrossRef]
133. Zhang, A.; Wang, F.; Chen, L.; Wei, X.; Xue, M.; Yang, F.; Jiang, S. 3D printing hydrogels for actuators: A review. *Chin. Chem. Lett.* **2021**, *32*, 2923–2932. [CrossRef]
134. Chen, L.; Duan, G.; Zhang, C.; Cheng, P.; Wang, Z. 3D printed hydrogel for soft thermo-responsive smart window. *Int. J. Extrem. Manuf.* **2022**, *4*, 025302. [CrossRef]
135. Guo, Y.; Lu, H.; Zhao, F.; Zhou, X.; Shi, W.; Yu, G. Biomass-derived hybrid hydrogel evaporators for cost-effective solar water purification. *Adv. Mater.* **2020**, *32*, 1907061. [CrossRef]
136. Wang, H.; Guo, J. Recent advances in 4D printing hydrogel for biological interfaces. *Int. J. Mater. Form.* **2023**, *16*, 55. [CrossRef]
137. Seo, J.W.; Shin, S.R.; Park, Y.J.; Bae, H. Hydrogel production platform with dynamic movement using photo-crosslinkable/temperature reversible chitosan polymer and stereolithography 4D printing technology. *Tissue Eng. Regen. Med.* **2020**, *17*, 423–431. [CrossRef]

Disclaimer/Publisher's Note: The statements, opinions and data contained in all publications are solely those of the individual author(s) and contributor(s) and not of MDPI and/or the editor(s). MDPI and/or the editor(s) disclaim responsibility for any injury to people or property resulting from any ideas, methods, instructions or products referred to in the content.

Review

Biopolymers for Tissue Engineering: Crosslinking, Printing Techniques, and Applications

David Patrocinio [1], Victor Galván-Chacón [1], J. Carlos Gómez-Blanco [1,*], Sonia P. Miguel [2,3], Jorge Loureiro [2], Maximiano P. Ribeiro [2,3], Paula Coutinho [2,3,*], J. Blas Pagador [1,4] and Francisco M. Sanchez-Margallo [4,5,6]

1. CCMIJU, Bioengineering and Health Technologies, Jesus Usón Minimally Invasive Surgery Center, 10071 Cáceres, Spain; dpatrocinio@ccmijesususon.com (D.P.); vpgalvan@ccmijesususon.com (V.G.-C.); jbpagador@ccmijesususon.com (J.B.P.)
2. CPIRN-IPG, Center of Potential and Innovation of Natural Resources, Polytechnic of Guarda, 6300-559 Guarda, Portugal; jorgeloureiro@ipg.pt (J.L.); mribeiro@ipg.pt (M.P.R.)
3. CICS-UBI, Health Science Research Center, University of Beira Interior, 6201-506 Covilhã, Portugal
4. CIBER CV, Centro de Investigación Biomédica en Red—Enfermedades Cardiovasculares, 28029 Madrid, Spain; msanchez@ccmijesususon.com
5. Scientific Direction, Jesus Usón Minimally Invasive Surgery Center, 10071 Cáceres, Spain
6. TERAV/ISCIII, Red Española de Terapias Avanzadas, Instituto de Salud Carlos III (RICORS, RD21/0017/0029), 28029 Madrid, Spain
* Correspondence: jcgomez@ccmijesususon.com (J.C.G.-B.); coutinho@ipg.pt (P.C.); Tel.: +34-927-18-10-32 (J.C.G.-B.); +351-96-55-44-187 (P.C.)

Citation: Patrocinio, D.; Galván-Chacón, V.; Gómez-Blanco, J.C.; Miguel, S.P.; Loureiro, J.; Ribeiro, M.P.; Coutinho, P.; Pagador, J.B.; Sanchez-Margallo, F.M. Biopolymers for Tissue Engineering: Crosslinking, Printing Techniques, and Applications. *Gels* **2023**, *9*, 890. https://doi.org/10.3390/gels9110890

Academic Editor: Enrique Aguilar

Received: 10 October 2023
Revised: 2 November 2023
Accepted: 7 November 2023
Published: 10 November 2023

Copyright: © 2023 by the authors. Licensee MDPI, Basel, Switzerland. This article is an open access article distributed under the terms and conditions of the Creative Commons Attribution (CC BY) license (https://creativecommons.org/licenses/by/4.0/).

Abstract: Currently, tissue engineering has been dedicated to the development of 3D structures through bioprinting techniques that aim to obtain personalized, dynamic, and complex hydrogel 3D structures. Among the different materials used for the fabrication of such structures, proteins and polysaccharides are the main biological compounds (biopolymers) selected for the bioink formulation. These biomaterials obtained from natural sources are commonly compatible with tissues and cells (biocompatibility), friendly with biological digestion processes (biodegradability), and provide specific macromolecular structural and mechanical properties (biomimicry). However, the rheological behaviors of these natural-based bioinks constitute the main challenge of the cell-laden printing process (bioprinting). For this reason, bioprinting usually requires chemical modifications and/or inter-macromolecular crosslinking. In this sense, a comprehensive analysis describing these biopolymers (natural proteins and polysaccharides)-based bioinks, their modifications, and their stimuli-responsive nature is performed. This manuscript is organized into three sections: (1) tissue engineering application, (2) crosslinking, and (3) bioprinting techniques, analyzing the current challenges and strengths of biopolymers in bioprinting. In conclusion, all hydrogels try to resemble extracellular matrix properties for bioprinted structures while maintaining good printability and stability during the printing process.

Keywords: natural polymers; hydrogels; crosslinking; bioprinting techniques; tissue engineering

1. Introduction

Tissue engineering (TE), as it is currently known, is a relatively recent field of innovation resulting from a complex merger of pre-existing lines of work from three different areas of knowledge: engineering, clinical medicine, and the biomedical field [1]. The TE area emerges as a consequence of clinical challenges related to the unavailability of tissue samples for the replacement of a damaged area when an accident or injury occurs and the immunological response associated with allograft transplants. In addition, autografts and allograft-based therapeutic approaches are associated with low availability, significant morbidity, and the requirement of immunosuppressive drugs for long-term treatment [2,3]. Driven by the pressing clinical need to develop alternative treatment therapies that can achieve the same outcomes as autografts and allografts but without their associated drawbacks, the field of tissue engineering emerged and has seen rapid progress over the past few decades [1].

In this way, TE aims to restore, maintain, improve, or replace different types of tissues and ultimately to incorporate or integrate the in vitro formed tissue into the body, contributing to the repair of injuries or the replacement of failing organs [1]. To accomplish these goals, TE proposes the production of provisional structures with structural and functional features that mimic the native tissue. In this field, different bioengineered constructs have been developed by employing different techniques, such as matrix decellularization [2,3], injectable hydrogels [4,5], and bioprinting [6,7]. In the first stage, the TE goal was to produce tissue constructs using biological or engineering techniques, but the evolution in this area has been contributing to the exploration of more integrated strategies involving the use of a biomaterial or biomimetic platform in combination with cells and biological factors that prompt the tissue regeneration process [6].

For the production of such biomimetic 3D supports, bioprinting is a promising technology that has revolutionized the TE and regenerative medicine fields, among others. Through this technology, it is possible to produce tissue models in a high-throughput manner, creating porous structures with controlled architecture using different biomaterials in combination with viable and physiologically relevant cells. Bioprinting enables the achievement of appropriate cell distribution, attachment, and growth within the biomaterial scaffold, which is complemented by improved nutrient and gas exchange as well as enhanced cell-cell communication due to the interconnected pores and large surface area [8,9]. In this way, bioprinting can contribute to overcoming the lack of dynamic and complex tissue structure in 2D cell culture as well as provide 3D scaffolds with spatial depth and more realistic cell-cell communication to better understand and simulate in vivo physiology [9]. Furthermore, the efficiency of 3D bioprinters and the availability of bioinks (polymer solutions and viable cells) with suitable structural, physicochemical, mechanical, thermal, and biological features are key factors for the success of bioprinting technology [9]. To be considered suitable for bioprinting, bioinks must have proper mechanical, rheological, chemical, and biological attributes [10]. This means that the materials used for the bioink formulation must ensure their printability, adequate mechanical properties according to the targeted tissue, high resolution and shape fidelity if needed, as well as biocompatibility, biodegradability, bioactivity, and reliability mimicking the microenvironment of tissues [11]. Thus, the choice of (bio)printable materials continues to be the bottleneck in bioprinting technology since it is required to use a polymer solution with suitable rheological and biological properties for bioprinting technique and cell incorporation, respectively. In addition, the printing techniques, such as photopolymerization, laser, extrusion, and droplets are another parameter that can be considered according to the properties of the polymeric bioink.

In this way, hydrogels arise as promising materials for TE and bioprinting, as they are characterized as highly biocompatible materials that allow the incorporation of cells and bioactive compounds. Moreover, their highly porous microstructures allow better cell internalization into the matrix and promote cell proliferation, owing to optimized oxygen and nutrient diffusion [12,13]. Biocompatibility, biodegradability, and specific structural and mechanical properties need to be ensured when developing a hydrogel formulation for TE, and more so when the formulation is intended for bioprinting, as bioprinter-specific requirements need to be considered. It is also important to note that natural materials are more sustainable than synthetic ones because not only are they degradable, but also, in most cases, they are obtained from renewal sources or by-products, such as proteins (e.g., gelatin, sericin), collagens [14], polysaccharides such as chitosan (CS) [15], alginate, or other natural-derived polysaccharides. Many hydrogel formulations contain polymers with hydrophilic groups in their structure, which confers enhanced interactions with biological tissues [16]. Similarly, a swollen-state hydrogel has a low interfacial contact angle with biological fluids, which decreases the chances of a negative immune response, leading to excessive inflammation, interference with healing, and implant rejection [17–19]. Another interesting characteristic of hydrogels is their stimuli-responsive nature due to their capacity to perform structural or mechanical changes in response to environmental signals. The stimuli-responsive nature can be classified into (i) physical, (ii) chemical, and (iii) biological, which allows the formulation of tunable bioinks for bioprinting [20].

To the best of our knowledge, this review highlights the characteristics and development of biological hydrogel-based bioinks for use in TE applications, describing the methodological modification/functionalization approaches, crosslinking techniques, and their respective applications on TE. Finally, the main challenges and future steps in using natural-based hydrogels in the 3D bioprinting area are also depicted.

2. Biopolymers for TE-Formulated Bioinks

Natural polymers are promising materials considering their intrinsic physicochemical and biological properties, mainly due to their large molecular chains and monomeric units. As previously outlined, natural polysaccharides and proteins are widely used in the development of scaffolds in regenerative medicine. Also, in the bioprinting field, biopolymers received special interest for their rheological and tribological features that allow them to generate complex geometries and structures of different tissues, as well as providing support for cell adhesion and proliferation. In this section, recent reports on proteins and polysaccharides used in 3D bioprinted hydrogels for TE are depicted.

2.1. Proteins

Collagen (Col) and fibrin (F) are proteins found in the Extracellular Matrix (ECM) of mammals and present specific bio-related features, such as biocompatibility and biodegradability [21], which is why they have been selected as bioinks to produce 3D bioprinted structures. In this way, several authors selected ECM structural proteins to produce their hydrogels and ensure provisional support for cell growth as well as mimicking the microenvironment available on the native tissue [22–30]. The rheological properties provided by these proteins are critical for bioprinting because of their weak mechanical behavior, which compromises the printability and consequently the integrity of the bioprinted material.

Among them, proteins such as Col, gelatin (Gel), silk fibroin (SF), and F have been used as bioinks to produce hydrogels for TE, as depicted in the following subsections and summarized in Table 1.

Table 1. Protein-based bioinks used for tissue engineering purposes.

Bioink Compounds [1]	Cells [2]	Printing Techniques	Applications	Reference
Col	HCCs	Extrusion	Cartilage tissue regeneration	[31]
Col + VEGF	NSCs	Droplet	Neural tissue regeneration	[32]
Gel + PU	MSCs	Extrusion	Tissue regeneration	[33]
Gel + GelMA + OCP	HUVECs	SLA	Bone regeneration	[34]
Gel + GelMA	Endothelial and ASCs	Extrusion	Bone regeneration	[35]
SF + GMA	NIH/3T3 and chondrocytes	DLP	Heart, vessel, brain, trachea, and ear regeneration	[36]
SF_GMA + GO	Neuro2a	DLP	Neural tissue engineering	[37]

[1] Biopolymer acronyms: Collagen (Col), Vascular Endothelial Growth Factor (VEGF), Gelatin (Gel), Polyurethane (PU), Methacrylate Gelatin (GelMA), Octacalcium Phosphate (OCP), Silk Fibroin (SF), Glycidyl Methacrylate (GMA), Glycidyl Methacrylate Silk Fibroin (SF_GMA), Graphene oxide (GO). [2] Cells acronyms: Human-derived Chondrocyte cells (HCCs), Neural Stem cells (NSCs), Mesenchymal Stem cells (MSCs), Human Umbilical Vein Endothelial cells (HUVECs), Adipose-derived Stem cells (ASCs), embryonic mouse fibroblast cells (NIH/3T3), and mouse neural crest-derived cells (Neuro2a).

2.1.1. Collagen

As the main structural protein of the ECM, Col has a high affinity for adherent cells [38], and Col hydrogels are widely used in 3D bioprinting purposes due to their good biocompatibility and bioactivity [39–42]. However, the weak mechanical properties and constraints in sterilization are major challenges [43]. To improve the material properties, a chemical modification can be performed by adding methacrylate (MA) moieties, allowing stronger crosslinking by intermolecular radical reactions [44].

It is important to note that these methacrylate groups can cause two undesirable effects on cells. First, free radical initiators could affect cell membranes under oxidative stress. Secondly, methacrylate and diacrylate chemical groups could provoke immunological responses in cells due to incomplete degradation of the scaffold, resulting in toxicity or disruption of the cell-cell interaction [45,46].

Also, it can be combined with other materials to generate a hydrogel that can be stabilized after printing [47]. In general, bioprinted Col hydrogels are used in microextrusion, inkjet, or laser-assisted bioprinting strategies [48]. So, this natural polymer can be used per se or combined with other compounds and cells to generate bioinks with promising properties for TE applications.

For example, Col was used as a bioink in the development of 3D skin substitutes by Yoon et al. The authors used extrusion bioprinting to construct Col 3D scaffolds seeded with primary human epidermal Gel keratinocytes (HEK) and human dermal fibroblasts (HDF). As the main findings, the results evidenced that the cell-laden 3D scaffolds implanted on a 1×1 cm^2 full-thickness excision mouse model were able to promote complete skin regeneration after one week of implantation [49].

On the other hand, Col combined with fibrinogen, autologous dermal fibroblasts, and epidermal keratinocytes were tested as bioinks on laser in situ bioprinting on murine full-thickness wound models (3×2.5 cm) and porcine full-thickness wound models (10×10 cm). In general, the results evidenced rapid wound closure, reduced contraction, and accelerated re-epithelialization [50].

In addition, Col was also used as bioink in the production of 3D bioprinted constructs for bone regeneration applications. For example, Kim and Kim combined Col with a bioceramic (β-TCP), preosteoblasts (MC3T3-E1), and human adipose stem cells (hASCs). The cell-laden structure was mechanically stable, and the cells remain viable, resulting in their proliferation and osteogenic differentiation [51].

2.1.2. Gelatin

Gelatin (Gel) is a low-cost biopolymer obtained by partial hydrolysis of Col, possessing a significant amount of arginine-glycine-aspartic acid (RGD), which will prompt cell attachment [52,53]. Furthermore, Gel is water-soluble and biodegradable through the action of enzymes such as collagenases (MMP-1 and MMP-8), having lower antigenicity than Col [54]. Indeed, different studies have evidenced the beneficial effects of Gel on cell adhesion, migration, and growth in the tissue regeneration process [55–57].

For bioprinting purposes, the solubility of Gel in warm water and its gelation at low temperatures allow the formation of physical crosslinked hydrogels [35,58]. Also, it is possible to increase the viscosity of bioinks for extrusion-based printing by decreasing the melting temperature of non-modified Gel. However, to obtain more stable scaffolds for TE, chemically crosslinked Gel hydrogels have been explored through the modification of Gel with methacryl groups [59,60].

In 2018, Yin et al. combined methacrylate gelatin with Gel to obtain stable structures through the extrusion-based bioprinting technique. So, the printed structures were mechanically stable, and the cell printing of bone marrow stem cells did not compromise their viability [61].

In addition, a composite biomaterial ink composed of Fish Scale and alginate dialdehyde (ADA)-Gel for the fabrication of 3D cell-laden hydrogels using mouse pre-osteoblast

MC3T3-E1 cells was revealed as a suitable biomaterial ink formulation with great potential for 3D bioprinting for bone TE applications [62].

A semi-synthetic bioink composed of 4.0% Gel, 0.75% alginate, and 1.4% carboxymethylated cellulose nanocrystal dissolved in 4.6% D-mannitol with normal human knee articular chondrocytes (NHAC-kn) was formulated and evaluated for 3D bioprinting. The results demonstrated the bioink as printable, stable under cell culture conditions, biocompatible, and able to maintain the native phenotype of chondrocytes, which, aside from meniscal tissue bioprinting, is suggested as a basis for the development of bioinks for various tissues [63].

Moreover, bioink hydrogel based on Gel blended with other polymers, such as phenylboronic acid-grafted hyaluronic acid (HA-PBA) and poly(vinyl alcohol), demonstrates that it can produce 3D bioprinted structures with anti-ROS ability. They try to improve cartilage regeneration in a chronic inflammatory and elevated ROS microenvironment [64].

Also, in Table 1, some recent works reporting the application of bioprinted Gel hydrogels for TE applications are summarized.

2.1.3. Silk Fibroin

Silk fibroin (SF) is extracted from silkworm silk and has been used as a potential biopolymer for TE due to its physicochemical and biological properties such as biocompatibility, biodegradability, low immunogenicity, and tunable mechanical properties [35,65,66]. On the other hand, silk biomaterials have received FDA approval for some medical products, such as sutures and surgical meshes, and can be processed into a variety of material formats [67,68]. Indeed, SF promotes the proliferation and adherence of different cells and avoids inflammation processes [69–71]. It can also be combined synergistically with other polymers by chemical interactions or covalent bonds, enabling physical crosslinking and thus avoiding harsh crosslinking chemicals.

An alternative way to use SF as bioink is based on its chemical modification with methacrylate groups that enable its photocrosslinking, resulting in a semi-synthetic SF. In this way, Yang et al. used methacrylated SF and metacrylated gelatin together in a new bioink to obtain rheological properties suitable for extrusion bioprinting. This bioink was able to encapsulate human umbilical vein endothelial and rat pheochromocytoma (PC12) cells, which maintain high viability and proliferation ability after the printing process. Also, the printed structures were implanted in the subcutaneous tissue of rats, revealing their promising potential to be used in TE applications [72].

In turn, a 3D-printed hydrogel based on alginate-silk nanofibril was introduced for soft tissue engineering, in which silk nanofibril reinforced Alg and was responsible for modulating its injectability by improving its shear-thinning behavior and shape retention before ionic crosslinking and supporting cell attachment and proliferation, with suitable physical and mechanical properties for 3D-printed structures in complex shapes such as ear cartilage [73]. Nevertheless, the low viscosity and frequent clogging during printing are some barriers described for SF use as bioink [74,75], despite the reports on the production of diverse 3D bioprinted structures for TE applications as described in Table 1.

2.1.4. Fibrin

Fibrin (F), as well as its precursor fibrinogen, belong to the family of glycoproteins that are crucial to induced blood clotting. Its three-dimensional network composed of randomly arranged fibers prevents blood leakage and contributes to the migration of cells towards the injury site, leading to the tissue regeneration process [76]. For these reasons, F has been widely used in the biomedical field due to its excellent biocompatibility and biodegradability, as well as the easily 3D processable capabilities of fibrin hydrogels [28,77].

Budharaju et al. reported a dual crosslinking strategy utilized towards 3D bioprinting of myocardial constructs by using a combination of alginate and fibrinogen, and it is based on a pre-crosslinking of the physically blended alginate-fibrinogen bioinks with $CaCl_2$ for improving shape fidelity and printability. This bioink was revealed to be cytocompatible and possesses the potential to be used for the biofabrication of thick myocardial constructs for regenerative medicine applications [78].

Additionally, other works reported a new droplet-based bioprinting system to integrate a cell-laden hydrogel with a microfibrous mesh produced by embedding human dermal fibroblasts in a collagen-alginate-fibrin hydrogel matrix. This study shows that cell-hydrogel-microfibre composites maintain high cell viability and promote cell–cell and cell–biomaterial interactions, offering an efficient way to create highly functional tissue precursors for laminar tissue engineering, particularly for wound healing and skin tissue engineering applications [79].

However, for bioprinting applications, Fibrin cannot withstand a stable 3D shape due to its low viscosity and Newtonian fluid behavior [80]. Therefore, different strategies have been developed through the combination of F with printable biomaterials, such as Gel and polyethylene glycol (PEG) [81,82], using a support bath to embed the printed structure [22,80], or crosslinking the hydrogel during the printing process [83]. Due to these reasons, the production of bioprinted fibrin-based hydrogels remains a challenge, and there are yet few works available in the literature, as listed in Table 1.

2.1.5. Other Proteins

In addition, some reports addressed the use of decellularized matrix and Matrigel as bioinks for TE purposes. The decellularized matrix (dECM) is obtained through physical and/or chemical methods to remove all cellular components while maintaining the ultrastructure and composition of the ECM [84]. As a bioink, dECM does not require the use of crosslinkers and allows to mimic tissue-specific characteristics into printed constructs [85–87], but any additional structural biopolymers are needed to recover its structural stability [84]. In fact, porcine-derived dECM bioink from different sources (liver, heart, skin, and cornea) was tested, and the authors verified that the tissue source of the dECM induced tissue-specific gene expression in human bone marrow mesenchymal stem cells [86]. On the other hand, dECM bioinks have been validated for skin regeneration applications, where a printed pre-vascularized skin patch promoted wound closure, epithelialization, and neovascularization [88]. In other work, the skin-derived dECM combined with a fibrinogen-based bioink incorporated with primary human skin fibroblasts improved the mechanical properties and viability of a bioprinted skin model [89]. Also, dECM as bioink was studied for the printing of human cardiac progenitor cells, demonstrating that the printed pre-vascularized stem cell patches promoted vascularization, maintained cell viability, and decreased cardiac remodeling and fibrosis [90,91].

Similarly, Matrigel consists of a composite, gelatinous mixture that mimics the human ECM and contains proteins such as laminin, Col, and entactin, which prompt cell growth and adhesion [92,93]. As recently reviewed, Matrigel applications in TE and bioprinting are based on the efficiency of cell differentiation, promoting angiogenesis and tissue regeneration both in vitro and in vivo [94].

In a work conducted by Li, the combination of Matrigel with sodium alginate was used to produce a hydrogel that was suitable for the growth of ectomesenchymal stem cells [95]. However, Matrigel has poor mechanical strength and requires a temperature-controlled system when used for extrusion printing [96,97], which limits its application in 3D bioprinting.

2.2. Polysaccharides

Among natural polymers, polysaccharides are described as the most promising macromolecules for biomedical applications. The most explored polysaccharides for biomedical applications are alginate, chitosan (CS), hyaluronic acid (HA), or cellulose [98,99], and they provide structural support in different cellular walls. These biopolymers are biosynthesized by living organisms, including plants, animals, algae, bacteria, and fungi [100], and present several important features crucial for 3D bioprinting applications in TE. The remarkable features of polysaccharides are their biocompatibility with mammalian cells and tissues and their biodegradability under physiological conditions, with the formation of nontoxic degradable products [20,101]. Moreover, these biopolymers are classified as eco-friendly materials due to their renewable and biodegradable nature [100], and they are considered safe (GRAS), being widely applied in the pharmaceutical industry as authorized excipients [102]. However, some studies concluded that alginate does not properly interact with mammalian cells, promoting minimal protein adsorption [103]. This feature allows researchers to develop new solutions that can be easily and quickly translated to the clinic. Nevertheless, the new 3D bioprinting technology implies a new challenge for biomedical sciences, namely regarding the development of bioinks able to provide adequate printability and simultaneously the mechanical and biological features needed for the TE field. In turn, the main weakness of biopolymer applications in 3D bioprinting is the difficulty in obtaining scaffolds with reproducible quality and properties and, to some extent, the lack of mechanical properties supporting their application as bioinks [104]. In this context, these polymers have been combined and blended with other natural and synthetic polymers to assure their application as bioinks for 3D bioprinting, as summarized in Table 2.

Table 2. Polysaccharide-based bioinks used for tissue engineering purposes.

Bioink Compounds [1]	Cells [2]	Printing Techniques	Applications	Reference
Alg + Aga	FBCs	Extrusion	Cartilage tissue engineering	[105]
Col + Fg	HUVECs and HDFs	Droplet	Vascular tissue regeneration	[106]
CS + PEGDA	hMSCs	SLA	Cartilage tissue engineering	[107]
HA + ChS	MSCs	Extrusion	Cartilage tissue engineering	[108]
Gel + XG	Fibroblasts and keratinocytes	Extrusion	Skin regeneration	[109]
XG + GOx + Glu + NHS	NIH/3T3	Extrusion	Soft tissue engineering	[110]
MA-κ-CA	NIH/3T3	DLP	Soft tissue engineering	[111]
MA-κ-CA + Alg	HUVECs	Extrusion	Vascular tissue regeneration	[112]
MA-κ-CA	HeLa and Fibroblasts	Extrusion	Soft tissue engineering	[113]
Pc + Plu + Alg	MIN6	Extrusion	Pancreatic tissue engineering	[114]
Pc + Plu	mBMSCs	Extrusion	Vascular tissue regeneration	[115]
Alg	NIH/3T3 and hMSCs	Laser	Skin regeneration	[116]
Alg + Fg	AC16	Extrusion	Myocardial regeneration	[78]

[1] Biopolymers acronyms: Alginate (Alg), Agarose (Aga), Collagen (Col), Fibrinogen (Fg), Chitosan (CS), Polyethylene Glycol Diacrylate (PEGDA), Hyaluronic Acid (HA), Chondroitin sulfate (ChS), Gelatin (Gel), Xanthan Gum (XG), Glucose Oxidase (GOx), Glucose (Glu), N-Hydroxy Sulfosuccinimide (NHS), Photocurable Methacrylate-κ-Carrageenan (MA-κ-CA), Pectin (Pc), Pluronic F127 (Plu). [2] Cells acronyms: Fetal Bovine Chondrocytes (FBCs), Human Umbilical Vein Endothelial cells (HUVECs), Human Dermal Fibroblasts (HDFs), Mesenchymal Stem cells (MSCs), Human Bone Marrow Mesenchymal Stem cells (hBMSCs), Embryonic Mouse Fibroblast cells (NIH/3T3), Mouse Insulinoma cells (MIN6), Mouse Bone Marrow Mesenchymal Stem cells (mBMSCs), AC16 cardiomyocytes (AC16).

2.2.1. Alginate

Alginate is a low-cost FDA-approved natural polysaccharide and probably the most used biomaterial in TE because it mimics the functions of the ECM and also has the possibility of being functionalized with cell adhesive links. Moreover, as a bioink, alginate has tunable degradation kinetics and can gelate easily [117–120].

Alginate solutions present the characteristic behavior of non-Newtonian fluids at low viscosities, which makes the production of a 3D geometrically defined structure difficult. Therefore, effective 3D printing alginate structures usually require crosslinking or the addition of thickening agents aiming to facilitate the solution extrusion as a filament [121]. For this reason, different approaches have been explored, from ionic crosslinking to combinations with other polymers.

Alginate is widely employed in vascular, cartilage, and bone tissue printing, as reviewed by [122,123]. The most successful applications of alginate as bioink are related to cartilage printing and bioprinting of vascularized tissues, for which the employment of coaxial (or triaxial) nozzle assemblies for printing alginate-based bioinks highlights excellent results. However, the mechanical performance of alginate compromises its application in bone tissue engineering, which can be overcome by combining alginate with other biomaterials such as Gel, hydroxyapatite, polycaprolactone, polyphosphate, or biosilica.

2.2.2. Chitosan

Chitosan (CS) is also highly used in TE due to its low price and some critical biological properties, such as biodegradability, biocompatibility, and antimicrobial activity [124]. Indeed, CS supports adequate cell proliferation and differentiation, mimicking the native tissue structure, and provides a suitable microenvironment for oxygen and nutrition exchanges [125,126]. In addition, the presence of CS in the composition of bioinks confers high antimicrobial activity due to the electrostatic interaction of the protonated NH_3^+ in CS with the negative cell wall of bacteria, which causes bacterial death or restricts their growth [127].

Further, the CS-containing solutions remain stable under physiological conditions, possessing appropriate viscosity values for bioprinting purposes [128,129]. However, as a natural polymer, CS presents a slow gelation rate and a weak mechanical strength [130–132]. Such a drawback has been overcome through combination with other polymers or the production of semi-synthetic CS derivatives. For example, the approach involves CS coupling with methacrylic anhydride (methacrylation of the backbone) to promote stronger crosslinking [133]. In Table 2, different CS-based bioinks produced for TE are listed.

In fact, the use of CS as bioink for TE applications has been widely explored, as evidenced in different reviews [134–136].

2.2.3. Hyaluronic Acid

Another remarkable polysaccharide is hyaluronic acid (HA), which is a linear polysaccharide with an important role in ECM formation, especially in connective tissues. HA plays key roles in cell proliferation and differentiation, immune modulation, and angiogenesis since it can mediate cell activity through receptor-ligand interactions with surface receptors such as CD44 and RHAMM [137,138]. In addition, HA structure is amenable to chemical modification through the grafting of active moieties or crosslinking, and diverse HA derivatives are available for clinical applications [138,139].

Two main limitations of using HA as a bioink are its limited mechanical performance and its viscous shear-thinning behavior, which hinders the interaction and fusion of printed filaments or droplets, preventing it from maintaining the desired shape throughout the printing process [140–142]. Because of this, HA is usually combined with other biomaterials, including natural or synthetic polymers, to obtain 3D bioprinted matrices for tissue regeneration applications, resulting in the creation of stable structures that can ensure cell viability during the extrusion process. In Table 2, the most recent works reporting the development of HA-based 3D bioprinted matrices for TE purposes are depicted.

Indeed, the intrinsic properties of HA make it an ideal choice of bioink for developing tissue constructs. Some reviews available in the literature describe the physicochemical properties, reaction chemistry involved in various cross-linking strategies, and biomedical applications of HA. Further, the features of HA bioinks, emerging strategies in HA bioink preparations, and their applications in 3D bioprinting have also been depicted [142,143].

2.2.4. Cellulose

Cellulose, mainly obtained from plants (Cellulose Nanofibril or CNF), but also produced by some bacteria (e.g., Cellulose Nanocrystals or CNC), is one of the most abundant biomaterials on earth [144] This biomaterial provides a stable matrix for TE with good mechanical properties. Another interesting feature is that inside the human organism, cellulose behaves as a non-degradable or very slowly degradable material [145,146]. Due to these reasons, cellulose is the main biopolymer in medical products, reinforcing its safety for humans.

For 3D bioprinting purposes, cellulose can support different process parameters, while its high viscosity, high surface area, good mechanical stability, tunable surface chemistry, and shear-thinning behavior allow the creation of 3D-printed freestanding and biocompatible constructs with favorable mechanical properties. However, it is necessary to use polymeric combinations or cellulose derivatives to improve the ink printability and shape fidelity after printing [147–149]. Nanocelluloses have attracted significant interest in the field of bioprinting, with previous research outlining the value of nanocellulose fibrils and bacterial nanocelluloses for 3D bioprinting tissues such as cartilage, augmenting the printability and chondrogenicity of bioinks [150], and as reviewed by [151], with important features for TE with different applications from vascular prosthesis to neural and bone regeneration. Furthermore, multicomponent bioinks based on pectin (Pc) and TEMPO-oxidized cellulose nanofibers (TOCNFs) for extrusion-based applications are used in complex-shaped scaffolds for tissue engineering [152] and in the production of hybrid hydrogels made of chemically modified pectin, Gel, and xanthan gum to create a supportive environment for cell adhesion and proliferation essential for wound healing and to incorporate cells in 3D bioprinting applications [153].

Moreover, novel, biodegradable, cost-effective, antimicrobial-loaded scaffolds are an emergent trend in biomedical devices and medical equipment manufacturing. So, hydrogel-based bioinks composed of cellulose and derivates revealed significant antibacterial and antibiofilm properties, suggesting the promising potential to be introduced in 3D bioprinting technology [154].

2.2.5. Gellan Gum

As a linear microbial exopolysaccharide, Gellan gum (GG) is a low-cost biopolymer that exhibits a variety of desirable properties, such as biocompatibility and biodegradability. Furthermore, according to their degrees of methylation, GG can show high transparency, thermo-responsive characteristics, flexible mechanical properties, efficient gelling, ease of manufacturing and crosslinking, and stability under physiological conditions that make it a compelling option as a bioink [155–157].

GG has been used to create bioprintable hydrogel when physically combined with lignin and crosslinked with magnesium ions, leading to improved in vitro chondrogenesis of hMSCs for cartilage regeneration [158]. Furthermore, when blended with Gel, it was used for vascularized bone regeneration, resulting in a hydrogel scaffold with a controlled release of deferoxamine that significantly promoted angiogenesis and osteogenic differentiation [159], as well as for adipose tissue engineering [155].

A recent review conducted by Cernencu described the various strategies that have been employed to propel the use of GG as bioink [160]. Also, a synopsis of the printable ink designs (e.g., compositions and fabrication approaches) that may be explored for tuning the properties of GG-based 3D-printed hydrogels for TE applications was discussed. Such work outlined the development of GG-based 3D printing inks, highlighting their possible biomedical applications.

2.2.6. Other Polysaccharides

Other polysaccharides of natural origin, such as xanthan, have been proposed as bioinks considering their biological activities. Xanthan gum (XG) is a heteropolysaccharide of high molecular weight produced by Xanthomonas campestris. The rheological properties of XG are attractive, as they enable the formation of pseudoplastic solutions at low concentrations. Additionally, XG exhibits good biocompatibility and is considered safe for various biomedical applications [161]. Moreover, it has been used in 3D bioprinting to produce a multilayered 3D construct by extrusion techniques on a hydrogel-based formulation containing alginate chemically crosslinked with $SrCl_2$, which can be conveniently sterilized via steam heat. Furthermore, its good rheological properties make it a cost-effective option to obtain a bioink able to produce biological tissues and other applications [162]. It can also be blended with Gel crosslinked with GTA to obtain a 3D-printed hydrogel that is biocompatible, maintains its print shape, resists swelling, and degrades easily [108]. Moreover, enzymatic polymerization or methacrylation has been proposed for the obtainment of semi-synthetic xanthan gum to modulate its printability [110] and biofunctionality [163].

Additionally, dextran (Dex) is a nontoxic and hydrophilic homopolysaccharide that can be used to produce biodegradable scaffolds due to its degradation by dextranase. As a bioink, dextran is usually combined with other biomaterials to modify its poor mechanical strength [164,165]. So, Dex was used for microextrusion hydrogels with chemically crosslinked Gel-varying ratios of Dex polyaldehyde. This material has the potential to: (1) form self-supporting structures in multiple layers that can be combined with living cells [166], and (2) produce a biocompatible and printable hydrogel using thermosensitive Gel and oxidized dextran with a tunable gelation time that can be easily post-reinforced through Schiff base crosslinking [167]. Dex exhibits remarkable properties in the TE field, specifically in skin and wound regeneration, when combined or modified with other biopolymers.

Starch is a neutral polysaccharide produced by plants such as rice, wheat, or maize. This polysaccharide is insoluble in water at room temperature, but its granules swell and gelatinize when heated. By cooling the starch suspension, the amylose phase separates, promoting the formation of a highly stable and biocompatible gel [168,169]. For bioprinting purposes, starch has been chemically modified or combined with other biomaterials to obtain hydrogels with good mechanical and rheological properties as well as enable the printing of structures without the need for any additional heat treatment. Thus, some authors have proposed starch as a highly desirable bio-ink to promote 3D TE by blending it with Gel nanoparticles and Col [170] or with CS in a 50/50 ratio for neural TE application [171].

Apart from alginates, other macroalgae-based polysaccharides have been evaluated for their suitability to be used in 3D bioprinting. Carrageenan (CA) is widely used in TE for its resemblance to the natural glycosaminoglycans of ECM. Despite the important properties of k-carrageenan (κ-CA) such as biocompatibility, biodegradability, shear-thinning, and ionic gelation, its gelation properties are challenging to control, which limits its use for applications in 3D bioprinting. To overcome this limitation, various modifications to the crosslinking process were tested to enhance the rheological properties of extrusion-based 3D-printed scaffolds. Several studies have reported the use of k-carrageenan-based bioinks in the extrusion-based 3D printing of cell-laden hydrogel structures with high cellular viability [172]. K-carrageenan has been blended with Gel [173], with alginate to obtain a bioink with suitable structural strength and printability without affecting cell viability [174], with methylcellulose [175] to be used on tissue regeneration, or with another bioink for cell encapsulation and 3D bioprinting in tubular tissue regeneration [112]. Moreover, other works reported the addition of methacrylate groups to obtain a hydrogel with later UV (ultraviolet) crosslinking. This approach was used for encapsulating NIH-3T3 cells [176] to obtain a photocurable bioink through visible light for soft tissue injuries in situ healing [113] or to produce tissue scaffolds using DLP-based 3D printing technology [111].

Pullulan (PUL) is a non-ionic exopolysaccharide with a molecular weight ranging from 10 to 400 kDa, resulting in low-viscosity solutions when dissolved in water. To obtain hydrogels, PUL can undergo chemical modifications with methacrylate groups. This modification led PUL to the synthesis of a photo-crosslinkable hydrogel suitable for 3D printing while allowing spatial and structural control over the porosity and shape of the printed structure [177]. Additionally, this modification of PUL did not alter its rheologic properties but assured its printability in an extrusion 3D printing process. To the best of our knowledge, the use of pullulan for the preparation of cell-laden bioinks for 3D bioprinting has not been reported yet [20].

Pectin (Pc), a natural heteropolysaccharide extracted from plant cell walls, exhibits shear-thinning behavior that makes the extrusion process easier [178]. However, the low viscosity values of pectin solutions hinder their printability. For this reason, partial crosslinking through cation addition, either during the printing or in the post-printing process, is reported as a useful strategy to obtain table 3D structures [179,180]. Additionally, pectin can also be combined with other biopolymers such as oligochitosan [181] or Pluronic (Plu) to obtain scaffold structures crosslinked with Ca^{2+} [182] and containing microspheres with bioactive molecules, such as estradiol, for the generation of vascularized tissue [115]. Further, 3D bioinks based on high-methoxylated pectin with Manuka honey have been developed to produce wound dressings [183]. However, Pc has been primarily used in the enhancement of bioinks by adding bioactivity, among other important biochemical properties. This enhanced capacity has been reported in bioprinted cell-laden devices made of an alginate-pluronic hydrogel that supports the viability of pancreatic β-cells while adding immunomodulating capacity [114] or as a Gel thickening agent and promoter of bioprintability [184,185].

Despite different works reporting the application of protein or polysaccharide-based bioink, the blended between them arises as an excellent option since it is possible to meet the bioactive and structural properties of proteins and polysaccharides, respectively, as listed in Table 3.

Table 3. Protein- and Polysaccharide-blended bioinks used for tissue engineering purposes.

Bioink Compounds [1]	Cells [2]	Printing Techniques	Applications	Reference
F + Alg + G	U87MG	Extrusion	Tissue construct	[186]
Col + Alg + TH	Fibroblasts	Extrusion	Vascular tissue regeneration	[187]
Col + GelMA	HUVECs and hMSCs	Droplet	Vascular tissue regeneration	[188]
SF + Gel	hMSCs	Extrusion	Cartilage tissue engineering	[189]
F + Alg + G	ASCs	Extrusion	Neural tissue engineering	[190]
F + G + Alg	hiPSCs	Extrusion	Neural tissue engineering	[191]
Alg + Gel + DEAE-C + Fg	HPFs and keratinocytes	Extrusion	Skin tissue engineering	[192]
Alg + TOCNF + PDA-NPs	Osteoblasts	Extrusion	Bone tissue regeneration	[193]
Alg + Gel + Soy	HUVECs	Extrusion	Vascular tissue regeneration	[194]
aAlg + aHA	MSCs	Extrusion	Cartilage tissue engineering	[195]
tCS + GHEC + CNCs	MC3T3-E1	Extrusion	Bone tissue regeneration	[196]
sCS + aDex + GelMA	hBMSCs and HUVECs	Extrusion	Vascular tissue regeneration	[165]
HAMA + Col	PC-12		Neural tissue engineering	[197]
HA + CMC	MC3T3	Extrusion	Bone tissue regeneration	[198]
HAMA + GelMA	CCs	Extrusion	Cartilage tissue engineering	[199]

Table 3. Cont.

Bioink Compounds [1]	Cells [2]	Printing Techniques	Applications	Reference
TOCNF + GelMA	ASCs	Extrusion	Vascular tissue engineering	[200]
TOCNF + Alg	hMF	Extrusion	Cartilage tissue engineering	[201]
GGMA + GelMA + DFO-Eth	HUVECs and hBMSCs	Extrusion	Bone regeneration	[159]
GG + Alg + LM	hiPSCs	Extrusion	3D neuromodeling	[202]
aDex + sCS + GelMA	hBMSCs and HUVECs	Extrusion	Skin regeneration	[165]

[1] Biopolymers acronyms: Fibrin (F), Alginate (Alg), Genipin (G), Collagen (Col), Tyramine Hydrochloride (TH), Methacrylate Gelatin (GelMA), Silk Fibroin (SF), Gelatin (Gel), Diethylaminoethyl Cellulose (DEAE-C), Fibrinogen (Fg), TEMPO-oxidized Cellulose Nanofibril (TOCNF), Polydopamine Nanoparticles (PDA-NP), Gelatin (Gel), Aldehyde Alginate (aAlg), Amine-Hyaluronic Acid (aHA), Thermogelling Chitosan (tCS), Glycerophosphate Hydroxyethyl Cellulose (GHEC), Cellulose Nanocrystals (CNC), Succinylated Chitosan (sCS), Dextran Aldehyde (aDex), Methacrylated Hyaluronic Acid (HAMA), Hyaluronic Acid (HA), Carboxymethylcellulose (CMC), Methacrylated Gellan Gum (GGMA), Deferoxamine-loaded ethosomes (DFO-Eth), Gellan Gum (GG), Laminin (LM). [2] Cells acronyms: Glioblastoma cells U87MG (U87MG), Human Umbilical Vein Endothelial cells (HUVECs), Human Mesenchymal Stem cells (hMSCs), Adipose-derived Stem cells (ASCs), Human induced Pluripotent Stem cells (hiPSCs), Human Dermal Fibroblasts (HPFs), Mesenchymal Stem cells (MSCs), Pre-osteoblast cells (MC3T3-E1), Human Bone Marrow Mesenchymal Stem cells (hBMSCs), Rat adrenal phaeochromocytoma cells (PC-12), Porcine Chondrocytes cells (CCs), Human Meniscus Fibrochondrocytes (hMF).

3. Biopolymers Requirements for Bioprinting

In bioprinting, bioink is the term used to refer to a solution composed of natural and/or synthetic polymers selected for their good biocompatibility and rheological properties. Typically, bioinks are cell-laden gels with specific properties designed to be used in biofabrication (Figure 1). The two main components of these inks are biopolymers and solvents. So, the fundamental function of biopolymers is to maintain the most favorable conditions for the cells during the formulation and processing stages, as well as to act as a safe supporting substrate for the cells. Meanwhile, solvents' main function is to provide appropriate rheological behavior and an adequate cell environment during the printing process. The concentration and the molecular weight of the biopolymer are critical in determining hydrogel viscosity behavior. Taking this into account, the most stable bioink formulations in the field of bioprinting are those using water as a solvent.

The combination of components and their proportion within the bioink formulation play a crucial role in establishing the printing conditions (e.g., pressure or temperature) and the overall quality of the printing outcome. For extrusion-based bioprinting, a high biomaterial concentration implies high viscosity, which usually has a negative impact on cell growth, migration, and differentiation, although it provides better fidelity control over the extruded material. Hydrogels with high molecular weight and low concentration biopolymers are commonly used and specifically formulated to determine the viscosity behavior, solubility shear rate, and working temperature of these hydrogels.

It is important to note that during the bioprinting process, bioinks undergo inner forces (compression and shear forces) that generate stress on the materials. The inner forces generated by each bioprinting technique must be considered during the setting because they have a major role in the bioink viscosity behavior, which controls important outcomes such as printability, fidelity, and cellular viability. This implies that bioinks must flow properly while maintaining their structural integrity to minimize cell stress and maintain the shape fidelity of the bioprinted structures.

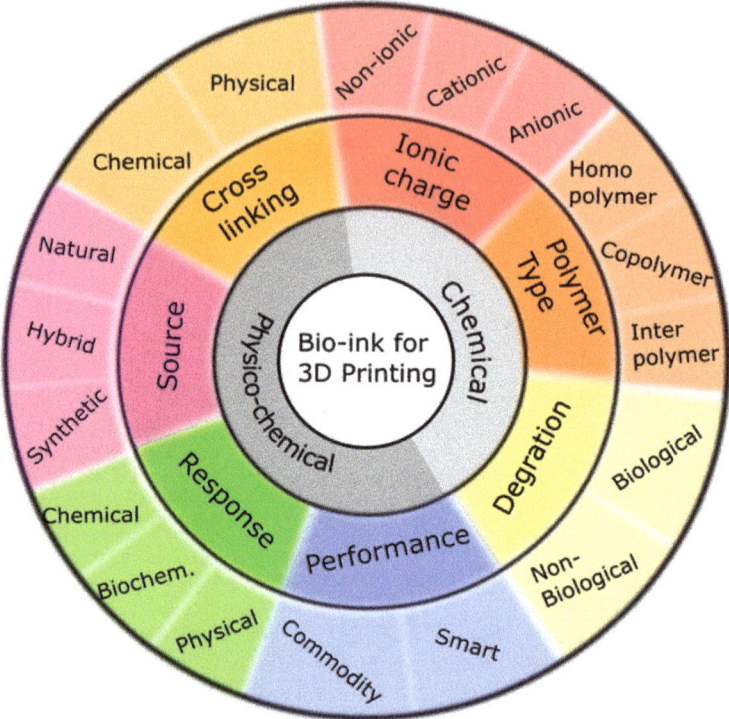

Figure 1. Classification of bioinks according to their properties.

The consolidation of the bioink is a crucial process that affects the qualities of the printed object and its application. The consolidation process determines the printing or bioprinting method to be used. During the consolidation process, the bioink undergoes a transition from its solvated (sol) state to its highly hydrophilic 3D reticulated macromolecule (Gel) state. The term "gelation" or "sol-gel transition" refers to this change of state. A polymer undergoes a transformation from its liquid or solvated state, exhibiting a pseudoplastic behavior (Sol), to a solid viscoelastic state (Gel), during the gelation process. In the case of bioinks, this consolidation process (sol-gel transition) is the result of the reactions of different macromolecular chains to gelate the bioink (crosslinking). The polymer's size increases as a result of the macromolecular chain combination, eventually forming a stable or pseudo-stable three-dimensional network.

Depending on the crosslinking process, hydrogels can be classified into those that crosslink by chemical (permanent) or physical (typically reversible) bonds. The former creates a stable covalent chemical bond, while the latter relies on non-permanent physical processes, as crosslinking occurs via dynamic and reversible non-covalent (hydrophilic, electrostatic, or hydrogen) bonding interactions. Due to the rapid formation rate, physical crosslinking is the most interesting method to maintain the shape of the printed material.

The bioink structure should be reticulated to promote the adhesion and growth of loaded cells and reconstruct tissue-specific structures. The bioink can form crosslinked structures that contribute to the mechanical effects on cells, increasing both mechanical strength and biological activity. Physical crosslinking can be achieved using light and/or heat under specific conditions, while chemical crosslinking can be achieved by the addition of chemical or natural crosslinking agents. Cell survival is directly impacted by the stiffness of the print substrate. The mechanical and rheological properties of the bioink can be adjusted by modifying its composition. So, the bioink components should facilitate cell attachment, growth, and proliferation inside the 3D construct. At the same time, components should facilitate the modification of the biopolymers' functional groups, promoting the inclusion and delivery of different biochemical signals or biomolecules.

It is common to apply a final fixing step or re-gelation process to maintain the final shape of the bioprinted object and keep a compromise between the different properties of the bioinks. Typically, this re-gelation stage is performed by incubating the bioprinted hydrogel in a solution with a physical or chemical crosslinking agent. Some common strategies utilized in bioprinting include adding photopolymerizing agents or agents that increase post-printing viscosity, facilitating crosslinking during these subsequent treatments. Another strategy is the addition of thermoplastic materials to mechanically reinforce and stabilize the structure of these printed objects [179,203–205].

4. Crosslinking of Biopolymers

The long-term stability and shape retention of a structure after printing are critical properties in bioprinting. The mechanism used by a hydrogel to achieve this is decisive when it comes to its use as a bioink. As previously mentioned, there are two types of crosslinking processes: chemical, which requires a crosslinking agent, and physical, which occurs through physical interactions between the biopolymer chemical groups and/or chains. Chemical crosslinking involves a molecule, the crosslinker, that reacts directly with the biomaterial and forms permanent interpolymeric bonds. Physical interactions are formed through various means, including electrostatic or macromolecule restructuration-type interactions, hydrogen bonds, the formation of stereogenic complexes, interpolymeric amphiphilic interactions, and reversible crystallization. Due to the lack of crosslinking agents, reversible hydrogels pose less toxicity, making them highly attractive for the development of bioinks [179,203,205–209].

4.1. Physical Crosslinking

Physically crosslinked polymers (Table 4) do not form any covalent bonds between chains, which implies that the bonds are formed through weak interactions. These bonds can be cyclically undone and re-done. Such behavior is typically seen in macromolecules with a specific preferred orientation and packaging, demonstrating a certain degree of order. Furthermore, these hydrogels have polar chemical groups that allow weak interactions, which can be broken with external stimuli such as temperature, pH, or similar, while exhibiting fast formation kinetics. Consequently, they are of great interest in the pharmaceutics and biomedicine fields, as they are not needed as crosslinking agents, which can be toxic/cytotoxic. Some of its counterparts have difficulty controlling the hydrogels' physical properties, gelation time, network pore size, chemical functionalities, and degradation time. Moreover, enhancing their mechanical properties is challenging [207].

Table 4. Different kinds of physical crosslinking with some examples of biopolymers and their chemical schema.

Physical Crosslinking	Biopolymers	Schema	References
Crystallization	Chitosan/PVA Gelatin/PVA Strach/PVA Dextran	Amorphous segment / Crystalline segment	[203,205–208]

Table 4. *Cont.*

Physical Crosslinking	Biopolymers	Schema	References
Stereo complexation	Dextran		[179,203,205,207,208,210]
Heating/Cooling	Gellan Gum, Gelatin, Carrageenan		[203,207,210]
Hydrogen bonding	Gelatin/Agar, Starch/Carboxymethyl Cellulose, Hyaluronic Acid/Methyl Cellulose		[203,205–208]
Ionic interaction	Alginate, Chitosan, Carrageenan		[203,205,207–209]
Hydrophobicity	Chitosan, Dextran, Pullulan, Carboxymethyl curdlan		[205,208]
Maturation	Alginate, Chitosan, Carrageenan, Arabic gum		[203,205]

4.1.1. Crystallization

The process of polymer crystallization consists of the formation of domains within the polymer matrix where different polymer sections are grouped together, increasing their arrangement, and clustering into pseudocrystalline structures (spherulites). This process affects the entire material, and multiple chain sections are involved in the formation of each domain. Therefore, the pseudocrystals reduce chain mobility and serve as crosslinking anchor points between consecutive chains. Polyvinyl alcohol (PVA) is one of the first hydrogels obtained using this type of crosslinking.

Polyvinyl alcohol (PVA) hydrogels, crosslinked by recrystallization, are gradually generated from an aqueous PVA solution at room temperature. The mechanical strength of these hydrogels manifests a significant dependence on the proceeding conditions. The molecular weight and the initial concentration are among those factors that influence the mechanical properties significantly. Additionally, the thermal history of the process and the gelation temperature have a significant impact. For example, the production of stable PVA hydrogel by crystallization at 37 °C may take up to 6 months. Hydrogels with this kind of crosslinking composition (PVA/CS, PVA/Starch, and PVA/Gel) are currently of interest in the fields of protein and peptide biotechnology as well as tissue engineering. Currently, there is a growing interest in the use of hydrogels with this crosslinking method in the fields of TE, protein, and peptide biotechnology. This includes the production of hydrogels with low-molecular-weight dextrin via recrystallization [203,205–208].

4.1.2. Stereocomplex Formation

Stereocomplexation is a process wherein the recrystallization process of optical isomeric polymers of racemic blends undergoes recrystallization. Complexation is a consequence of the interactions generated by the stereoselective van der Waals forces. So, hydrogels crosslinked via this method show improved thermal and mechanical properties compared to their homocrystal counterparts. However, it is important to take into consideration the limited number of polymers compatible with this crosslinking method, such as mixtures of high molecular weight such as PDLA or PLLA. This mixture exhibits a phase transition or melting temperature (Tm) at higher temperatures (230 °C). In this sense, the modification of the melting temperature is attributed to the formation of stereocomplexes. Additionally, protein-based hydrogels are often formed from dextran-oligolactate suspensions. The degradation kinetics of these hydrogels are highly dependent on the number, length, and polydispersity of the lactate grafts, as well as the initial water content [179,203,205,207,208,210].

4.1.3. Heating/Cooling

Heating/cooling crosslinked gels are generated through the processes of macromolecular chain recombination and conformational change, such as the restructuring of alpha-helix sections. For polymer solution temperatures above the melting point (Tm), the conformation of the macromolecular helix is lost. Posterior cooling redoes the helix structure due to intermolecular forces. This process involves the association of different macromolecule structures, leading to the formation of stronger junction zones, such as double helices. Furthermore, when K^+ or Na^+ cations are present, the double helices further aggregate, forming more stable gels [203,207,210].

4.1.4. Hydrogen Bonding

Hydrogen bond crosslinking is caused by the interaction between an electron-deficient hydrogen atom in a molecule and another functional group possessing a higher electron density in a closer molecule. In hydrogen bonds, the presence of protonated carboxylic acid groups is prevalent, and thus, the swelling of these gels is dependent on the environmental pH. This crosslinking process is usually found on gelatin-based hydrogels, such as blends of gelatin-agar, starch-carboxymethyl cellulose, or hyaluronic acid-methacrylate (HAMA). It is noteworthy that the hydrogel crosslinking process can be reversed with temperature when this hydrogen bridge formation does not happen at the same time as a macromolecular restructuring. Furthermore, other studies have explored hydrogel systems in which crosslinking is a result of the hybridization of hydrogen bonding and the stacking of bases [203,205–208].

4.1.5. Ionic Interaction

One of the most widespread gelation processes is ionic interaction, which can be achieved under regular conditions such as room temperature and physiological pH. Usually, hydrogels based on ionic crosslinking are polysaccharide-based, with alginate-based hydrogels being the most commonly used. The crosslinking of alginate with polycationic solutions (Ca^{2+}, Fe^{2+}, and Fe^{3+}) is feasible due to the presence of mannuronic and glucuronic acids. These gels are frequently used as matrices to encapsulate living cells and for protein delivery. Similarly, hydrogels can be produced by crosslinking CS, which contains 1,4-linked-β-glucosamine units, with glycerol-phosphate disodium salt. CS solutions below room temperature remain in a liquid phase when this salt is present but solidify quickly upon heating. Therefore, hydrogels that are in liquid form at room temperature but solidify at physiological temperature can be easily formulated. These gels can deliver active proteins, previously included in the hydrogel formulation before the gelation process, that stimulate bone and cartilage formation.

The addition of ions to produce hydrogels via ionic crosslinking does not require the presence of ionic groups in the polymer. Carrageenan is a polysaccharide containing 1,4-α-D-galactose and 1,3-β-D-galactose, as well as a few sulphate groups on its polymeric chain, that can form a gel either with potassium ions or under salt-free conditions. Furthermore, very rigid hydrogels can be produced in the presence of metallic cations. Dextran is another biopolymer that forms a hydrogel in the presence of potassium ions by encasing the potassium ions within the macromolecule structure. However, this hydrogel is not suitable for drug delivery due to its instability when in contact with water. Hydrogels can also be produced by the complexation of polyanions with polycations. The most frequently used ionically crosslinked CS hydrogels are the result of the complexation between CS and polyanions, such as dextran sulphate or polyphosphoric acid [203,205,207–209].

4.1.6. Hydrophobicity

Hydrogels with a high water content, typically around 80 wt%, are physically crosslinked by hydrophobic interaction. This is achieved using compounds such as hydrophobic modified CS, dextran, pullulan, and carboxymethyl curdlan, which interact in the presence of water. These specific compounds are utilized in the formation of monodisperse hydrogel nanoparticles as a means of creating a microenvironment to safeguard proteins from thermal denaturation, aggregation, and enzymatic degradation. Furthermore, highly porous and low-swelling solid materials can be obtained via freeze drying and subsequent hydration with an alkaline buffer. This method can be used to obtain precise control over material properties and structure, producing materials that can increase their weight about 20 times without swelling [205,208].

4.1.7. Maturation

The process of maturation consists of an increase in molecular weight, resulting in the production of a precisely structured hydrogel with controlled, structured molecular dimensions. An example of heat-induced gelation is the thermal treatment of gum Arabic, which contains about 2–3% protein [arabinogalactan protein (AGP), arabinogalactan (AG), and glycoprotein (GP)] as an integral part of its structure. So, these proteins are responsible for aggregation, leading to the improvement of their mechanical properties and water-binding capability [203,205].

4.2. Chemical Crosslinking

The formation of chemically crosslinked hydrogels, as represented in Table 5, is the result of a reaction between multifunctional monomers or groups and the biopolymer. The resulting hydrogel becomes insoluble in any solvent because of the formation of covalent bonds, which remain intact until breakage. Compared to physically crosslinked hydrogels, these hydrogels show better mechanical stability, resistance, longer degradation times, and the ability to regulate hydrogel formation. A wide range of chemical processes are available to form these hydrogels, such as radical polymerization, Michael reactions, Schiff based reactions, and enzymatic crosslinking reactions, based on the functional groups present (such as OH, COOH, NH_2, and others).

Table 5. Different kinds of chemical crosslinking, with some examples of biopolymers and their chemical schema.

Chemical Crosslinking	Biopolymers	Schema	Reference
Complementary Groups			
1. Aldehyde, Dihydrazide, and Schiff's base	Hyaluronic acid-based Dextran Chitosan Alginate		[210]
2. Thiol-ene Michael addition	Hyaluronic acid-based Dextran Chitosan Alginate	k_i = First initiation etape, thiol deprotonation stage. k_{p1} = First propagation etape constant (n = 0). k_{p2} = First propagation etape constant (n = 1,2,3, …). k_{trp} = Termination constant by transference to tiol initiator.	[210]
3. Condensation	Hyaluronic acid-based Dextran Chitosan Alginate	Passerini Condensation; Ugi Condensation	[205,207,208,211]
Radical polymerization	Hyaluronic acid Dextran Chitosan	dex-GMA; dex-MA	[205,207–209,211]
Enzymatic	Gellan gum Polypeptides Hyaluronic acid Dextran Cellulose Alginate		[180,204–206,210,211]

4.2.1. Complementary Groups

Chemical groups in hydrophilic polymers comprise atoms with varying electronegativities. As a result, some of these chemical groups, such as COOH, will be electron-deficient, while others, such as NH_2 or OH, will be electron-dense. The presence of these groups in the polymer enables the creation of hydrogels through their reactivity, such as in the cases of amino-carboxylic acid, isocyanate-OH/NH_2, or the synthesis of Schiff bases. Covalent bonds are created between the polymer chains during this process. A bifunctional molecule with the appropriate chemical group may be necessary as a crosslinking agent under certain circumstances. The dilution of the polymer's aqueous starting solution will start hydrogel production. On the other hand, resins will be produced if the concentration is very high.

- Aldehyde, dihydrazide, and Schiff's base

Aldehydes and dihydrazides (such as glutaraldehyde and adipic acid dihydrazide) are two important crosslinking agents. Crosslinking usually begins with low pH, high temperature, and methanol addition as a quencher. When preparing hydrogels, the polymer's amino group can react with the aldehyde to form a Schiff base. The resulting imine link can be hydrolyzed, resulting in the breakdown of Schiff bases at low pH levels. So, using dialdehyde or formaldehyde as a crosslinking agent makes it possible to form gelled hydrogels based on proteins such as Gel, albumin, and amine-containing polysaccharides. In this sense, some hydrogel films are made by modifying HA with adipic dihydrazide and then crosslinked with poly (ethylene glycol)-propiondialdehyde. These hydrogels are commonly used for controlled medication delivery [210].

- Thiol-ene Michael addition

The thiol-based Michael-type reaction has been investigated as a viable method for producing chemically crosslinked hydrogels due to its mild reaction conditions, well-characterized reaction processes, and capacity to create homogeneous constructions. This chemical crosslinking is based on nucleophilic addition to unsaturated carbonyl compounds. Typical nucleophiles include thiol and amine-bearing molecules, while unsaturated carbonyl can be found in acrylate, methacrylate, and vinyl sulfone groups. Photoinitiated thiol-ene crosslinking has recently been investigated as a desirable method for cell encapsulation. This photopolymerization mode requires highly reactive components and a short exposure time. So, cells can benefit from lower exposure to reactive radicals. Additionally, the thiol-ene chemistry has a reduced susceptibility to oxygen inhibition, allowing cell encapsulation in an oxygen-rich environment, which is a desirable attribute for 3D bioprinting technologies [210].

- Condensation

Condensation reactions are an effective method for crosslinking water-soluble polymers. Therefore, their application in hydrogel creation is being encouraged. The Passerini and Ugi condensation processes are extensively used in the synthesis of polymers, specifically in the production of polyesters and polyamides. For this reason, biopolymers that present hydroxyl groups or amines with carboxylic acids or derivatives are necessary, respectively. To facilitate the sol-gel transition, the crosslinking reaction can be conducted in slightly acidic water at room temperature. A few molecules are commonly included in this process to prevent any unwanted side reactions and to achieve tighter control of the gel's crosslinking density. The rate of gel degradation at room temperature and pH 9.5 can be adjusted based on the crosslinking density. The inclusion of bioactive molecules, such as antimicrobial proteins, lysozymes, or negatively charged polysaccharides, to increase loading capacity is another intriguing alteration [205,207,208,211].

4.2.2. Radical Polymerization

Radical polymerization has become the most widely used technique for creating chemically crosslinked hydrogels. Specifically, two reaction subclasses can be further classified under this category: random and controlled polymerization. To produce hydrogels through free radical processes, a monomer, oligomer, or macromer with unsaturated vinyl functional groups must undergo polymerization. The macromolecular backbones are activated in chemical grafting via radical polymerization, which occurs through the reaction of a chemical reagent (visible or UV light, heat, or redox initiators). This crosslinking technique is highly effective and frequently utilized, even in modest conditions. The characteristics of the hydrogel and its swelling capacity depend on the proportion of crosslinker used. The most typical natural polymers used in radical crosslinking are HA, dextran, and CS, as well as biopolymers functionalized with acrylate groups. Therefore, photocrosslinking techniques are typically the most promising for using cell-laden bioinks in bioprinting [205,207–209,211].

Additionally, the classification of a radical initiation method can be extended to include thermal and photoinitiation processes, as well as polymerization induced by heat and/or light. This crosslinking by light-induced polymerization is an incredibly quick hydrogel process that is particularly attractive for the creation of patterned structures. The selection of the appropriate photoinitiator type and solvent is crucial, as they may leak out of the hydrogel post-production. The radicals produced during polymerization in the presence of proteins can harm the protein's structure. Additionally, the UV light's intensity is restricted to prevent cell damage because the heat released during the crosslinking process may result in cell death [179,205,207].

4.2.3. Enzymatic

Enzymatic crosslinking offers excellent substrate selectivity and avoids adverse responses. The peptide ratio can be changed to customize the gel's properties. The reaction kinetics of the gel are influenced by the ratio of reactants, enzyme concentration, and macromer shape and composition. This qualifies these hydrogels as suitable for in situ gelling systems.

Transglutaminase (TG), a Ca^{2+}-dependent enzyme, is capable of catalyzing the formation of polypeptide-like hydrogels rapidly at biological temperature (37 °C). The TG forms these hydrogels by amide linkage between the carboxamide and primary amines on polymers or polypeptides. Horseradish peroxidase and H_2O_2 have been used in similar proceedings to crosslink hydrogels based on HA, dextran, cellulose, and alginate [180,204–206,210,211].

5. Printing Techniques for Biopolymers

Bioprinting is defined by the Cambridge English Dictionary as 'the process of producing tissue or organs similar to natural body parts and containing living cells using 3D printing'. The field of bioprinting has experienced tremendous growth over the last decade as it allows the deposition of biomaterials with cells following precise geometries that aim to mimic the tissue or organ structures present in the body. One of the current and potential uses of bioprinting is the fabrication of viable tissues and organs for application in regenerative medicine, the creation of biomimetic in vitro models for drug screening, and the use of bioprinting in personalized medicine through a collection of tissues obtained from the same patients receiving the treatment.

The cells used in bioprinting require an environment capable of providing, at the same time, adequate mechanical support for shape retention and a biocompatible environment to allow cell growth while maintaining their functionality as closely to their native physiological environment as possible. For this purpose, printing typically follows the structure of porous scaffolds, with pores providing an area for cell growth and nutrients and oxygen penetration within the printed construct. Natural polymers, as opposed to synthetic ones, offer advantages in terms of better biocompatibility, biodegradability, and mimicking of the physiological properties of extracellular matrix (ECM) [212] e.g., stiffness.

Different biopolymers and gelation techniques have been used to generate bioinks. Therefore, printing techniques have been developed based on different principles, printing parameters (Table 6), and foundations adapted to each hydrogel and cell type. In this sense, these printing technologies can be divided into photopolymerization, laser-assisted, and deposition printing approaches depending on the underlying concepts upon which they are founded (Figure 2). Therefore, the properties of these scaffolds and their ultimate application hinge on the manufacturing procedure used to create them.

Table 6. Bioprinting techniques and their printing parameters.

Bioprinting Technique	Parameters	Reference
Photopolymerization	Wavelength Light/laser power Exposure time Repetition rate Pulse width Environmental temperature	[212–214]
Laser-based	Wavelength Laser intensity/power Pulse rate/frequency Laser fluence Hydrogel viscosity Thickness of the absorbing layer Thickness of the bioink layer Travel distance Printing velocity Printing pressure	[215–217]
Extrusion	Printing velocity Nozzle height Flow rate Printing temperature Hydrogel viscosity Extrusion gauge Environmental temperature Pressure	[217–219]
Droplet	Droplet rate Droplet volume Hydrogel viscosity Kinetic momentum Printing speed Printing time Voltage pulse (amplitude, rise and fall times, dwell time, echo time, and frequency) Substrate hydrophobicity/hydrophilicity	[212,220,221]

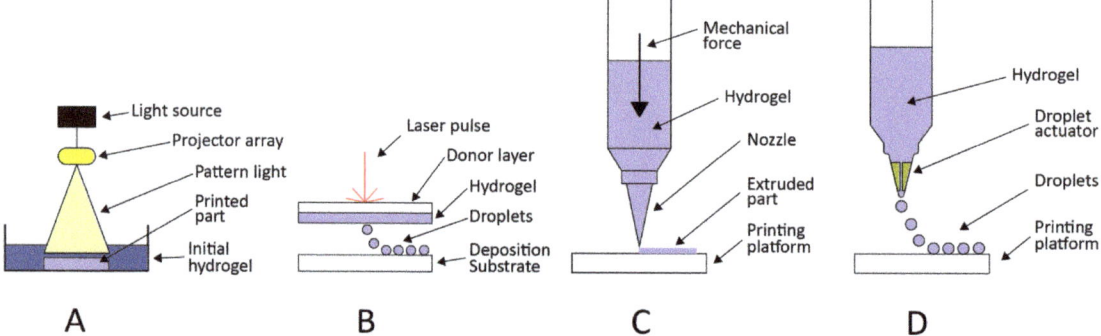

Figure 2. Different printing techniques used with biopolymers: (**A**) photopolymerization, (**B**) laser, (**C**) extrusion and (**D**) droplet.

5.1. Photopolymerization-Based

In photopolymerization-based printing, resins contain a photoinitiator activated by laser to produce the photopolymerization of a small volume around the focus spot. For this reason, the selection of an adequate photoinitiator agent is critical for biocompatibility since some have cytotoxic effects [222]. Furthermore, the photoinitiator will affect the final resolution, and occasionally a photoinitiator and a photoblocker are combined to achieve a smaller spot size.

Several techniques using photopolymerization principles have been developed, including stereolithography (SLA), digital light printing (DLP), and two-photon polymerization (2PP).

SLA works by using a mirror (galvanometer) to focus a laser beam on the desired spot, typically inside a resin tank, and achieves high resolution at a relatively low printing speed. Each layer of the object is printed by scanning dot by dot with the mirror (Figure 2A). Improvements in SLA technology have led to DLP, a more efficient method in which a digital light projector replaces the laser beam, curing each full layer at a time. Resolution in SLA and DLP printers is typically within 10–75 μm and rapidly improving. 2PP is based on photopolymer crosslinking induced by high-intensity femtosecond laser pulses. Due to the non-linear nature of the two-photon absorption, proportional to the square of the light intensity, it is possible to achieve very small voxel sizes, leading to submicrometer resolution [212], albeit with longer printing times.

The range of materials available for photopolymerization was initially restricted to photocurable synthetic resins, including hydrogels made from modified biopolymers. In this sense, it should be noted that the materials used in photopolymerization are not strictly natural polymers but modified polymers including acrylate or diacrylate groups, allowing the polymerization process to occur after the photoinitiator has been excited. Vinyl chemical groups are often incorporated through modification with acrylates, such as methacrylic anhydrides. Thus, polymers such as gelatin methacrylate (GelMA), hyaluronic acid methacrylate (HAMA), or Col methacrylate (ColMA) are among those commonly used.

The biocompatibility of constructs produced by photopolymerization may be limited by several factors. Photoinitiators are required in the hydrogel, which could harm biocompatibility through possible residual toxic photocuring agents [222]. Additionally, UV and shorter wavelengths of light may cause mutations in DNA, and intense light can decrease cell viability. Therefore, caution is needed while selecting and utilizing photoinitiators, as well as the wavelength and intensity of the light. Printing time is another crucial consideration because long print durations will result in insufficient immersion of cells in the medium and a lack of necessary CO_2 supply and humidity [214]. The porosity of the printing scaffold and the rheological properties of the hydrogel will determine whether the cells can obtain enough oxygen and nutrients, further spread, and migrate through the material, which is another important consideration.

Postprocessing in photopolymerization bioprinting affects both the structural properties of the construct and cell viability. Typically, these samples need to be immersed in a bath with a solvent (e.g., water or isopropanol) that removes the non-crosslinked volume, and in some cases, the extra support material must be mechanically removed. Furthermore, inks made of synthetic polymers commonly contain organic solvents that are usually cytotoxic. Thus, post-printing cell seeding is more prevalent than cell-laden photopolymerization-based bioprinting.

Examples of photopolymerization-based bioprinting are numerous. Both SLA and DLP have been used to produce vascular grafts and vascularized tissue constructs [213]. These techniques allow the building of cylindrical structures with a very narrow lumen due to their high resolution. For the same reason, they are also used for printing other tubule-like structures, such as nerve conduits [223,224]. 2PP lithography was initially limited to resin-like materials [225], resulting in biomimetic models that are particularly useful for tissue engineering, such as trabecular bone [226,227], cardiac tissue [200], or blood-brain barrier [228]. Nowadays, natural polymers find application in various fields, such as printing microvasculature channels in Col hydrogels [229,230], creating structures for cell encapsulation in GelMA [231], or reproducing the microstructure of a lung's alveolar tissue parenchyma in gelatin methacryloyl-based resin [232].

5.2. Laser-Based

In laser-assisted bioprinting (LAB), a laser beam is focused on the upper surface of a bioink layer affixed to a transparent support. In response to the laser, a bubble forms at the focus spot, making the ink stream along the vacuolar membrane. After the bubble collapses, the ink forms a jet or droplet that is deposited on the substrate [220], which typically contains a biopolymer or cell culture medium to ensure cell adhesion (Figure 2B) [216]. The small size of the droplets guarantees a higher resolution than extrusion-based bioprinting. Despite the numerous factors potentially affecting cell viability in this technique (laser radiation, thermal changes, mechanical forces during the transfer and landing processes), LAB has been reported to achieve viabilities of over 90% [215,217]. LAB allows printing using more viscous bioinks than extrusion-based systems, as the ink does not pass through a nozzle [220].

LAB has been used to print cellularized skin grafts in a mouse skin model using fibroblasts and keratinocytes positioned in a Col solution on the surface of a commercially available Col/elastin matrix (Matriderm®) [116]. After 11 days of implantation, the grafts were explanted and analyzed, revealing that the printed cells had formed a tissue similar to the native mice skin. This formation included a dense epidermis and the formation of a microvascularization network.

5.3. Extrusion-Based

In this bioprinting technique, hydrogels are extruded through a nozzle or printing head and deposited over a substrate to construct layer-by-layer through a mechanical force (Figure 2C). Due to its working mechanism, the bioink must have adequate rheological properties, provide good extrudability, and maintain structural integrity after printing. These requirements, often referred to as printability, contribute significantly to the regulation of cell functioning [233,234]. Thus, the design of a hydrogel for extrusion bioprinting must consider both rheological and biological factors. Bioinks are composed of one or more biomaterials mixed with living cells and eventually other biologically active components that impact their flow behavior. The viscosity of bioinks must be low enough to allow the extrusion process without requiring high pressure. This high pressure causes excessive shear stress, which can dramatically affect cell viability [235]. On the other hand, the retention of the shape of the construct after the extrusion process is primarily influenced by the material yield stress and, secondly, by its viscosity.

Crosslinking/gelation can also be of crucial importance in controlling both printability and cell function in extrusion-based bioprinting. Hydrogel gelation can be performed before, during, and after the extrusion printing procedure. So, pre-printing crosslinking can be used as a printability enhancer to obtain gel-like materials adequate for printing. This procedure must be carefully considered due to the possibility of provoking nozzle clogging, and thus, decreasing the material's extrudability [234] or even inducing the total gelation of the bioink, which is then no longer extrudable. Additionally, crosslinking while the printing process is ongoing, such as immediate photopolymerization after the bioink exits the nozzle, helps in maintaining the structural integrity of the printed part. However, crosslinking agents activated by visible light are preferred compared to those activated by UV light, as they may damage cells, and the presence of photoinitiators may also have deleterious effects on the cells [223]. Finally, post-printing crosslinking is typically accomplished by a light source, thermal treatment, or the addition of a metal-ion source such as $CaCl_2$. However, thermal post-printing treatment in cell-laden hydrogels is limited by the presence of the cells and the range of temperatures they can stand. The addition of $CaCl_2$, which is generally well-tolerated by most cell types, could affect the survival of those cells strongly affected by the presence of X^{2+} ions, such as cardiomyocytes.

Alginate/gelatin (AlgGel) hydrogel is a widespread example of a natural polymeric bioink for extrusion bioprinting due to its excellent printability, biocompatibility, and biodegradability. Alginate is renowned for its printability properties, but it lacks biological activity, while Gel enhances biological activity, resulting in a biocompatible and biodegradable easy-to-print bioink [207]. This combination has been used, for example, to print cardiac patches, mixing the hydrogel with human coronary artery endothelial cells (HCAECs) and separately with mouse vascularized cardiac spheroids (VCSs) [236]. The results showed high viability at 28 days, with the HCAECs successfully establishing a vascular network and the VCSs showing contractility. Another example is the printing of bilayer renal tubular tissue with a decellularized kidney ECM mixed with alginate using a coaxial extruder [237].

5.4. Droplet-Based

In droplet-based bioprinting (DBB), the structure is built by small bioink droplets deposited on a substrate. There are different approaches to this technique: inkjet, acoustic, and microvalve bioprinting (Figure 2D) [238]. Similar to extrusion bioprinting, bioinks used in DBB require low viscosity, biocompatibility, and biodegradability.

Printing hydrogel droplets on a substrate is the most frequent approach, but an alternative approach involving printing droplets containing cells or other biologicals on top of hydrogel substrates has also been reported [238]. For instance, skin graft substitutes have been fabricated by printing a solution of human dermal microvascular endothelial cells mixed with thrombin onto a substrate made of Col type I with neonatal human fibroblasts [239]. These skin grafts were implanted in a mouse model, demonstrating improvements in wound contraction.

Besides Col, several natural polymeric hydrogels have been reported as effective options for DBB. Alginate is frequently used in DBB and is often preferred due to its excellent printability, for example, in printing microvasculature constructs [240]. GelMA is also one of the most commonly used natural polymer hydrogels due to its degradability and tunable mechanical properties. GelMA has been used to recreate tumor microenvironments by mixing co-cultured CAL27-CALF tumor spheroids with a GelMA-based bioink and printing them using acoustic droplet printing [241]. F has been used to produce microvasculature by inkjet printing droplets of thrombin loaded with human dermal microvascular endothelial cells on top of a fibrinogen substrate, forming a microcapillary network after 21 days of culture [241].

6. Discussion

3D bioprinting is a recent and promising technology for TE but still presents several challenges to solve. The most relevant ones are: (1) the identification of biodegradable and biomimetic printable biopolymers that enable prompt cell adhesion and proliferation; (2) the need for vascularization at the single-cell level; (3) the development of complex patterning of heterocellular tissues; and (4) the conservation of cell viability and long-term post-printing functionality until the remodeling and regeneration of the tissue are completed [179,206,209,228].

Figure 3 presents the most relevant data extracted in this review, showing that biopolymers are commonly used in six specific TE applications: bone, cartilage, neural, vascular, skin, and pancreatic tissue regeneration. Other applications are included under a general concept named other tissue regeneration. This figure represents the interaction of commonly used biopolymers, their bonding into new hydrogels, and the printing and crosslinking techniques used for each application.

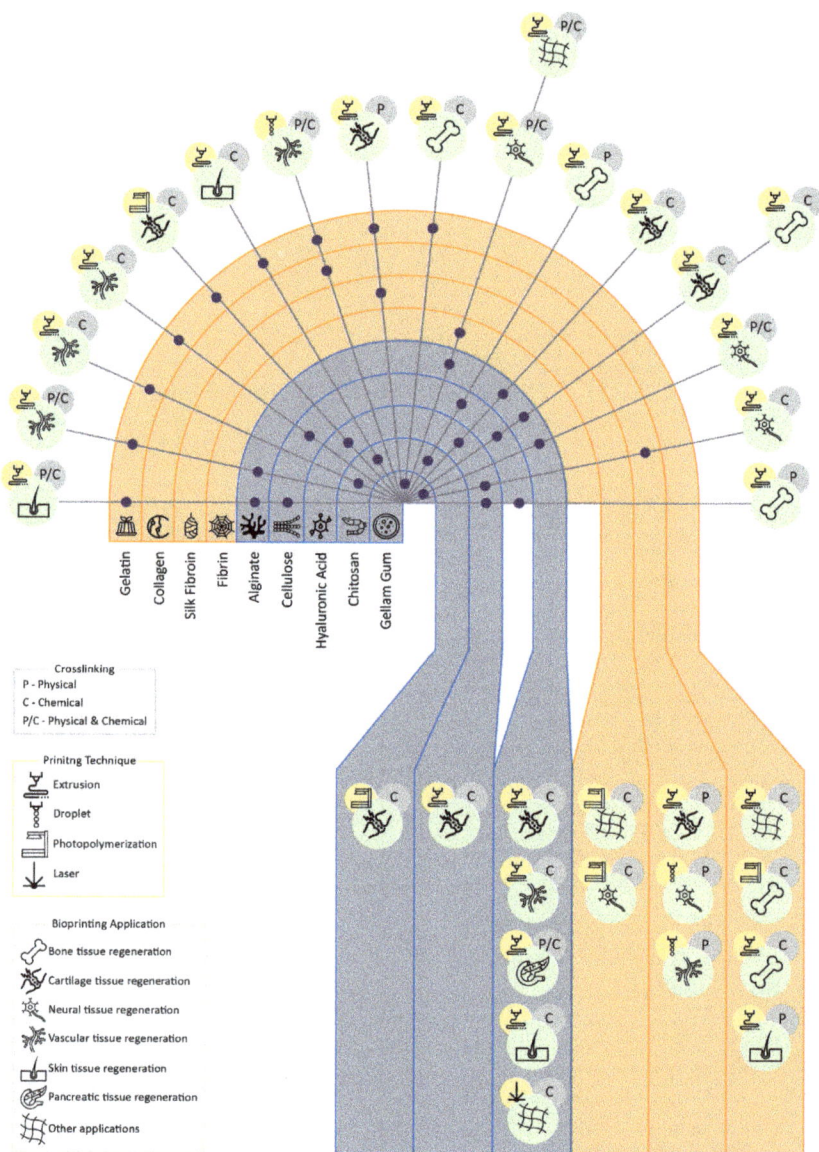

Figure 3. General overview of the data analysis where the most used biopolymers are classified with circles according to: (1) their application (green), (2) the printing procedure (yellow), and (3) the crosslinking type (grey). Hydrogel's blend is represented in the upper part (purple dots), while one-biopolymer hydrogel appears in the lower part of the figure, although other materials can be part of the blend (not included for simplicity).

7. Conclusions and Future Perspectives

In conclusion, hydrogels based on biological source compounds, namely, polysaccharides and proteins, are widely used in the composition of bioinks due to their biocompatibility, biodegradability, and resemblance to some properties of the ECM (e.g., stiffness). The potential of natural-based bioinks has been evaluated in several TE applications, such as bone, cartilage, neural, vascular, skin, and pancreatic tissue regeneration. These types of bioinks have been studied due to their biocompatibility, flexibility to be chemically modified or blended to improve their printability and mechanical stability during and after the printing process, and also to augment the biological or mechanical features of the 3D-printed structures. However, the rheological properties of natural compounds are unsuitable to be used alone as bioinks. So, several procedures to improve their printability are commonly used, such as blending with other natural or synthetic materials or chemical modification. In general, alginate and gelatin are biopolymers widely used in TE for a variety of applications, together with other methacrylated biopolymers (GelMA, ColMA, and HAMA). However, new alternative biopolymers are currently under study to avoid undesirable effects that compromise cell proliferation and maturation using Michael reactions for crosslinking.

Furthermore, future research may also attempt to optimize the printing techniques and printing process parameters to develop specific multifunctional bioprinted constructs. By combining state-of-the-art TE strategies and achievements made by current and ongoing research, it aims to propel the development of ideal 3D bioprinted materials based on biopolymers and consequently their successful translation into clinical applications in the future.

Author Contributions: Conceptualization, D.P., J.L., S.P.M., V.G.-C., P.C., M.P.R. and J.B.P.; methodology, D.P., J.L., S.P.M., V.G.-C., P.C., M.P.R. and J.B.P.; data curation, D.P., J.L., S.P.M., V.G.-C., P.C., M.P.R. and J.B.P.; formal analysis, D.P., J.L., S.P.M., V.G.-C., J.C.G.-B., P.C., M.P.R. and J.B.P.; writing—original draft preparation, D.P., J.L., S.P.M. and V.G.-C.; writing—review and editing, D.P., J.L., S.P.M., V.G.-C., J.C.G.-B., P.C., M.P.R. and J.B.P.; visualization, D.P., J.L., S.P.M., J.C.G.-B. and J.B.P.; supervision, F.M.S.-M., P.C., M.P.R. and J.B.P.; project administration, F.M.S.-M., P.C., M.P.R. and J.B.P.; funding acquisition, F.M.S.-M., J.C.G.-B., P.C., M.P.R. and J.B.P. All authors have read and agreed to the published version of the manuscript.

Funding: This work was supported by the project BioimpACE (0633_BIOIMP_ACE_4_E), co-funded by the European Union/ERDF, ESF, European Regional Development Fund ERDF under the Interreg V-A Spain-Portugal (POCTEP) 2014–2020 program. This work was supported by the Instituto de Salud Carlos III (RD21/0017/0014) and co-granted by the European Union, NextGenerationEU, Plan de Recuperación Transformación y Resiliencia.

Institutional Review Board Statement: Not applicable.

Informed Consent Statement: Not applicable.

Data Availability Statement: No new data was created or analyzed in this study. Data sharing is not applicable to this article.

Acknowledgments: The authors acknowledge Luisa F. Sánchez-Peralta for her help in the synthesis of data and definition of figures.

Conflicts of Interest: The authors declare no conflict of interest.

References

1. Viola, J.; Lal, B.; Grad, O. *The Emergence of Tissue Engineering as a Research Field*; The National Science Foundation Inc.: Alexandria, VA, USA, 2004.
2. Badylak, S.F.; Tullius, R.; Kokini, K.; Shelbourne, K.D.; Klootwyk, T.; Voytik, S.L.; Kraine, M.R.; Simmons, C. The Use of Xenogeneic Small Intestinal Submucosa as a Biomaterial for Achille's Tendon Repair in a Dog Model. *J. Biomed. Mater. Res.* **1995**, *29*, 977–985. [CrossRef]
3. Guyette, J.P.; Charest, J.M.; Mills, R.W.; Jank, B.J.; Moser, P.T.; Gilpin, S.E.; Gershlak, J.R.; Okamoto, T.; Gonzalez, G.; Milan, D.J.; et al. Bioengineering Human Myocardium on Native Extracellular Matrix. *Circ. Res.* **2016**, *118*, 56–72. [CrossRef] [PubMed]
4. Jongpaiboonkit, L.; King, W.J.; Lyons, G.E.; Paguirigan, A.L.; Warrick, J.W.; Beebe, D.J.; Murphy, W.L. An Adaptable Hydrogel Array Format for 3-Dimensional Cell Culture and Analysis. *Biomaterials* **2008**, *29*, 3346–3356. [CrossRef] [PubMed]
5. Rane, A.A.; Chuang, J.S.; Shah, A.; Hu, D.P.; Dalton, N.D.; Gu, Y.; Peterson, K.L.; Omens, J.H.; Christman, K.L. Increased Infarct Wall Thickness by a Bio-Inert Material Is Insufficient to Prevent Negative Left Ventricular Remodeling after Myocardial Infarction. *PLoS ONE* **2011**, *6*, e21571. [CrossRef] [PubMed]

6. Jorgensen, A.M.; Varkey, M.; Gorkun, A.; Clouse, C.; Xu, L.; Chou, Z.; Murphy, S.V.; Molnar, J.; Lee, S.J.; Yoo, J.J.; et al. Bioprinted Skin Recapitulates Normal Collagen Remodeling in Full-Thickness Wounds. *Tissue Eng. Part A* **2020**, *26*, 512–526. [CrossRef]
7. Yang, H.; Sun, L.; Pang, Y.; Hu, D.; Xu, H.; Mao, S.; Peng, W.; Wang, Y.; Xu, Y.; Zheng, Y.-C.; et al. Three-Dimensional Bioprinted Hepatorganoids Prolong Survival of Mice with Liver Failure. *Gut* **2021**, *70*, 567–574. [CrossRef]
8. Knowlton, S.; Anand, S.; Shah, T.; Tasoglu, S. Bioprinting for Neural Tissue Engineering. *Trends Neurosci.* **2018**, *41*, 31–46. [CrossRef]
9. Matai, I.; Kaur, G.; Seyedsalehi, A.; McClinton, A.; Laurencin, C.T. Progress in 3D Bioprinting Technology for Tissue/Organ Regenerative Engineering. *Biomaterials* **2020**, *226*, 119536. [CrossRef]
10. Lee, H.J.; Kim, Y.B.; Ahn, S.H.; Lee, J.-S.; Jang, C.H.; Yoon, H.; Chun, W.; Kim, G.H. A New Approach for Fabricating Collagen/ECM-Based Bioinks Using Preosteoblasts and Human Adipose Stem Cells. *Adv. Healthc. Mater.* **2015**, *4*, 1359–1368. [CrossRef]
11. Loo, Y.; Lakshmanan, A.; Ni, M.; Toh, L.L.; Wang, S.; Hauser, C.A.E. Peptide Bioink: Self-Assembling Nanofibrous Scaffolds for Three-Dimensional Organotypic Cultures. *Nano Lett.* **2015**, *15*, 6919–6925. [CrossRef]
12. Grenier, J.; Duval, H.; Barou, F.; Lv, P.; David, B.; Letourneur, D. Mechanisms of Pore Formation in Hydrogel Scaffolds Textured by Freeze-Drying. *Acta Biomater.* **2019**, *94*, 195–203. [CrossRef] [PubMed]
13. Rezwan, K.; Chen, Q.Z.; Blaker, J.J.; Boccaccini, A.R. Biodegradable and Bioactive Porous Polymer/Inorganic Composite Scaffolds for Bone Tissue Engineering. *Biomaterials* **2006**, *27*, 3413–3431. [CrossRef] [PubMed]
14. Sparkes, J.; Holland, C. The Rheological Properties of Native Sericin. *Acta Biomater.* **2018**, *69*, 234–242. [CrossRef] [PubMed]
15. Cadano, J.R.; Jose, M.; Lubi, A.G.; Maling, J.N.; Moraga, J.S.; Shi, Q.Y.; Vegafria, H.M.; VinceCruz-Abeledo, C.C. A Comparative Study on the Raw Chitin and Chitosan Yields of Common Bio-Waste from Philippine Seafood. *Environ. Sci. Pollut. Res.* **2021**, *28*, 11954–11961. [CrossRef] [PubMed]
16. Peppas, N.A. *Hydrogels in Medicine and Pharmacy*; CRC Press: Boca Raton, FL, USA, 1986; Volume 1.
17. Bashir, S.; Hina, M.; Iqbal, J.; Rajpar, A.H.; Mujtaba, M.A.; Alghamdi, N.A.; Wageh, S.; Ramesh, K.; Ramesh, S. Fundamental Concepts of Hydrogels: Synthesis, Properties, and Their Applications. *Polymers* **2020**, *12*, 2702. [CrossRef]
18. Negrescu, A.-M.; Cimpean, A. The State of the Art and Prospects for Osteoimmunomodulatory Biomaterials. *Materials* **2021**, *14*, 1357. [CrossRef]
19. Rana, D.; Desai, N.; Salave, S.; Karunakaran, B.; Giri, J.; Benival, D.; Gorantla, S.; Kommineni, N. Collagen-Based Hydrogels for the Eye: A Comprehensive Review. *Gels* **2023**, *9*, 643. [CrossRef]
20. Teixeira, M.C.; Lameirinhas, N.S.; Carvalho, J.P.F.; Silvestre, A.J.D.; Vilela, C.; Freire, C.S.R. A Guide to Polysaccharide-Based Hydrogel Bioinks for 3D Bioprinting Applications. *Int. J. Mol. Sci.* **2022**, *23*, 6564. [CrossRef]
21. Cheng, L.; Yao, B.; Hu, T.; Cui, X.; Shu, X.; Tang, S.; Wang, R.; Wang, Y.; Liu, Y.; Song, W.; et al. Properties of an Alginate-Gelatin-Based Bioink and Its Potential Impact on Cell Migration, Proliferation, and Differentiation. *Int. J. Biol. Macromol.* **2019**, *135*, 1107–1113. [CrossRef]
22. Beketov, E.E.; Isaeva, E.V.; Yakovleva, N.D.; Demyashkin, G.A.; Arguchinskaya, N.V.; Kisel, A.A.; Lagoda, T.S.; Malakhov, E.P.; Kharlov, V.I.; Osidak, E.O.; et al. Bioprinting of Cartilage with Bioink Based on High-Concentration Collagen and Chondrocytes. *Int. J. Mol. Sci.* **2021**, *22*, 11351. [CrossRef]
23. Costa, J.B.; Silva-Correia, J.; Oliveira, J.M.; Reis, R.L. Fast Setting Silk Fibroin Bioink for Bioprinting of Patient-Specific Memory-Shape Implants. *Adv. Healthc. Mater.* **2017**, *6*, 1701021. [CrossRef]
24. Hong, S.; Sycks, D.; Chan, H.F.; Lin, S.; Lopez, G.P.; Guilak, F.; Leong, K.W.; Zhao, X. 3D Printing of Highly Stretchable and Tough Hydrogels into Complex, Cellularized Structures. *Adv. Mater.* **2015**, *27*, 4035–4040. [CrossRef] [PubMed]
25. Li, Z.; Huang, S.; Liu, Y.; Yao, B.; Hu, T.; Shi, H.; Xie, J.; Fu, X. Tuning Alginate-Gelatin Bioink Properties by Varying Solvent and Their Impact on Stem Cell Behavior. *Sci. Rep.* **2018**, *8*, 8020. [CrossRef]
26. Osidak, E.O.; Karalkin, P.A.; Osidak, M.S.; Parfenov, V.A.; Sivogrivov, D.E.; Pereira, F.D.A.S.; Gryadunova, A.A.; Koudan, E.V.; Khesuani, Y.D.; Kasyanov, V.A.; et al. Viscoll Collagen Solution as a Novel Bioink for Direct 3D Bioprinting. *J. Mater. Sci. Mater. Med.* **2019**, *30*, 31. [CrossRef] [PubMed]
27. Sharma, R.; Smits, I.P.M.; De La Vega, L.; Lee, C.; Willerth, S.M. 3D Bioprinting Pluripotent Stem Cell Derived Neural Tissues Using a Novel Fibrin Bioink Containing Drug Releasing Microspheres. *Front. Bioeng. Biotechnol.* **2020**, *8*, 57. [CrossRef] [PubMed]
28. Wang, X.; Liu, C. Fibrin Hydrogels for Endothelialized Liver Tissue Engineering with a Predesigned Vascular Network. *Polymers* **2018**, *10*, 1048. [CrossRef]
29. Yan, W.-C.; Davoodi, P.; Vijayavenkataraman, S.; Tian, Y.; Ng, W.C.; Fuh, J.Y.H.; Robinson, K.S.; Wang, C.-H. 3D Bioprinting of Skin Tissue: From Pre-Processing to Final Product Evaluation. *Adv. Drug Deliv. Rev.* **2018**, *132*, 270–295. [CrossRef]
30. Zheng, Z.; Wu, J.; Liu, M.; Wang, H.; Li, C.; Rodriguez, M.J.; Li, G.; Wang, X.; Kaplan, D.L. 3D Bioprinting of Self-Standing Silk-Based Bioink. *Adv. Healthc. Mater.* **2018**, *7*, 1701026. [CrossRef]
31. Lan, X.; Liang, Y.; Erkut, E.J.N.; Kunze, M.; Mulet-Sierra, A.; Gong, T.; Osswald, M.; Ansari, K.; Seikaly, H.; Boluk, Y.; et al. Bioprinting of Human Nasoseptal Chondrocytes-laden Collagen Hydrogel for Cartilage Tissue Engineering. *FASEB J.* **2021**, *35*, 21191. [CrossRef]
32. Lee, Y.-B.; Polio, S.; Lee, W.; Dai, G.; Menon, L.; Carroll, R.S.; Yoo, S.-S. Bio-Printing of Collagen and VEGF-Releasing Fibrin Gel Scaffolds for Neural Stem Cell Culture. *Exp. Neurol.* **2010**, *223*, 645–652. [CrossRef]

33. Hsieh, C.-T.; Hsu, S. Double-Network Polyurethane-Gelatin Hydrogel with Tunable Modulus for High-Resolution 3D Bioprinting. *ACS Appl. Mater. Interfaces* **2019**, *11*, 32746–32757. [CrossRef] [PubMed]
34. Anada, T.; Pan, C.-C.; Stahl, A.; Mori, S.; Fukuda, J.; Suzuki, O.; Yang, Y. Vascularized Bone-Mimetic Hydrogel Constructs by 3D Bioprinting to Promote Osteogenesis and Angiogenesis. *Int. J. Mol. Sci.* **2019**, *20*, 1096. [CrossRef] [PubMed]
35. Leucht, A.; Volz, A.-C.; Rogal, J.; Borchers, K.; Kluger, P.J. Advanced Gelatin-Based Vascularization Bioinks for Extrusion-Based Bioprinting of Vascularized Bone Equivalents. *Sci. Rep.* **2020**, *10*, 5330. [CrossRef] [PubMed]
36. Kim, S.H.; Yeon, Y.K.; Lee, J.M.; Chao, J.R.; Lee, Y.J.; Seo, Y.B.; Sultan, M.T.; Lee, O.J.; Lee, J.S.; Yoon, S.; et al. Precisely Printable and Biocompatible Silk Fibroin Bioink for Digital Light Processing 3D Printing. *Nat. Commun.* **2018**, *9*, 1620. [CrossRef]
37. Ajiteru, O.; Sultan, M.T.; Lee, Y.J.; Seo, Y.B.; Hong, H.; Lee, J.S.; Lee, H.; Suh, Y.J.; Ju, H.W.; Lee, O.J.; et al. A 3D Printable Electroconductive Biocomposite Bioink Based on Silk Fibroin-Conjugated Graphene Oxide. *Nano Lett.* **2020**, *20*, 6873–6883. [CrossRef]
38. Lee, C.H.; Singla, A.; Lee, Y. Biomedical Applications of Collagen. *Int. J. Pharm.* **2001**, *221*, 1–22. [CrossRef]
39. Hinton, T.J.; Jallerat, Q.; Palchesko, R.N.; Park, J.H.; Grodzicki, M.S.; Shue, H.-J.; Ramadan, M.H.; Hudson, A.R.; Feinberg, A.W. Three-Dimensional Printing of Complex Biological Structures by Freeform Reversible Embedding of Suspended Hydrogels. *Sci. Adv.* **2015**, *1*, 1500758. [CrossRef]
40. Lee, A.; Hudson, A.R.; Shiwarski, D.J.; Tashman, J.W.; Hinton, T.J.; Yerneni, S.; Bliley, J.M.; Campbell, P.G.; Feinberg, A.W. 3D Bioprinting of Collagen to Rebuild Components of the Human Heart. *Science* **2019**, *365*, 482–487. [CrossRef]
41. Roth, E.A.; Xu, T.; Das, M.; Gregory, C.; Hickman, J.J.; Boland, T. Inkjet Printing for High-Throughput Cell Patterning. *Biomaterials* **2004**, *25*, 3707–3715. [CrossRef]
42. Xu, F.; Moon, S.; Emre, A.E.; Lien, C.; Turali, E.S.; Demirci, U. Cell Bioprinting as a Potential High-Throughput Method for Fabricating Cell-Based Biosensors (CBBs). In Proceedings of the 2009 IEEE Sensors, Christchurch, New Zealand, 25–28 October 2009; IEEE: Piscataway, NJ, USA, 2009; pp. 387–391.
43. Stepanovska, J.; Supova, M.; Hanzalek, K.; Broz, A.; Matejka, R. Collagen Bioinks for Bioprinting: A Systematic Review of Hydrogel Properties, Bioprinting Parameters, Protocols, and Bioprinted Structure Characteristics. *Biomedicines* **2021**, *9*, 1137. [CrossRef]
44. Lee, V.; Singh, G.; Trasatti, J.P.; Bjornsson, C.; Xu, X.; Tran, T.N.; Yoo, S.-S.; Dai, G.; Karande, P. Design and Fabrication of Human Skin by Three-Dimensional Bioprinting. *Tissue Eng. Part C Methods* **2014**, *20*, 473–484. [CrossRef] [PubMed]
45. Chan, K.; O'Brien, P.J. Structure-Activity Relationships for Hepatocyte Toxicity and Electrophilic Reactivity of α, β-Unsaturated Esters, Acrylates and Methacrylates. *J. Appl. Toxicol.* **2008**, *28*, 1004–1015. [CrossRef] [PubMed]
46. LoPachin, R.M.; Gavin, T. Molecular Mechanisms of Aldehyde Toxicity: A Chemical Perspective. *Chem. Res. Toxicol.* **2014**, *27*, 1081–1091. [CrossRef]
47. Liu, C.Z.; Xia, Z.D.; Han, Z.W.; Hulley, P.A.; Triffitt, J.T.; Czernuszka, J.T. Novel 3D Collagen Scaffolds Fabricated by Indirect Printing Technique for Tissue Engineering. *J. Biomed. Mater. Res. B Appl. Biomater.* **2008**, *85B*, 519–528. [CrossRef] [PubMed]
48. Gungor-Ozkerim, P.S.; Inci, I.; Zhang, Y.S.; Khademhosseini, A.; Dokmeci, M.R. Bioinks for 3D Bioprinting: An Overview. *Biomater. Sci.* **2018**, *6*, 915–946. [CrossRef]
49. Yoon, H.; Lee, J.-S.; Yim, H.; Kim, G.; Chun, W. Development of Cell-Laden 3D Scaffolds for Efficient Engineered Skin Substitutes by Collagen Gelation. *RSC Adv.* **2016**, *6*, 21439–21447. [CrossRef]
50. Albanna, M.; Binder, K.W.; Murphy, S.V.; Kim, J.; Qasem, S.A.; Zhao, W.; Tan, J.; El-Amin, I.B.; Dice, D.D.; Marco, J.; et al. In Situ Bioprinting of Autologous Skin Cells Accelerates Wound Healing of Extensive Excisional Full-Thickness Wounds. *Sci. Rep.* **2019**, *9*, 1856. [CrossRef] [PubMed]
51. Kim, W.; Kim, G. Collagen/Bioceramic-Based Composite Bioink to Fabricate a Porous 3D HASCs-Laden Structure for Bone Tissue Regeneration. *Biofabrication* **2019**, *12*, 015007. [CrossRef]
52. Harrington, W.F.; Von Hippel, P.H. The Structure of Collagen And Gelatin. *Adv. Protein Chem.* **1962**, *16*, 1–138.
53. Liu, D.; Nikoo, M.; Boran, G.; Zhou, P.; Regenstein, J.M. Collagen and Gelatin. *Annu. Rev. Food Sci. Technol.* **2015**, *6*, 527–557. [CrossRef]
54. Singh, S.; Prakash, C.; Ramakrishna, S. *Additive Manufacturing*; World Scientific: Singapore, 2020; ISBN 978-981-12-2481-2.
55. Li, J.; Sun, H.; Jiang, L.; Zhang, K.; Liu, W.; Zhu, Y.; Fangteng, J.; Shi, C.; Zhao, L.; Sun, H.; et al. Enhanced Biocompatibility of PLGA Nanofibers with Gelatin/Nano-Hydroxyapatite Bone Biomimetics Incorporation. *ACS Appl. Mater. Interfaces* **2014**, *6*, 9402–9410. [CrossRef] [PubMed]
56. Pezeshki-Modaress, M.; Zandi, M.; Mirzadeh, H. Fabrication of Gelatin/Chitosan Nanofibrous Scaffold: Process Optimization and Empirical Modeling. *Polym. Int.* **2015**, *64*, 571–580. [CrossRef]
57. Wu, S.; Deng, L.; Hsia, H.; Xu, K.; He, Y.; Huang, Q.; Peng, Y.; Zhou, Z.; Peng, C. Evaluation of Gelatin-Hyaluronic Acid Composite Hydrogels for Accelerating Wound Healing. *J. Biomater. Appl.* **2017**, *31*, 1380–1390. [CrossRef] [PubMed]
58. Nguyen, B.B.; Moriarty, R.A.; Kamalitdinov, T.; Etheridge, J.M.; Fisher, J.P. Collagen Hydrogel Scaffold Promotes Mesenchymal Stem Cell and Endothelial Cell Coculture for Bone Tissue Engineering. *J. Biomed. Mater. Res.* **2017**, *105*, 1123–1131. [CrossRef]
59. Claaßen, C.; Claaßen, M.H.; Truffault, V.; Sewald, L.; Tovar, G.E.M.; Borchers, K.; Southan, A. Quantification of Substitution of Gelatin Methacryloyl: Best Practice and Current Pitfalls. *Biomacromolecules* **2018**, *19*, 42–52. [CrossRef]
60. Hoch, E.; Hirth, T.; Tovar, G.E.M.; Borchers, K. Chemical Tailoring of Gelatin to Adjust Its Chemical and Physical Properties for Functional Bioprinting. *J. Mater. Chem. B* **2013**, *1*, 5675. [CrossRef]

61. Yin, J.; Yan, M.; Wang, Y.; Fu, J.; Suo, H. 3D Bioprinting of Low-Concentration Cell-Laden Gelatin Methacrylate (GelMA) Bioinks with a Two-Step Cross-Linking Strategy. *ACS Appl. Mater. Interfaces* **2018**, *10*, 6849–6857. [CrossRef]
62. Kara Özenler, A.; Distler, T.; Tihminlioglu, F.; Boccaccini, A.R. Fish Scale Containing Alginate Dialdehyde-Gelatin Bioink for Bone Tissue Engineering. *Biofabrication* **2023**, *15*, 025012. [CrossRef]
63. Semba, J.A.; Mieloch, A.A.; Tomaszewska, E.; Cywoniuk, P.; Rybka, J.D. Formulation and Evaluation of a Bioink Composed of Alginate, Gelatin, and Nanocellulose for Meniscal Tissue Engineering. *Int. J. Bioprinting* **2022**, *9*, 621. [CrossRef]
64. Shi, W.; Fang, F.; Kong, Y.; Greer, S.E.; Kuss, M.; Liu, B.; Xue, W.; Jiang, X.; Lovell, P.; Mohs, A.M.; et al. Dynamic Hyaluronic Acid Hydrogel with Covalent Linked Gelatin as an Anti-Oxidative Bioink for Cartilage Tissue Engineering. *Biofabrication* **2022**, *14*, 014107. [CrossRef]
65. Song, W.; Muthana, M.; Mukherjee, J.; Falconer, R.J.; Biggs, C.A.; Zhao, X. Magnetic-Silk Core–Shell Nanoparticles as Potential Carriers for Targeted Delivery of Curcumin into Human Breast Cancer Cells. *ACS Biomater. Sci. Eng.* **2017**, *3*, 1027–1038. [CrossRef] [PubMed]
66. Zhang, C.; Zhang, Y.; Shao, H.; Hu, X. Hybrid Silk Fibers Dry-Spun from Regenerated Silk Fibroin/Graphene Oxide Aqueous Solutions. *ACS Appl. Mater. Interfaces* **2016**, *8*, 3349–3358. [CrossRef] [PubMed]
67. Tao, H.; Kaplan, D.L.; Omenetto, F.G. Silk Materials—A Road to Sustainable High Technology. *Adv. Mater.* **2012**, *24*, 2824–2837. [CrossRef] [PubMed]
68. Tao, H.; Marelli, B.; Yang, M.; An, B.; Onses, M.S.; Rogers, J.A.; Kaplan, D.L.; Omenetto, F.G. Inkjet Printing of Regenerated Silk Fibroin: From Printable Forms to Printable Functions. *Adv. Mater.* **2015**, *27*, 4273–4279. [CrossRef] [PubMed]
69. Lee, O.J.; Lee, J.M.; Kim, J.H.; Kim, J.; Kweon, H.; Jo, Y.Y.; Park, C.H. Biodegradation Behavior of Silk Fibroin Membranes in Repairing Tympanic Membrane Perforations. *J. Biomed. Mater. Res. A* **2012**, *100A*, 2018–2026. [CrossRef]
70. Park, H.J.; Lee, O.J.; Lee, M.C.; Moon, B.M.; Ju, H.W.; Lee, J.M.; Kim, J.-H.; Kim, D.W.; Kim, J.H.; Park, C.H. Fabrication of 3D Porous Silk Scaffolds by Particulate (Salt/Sucrose) Leaching for Bone Tissue Reconstruction. *Int. J. Biol. Macromol.* **2015**, *78*, 215–223. [CrossRef]
71. Park, W.H.; Kim, M.H. Chemically Cross-Linked Silk Fibroin Hydrogel with Enhanced Elastic Properties, Biodegradability, and Biocompatibility. *Int. J. Nanomed.* **2016**, *11*, 2967. [CrossRef]
72. Yang, J.; Li, Z.; Li, S.; Zhang, Q.; Zhou, X.; He, C. Tunable Metacrylated Silk Fibroin-Based Hybrid Bioinks for the Bioprinting of Tissue Engineering Scaffolds. *Biomater. Sci.* **2023**, *11*, 1895–1909. [CrossRef]
73. Mohammadpour, Z.; Kharaziha, M.; Zarrabi, A. 3D-Printing of Silk Nanofibrils Reinforced Alginate for Soft Tissue Engineering. *Pharmaceutics* **2023**, *15*, 763. [CrossRef]
74. Rodriguez, M.J.; Brown, J.; Giordano, J.; Lin, S.J.; Omenetto, F.G.; Kaplan, D.L. Silk Based Bioinks for Soft Tissue Reconstruction Using 3-Dimensional (3D) Printing with in Vitro and in Vivo Assessments. *Biomaterials* **2017**, *117*, 105–115. [CrossRef]
75. Singh, Y.P.; Bandyopadhyay, A.; Mandal, B.B. 3D Bioprinting Using Cross-Linker-Free Silk–Gelatin Bioink for Cartilage Tissue Engineering. *ACS Appl. Mater. Interfaces* **2019**, *11*, 33684–33696. [CrossRef] [PubMed]
76. Undas, A.; Ariëns, R.A.S. Fibrin Clot Structure and Function. *Arterioscler. Thromb. Vasc. Biol.* **2011**, *31*, 88–99. [CrossRef] [PubMed]
77. de Melo, B.A.G.; Jodat, Y.A.; Cruz, E.M.; Benincasa, J.C.; Shin, S.R.; Porcionatto, M.A. Strategies to Use Fibrinogen as Bioink for 3D Bioprinting Fibrin-Based Soft and Hard Tissues. *Acta Biomater.* **2020**, *117*, 60–76. [CrossRef] [PubMed]
78. Budharaju, H.; Sundaramurthi, D.; Sethuraman, S. Efficient Dual Crosslinking of Protein-in–Polysaccharide Bioink for Biofabrication of Cardiac Tissue Constructs. *Biomater. Adv.* **2023**, *152*, 213486. [CrossRef]
79. Kotlarz, M.; Ferreira, A.M.; Gentile, P.; Russell, S.J.; Dalgarno, K. Droplet-Based Bioprinting Enables the Fabrication of Cell–Hydrogel–Microfibre Composite Tissue Precursors. *Bio-Des. Manuf.* **2022**, *5*, 512–528. [CrossRef]
80. de Melo, B.A.G.; Jodat, Y.A.; Mehrotra, S.; Calabrese, M.A.; Kamperman, T.; Mandal, B.B.; Santana, M.H.A.; Alsberg, E.; Leijten, J.; Shin, S.R. 3D Printed Cartilage-Like Tissue Constructs with Spatially Controlled Mechanical Properties. *Adv. Funct. Mater.* **2019**, *29*, 1906330. [CrossRef]
81. Anil Kumar, S.; Alonzo, M.; Allen, S.C.; Abelseth, L.; Thakur, V.; Akimoto, J.; Ito, Y.; Willerth, S.M.; Suggs, L.; Chattopadhyay, M.; et al. A Visible Light-Cross-Linkable, Fibrin–Gelatin-Based Bioprinted Construct with Human Cardiomyocytes and Fibroblasts. *ACS Biomater. Sci. Eng.* **2019**, *5*, 4551–4563. [CrossRef]
82. Rutz, A.L.; Hyland, K.E.; Jakus, A.E.; Burghardt, W.R.; Shah, R.N. A Multimaterial Bioink Method for 3D Printing Tunable, Cell-Compatible Hydrogels. *Adv. Mater.* **2015**, *27*, 1607–1614. [CrossRef]
83. de la Vega, L.; Gómez, D.A.R.; Abelseth, E.; Abelseth, L.; Allisson da Silva, V.; Willerth, S. 3D Bioprinting Human Induced Pluripotent Stem Cell-Derived Neural Tissues Using a Novel Lab-on-a-Printer Technology. *Appl. Sci.* **2018**, *8*, 2414. [CrossRef]
84. Keane, T.J.; Swinehart, I.T.; Badylak, S.F. Methods of Tissue Decellularization Used for Preparation of Biologic Scaffolds and in Vivo Relevance. *Methods* **2015**, *84*, 25–34. [CrossRef]
85. Dzobo, K.; Motaung, K.S.C.M.; Adesida, A. Recent Trends in Decellularized Extracellular Matrix Bioinks for 3D Printing: An Updated Review. *Int. J. Mol. Sci.* **2019**, *20*, 4628. [CrossRef] [PubMed]
86. Han, W.; Singh, N.K.; Kim, J.J.; Kim, H.; Kim, B.S.; Park, J.Y.; Jang, J.; Cho, D.-W. Directed Differential Behaviors of Multipotent Adult Stem Cells from Decellularized Tissue/Organ Extracellular Matrix Bioinks. *Biomaterials* **2019**, *224*, 119496. [CrossRef] [PubMed]
87. Pati, F.; Jang, J.; Ha, D.-H.; Won Kim, S.; Rhie, J.-W.; Shim, J.-H.; Kim, D.-H.; Cho, D.-W. Printing Three-Dimensional Tissue Analogues with Decellularized Extracellular Matrix Bioink. *Nat. Commun.* **2014**, *5*, 3935. [CrossRef]

88. Kim, B.S.; Kwon, Y.W.; Kong, J.-S.; Park, G.T.; Gao, G.; Han, W.; Kim, M.-B.; Lee, H.; Kim, J.H.; Cho, D.-W. 3D Cell Printing of in Vitro Stabilized Skin Model and in Vivo Pre-Vascularized Skin Patch Using Tissue-Specific Extracellular Matrix Bioink: A Step towards Advanced Skin Tissue Engineering. *Biomaterials* **2018**, *168*, 38–53. [CrossRef] [PubMed]
89. Jorgensen, A.M.; Chou, Z.; Gillispie, G.; Lee, S.J.; Yoo, J.J.; Soker, S.; Atala, A. Decellularized Skin Extracellular Matrix (DsECM) Improves the Physical and Biological Properties of Fibrinogen Hydrogel for Skin Bioprinting Applications. *Nanomaterials* **2020**, *10*, 1484. [CrossRef]
90. Jang, J.; Kim, T.G.; Kim, B.S.; Kim, S.-W.; Kwon, S.-M.; Cho, D.-W. Tailoring Mechanical Properties of Decellularized Extracellular Matrix Bioink by Vitamin B2-Induced Photo-Crosslinking. *Acta Biomater.* **2016**, *33*, 88–95. [CrossRef]
91. Jang, J.; Park, H.-J.; Kim, S.-W.; Kim, H.; Park, J.Y.; Na, S.J.; Kim, H.J.; Park, M.N.; Choi, S.H.; Park, S.H.; et al. 3D Printed Complex Tissue Construct Using Stem Cell-Laden Decellularized Extracellular Matrix Bioinks for Cardiac Repair. *Biomaterials* **2017**, *112*, 264–274. [CrossRef]
92. Hiller, T.; Berg, J.; Elomaa, L.; Röhrs, V.; Ullah, I.; Schaar, K.; Dietrich, A.-C.; Al-Zeer, M.; Kurtz, A.; Hocke, A.; et al. Generation of a 3D Liver Model Comprising Human Extracellular Matrix in an Alginate/Gelatin-Based Bioink by Extrusion Bioprinting for Infection and Transduction Studies. *Int. J. Mol. Sci.* **2018**, *19*, 3129. [CrossRef]
93. Nerger, B.A.; Brun, P.-T.; Nelson, C.M. Microextrusion Printing Cell-Laden Networks of Type I Collagen with Patterned Fiber Alignment and Geometry. *Soft Matter* **2019**, *15*, 5728–5738. [CrossRef]
94. Duarte, A.C.; Costa, E.C.; Filipe, H.A.L.; Saraiva, S.M.; Jacinto, T.; Miguel, S.P.; Ribeiro, M.P.; Coutinho, P. Animal-Derived Products in Science and Current Alternatives. *Biomater. Adv.* **2023**, *151*, 213428. [CrossRef]
95. Li, Y.; Cao, X.; Deng, W.; Yu, Q.; Sun, C.; Ma, P.; Shao, F.; Yusif, M.M.; Ge, Z.; Wang, K.; et al. 3D Printable Sodium Alginate-Matrigel (SA-MA) Hydrogel Facilitated Ectomesenchymal Stem Cells (EMSCs) Neuron Differentiation. *J. Biomater. Appl.* **2021**, *35*, 709–719. [CrossRef] [PubMed]
96. Fan, R.; Piou, M.; Darling, E.; Cormier, D.; Sun, J.; Wan, J. Bio-Printing Cell-Laden Matrigel–Agarose Constructs. *J. Biomater. Appl.* **2016**, *31*, 684–692. [CrossRef] [PubMed]
97. Snyder, J.E.; Hamid, Q.; Wang, C.; Chang, R.; Emami, K.; Wu, H.; Sun, W. Bioprinting Cell-Laden Matrigel for Radioprotection Study of Liver by pro-Drug Conversion in a Dual-Tissue Microfluidic Chip. *Biofabrication* **2011**, *3*, 034112. [CrossRef] [PubMed]
98. Maity, M.; Hasnain, M.S.; Nayak, A.K.; Aminabhavi, T.M. Biomedical Applications of Polysaccharides. In *Tailor-Made Polysaccharides in Biomedical Applications*; Nayak, A., Hasnain, M.S.T., Eds.; Elsevier: Boston, MA, USA, 2020; pp. 1–34.
99. Souza, P.R.; de Oliveira, A.C.; Vilsinski, B.H.; Kipper, M.J.; Martins, A.F. Polysaccharide-Based Materials Created by Physical Processes: From Preparation to Biomedical Applications. *Pharmaceutics* **2021**, *13*, 621. [CrossRef]
100. Chaabouni, E.; Gassara, F.; Brar, S.K. Biopolymers Synthesis and Application. In *Biotransformation of Waste Biomass into High Value Biochemicals*; Brar, S.K., Dhillon, G.S., Soccol, C.R., Eds.; Springer: New York, NY, USA, 2014; pp. 415–443.
101. Stanton, M.M.; Samitier, J.; Sánchez, S. Bioprinting of 3D Hydrogels. *Lab Chip* **2015**, *15*, 3111–3115. [CrossRef]
102. Abdelhak, M. A Review: Application of Biopolymers in the Pharmaceutical Formulation. *J. Adv. Biopharm. Pharmacovigil.* **2019**, *1*, 11. [CrossRef]
103. Rowley, J.A.; Madlambayan, G.; Mooney, D.J. Alginate Hydrogels as Synthetic Extracellular Matrix Materials. *Biomaterials* **1999**, *20*, 45–53. [CrossRef]
104. Benwood, C.; Chrenek, J.; Kirsch, R.L.; Masri, N.Z.; Richards, H.; Teetzen, K.; Willerth, S.M. Natural Biomaterials and Their Use as Bioinks for Printing Tissues. *Bioengineering* **2021**, *8*, 27. [CrossRef]
105. López-Marcial, G.R.; Zeng, A.Y.; Osuna, C.; Dennis, J.; García, J.M.; O'Connell, G.D. Agarose-Based Hydrogels as Suitable Bioprinting Materials for Tissue Engineering. *ACS Biomater. Sci. Eng.* **2018**, *4*, 3610–3616. [CrossRef]
106. Kreimendahl, F.; Köpf, M.; Thiebes, A.L.; Duarte Campos, D.F.; Blaeser, A.; Schmitz-Rode, T.; Apel, C.; Jockenhoevel, S.; Fischer, H. Three-Dimensional Printing and Angiogenesis: Tailored Agarose-Type I Collagen Blends Comprise Three-Dimensional Printability and Angiogenesis Potential for Tissue-Engineered Substitutes. *Tissue Eng. Part C Methods* **2017**, *23*, 604–615. [CrossRef]
107. Morris, V.B.; Nimbalkar, S.; Younesi, M.; McClellan, P.; Akkus, O. Mechanical Properties, Cytocompatibility and Manufacturability of Chitosan:PEGDA Hybrid-Gel Scaffolds by Stereolithography. *Ann. Biomed. Eng.* **2017**, *45*, 286–296. [CrossRef] [PubMed]
108. Mihajlovic, M.; Rikkers, M.; Mihajlovic, M.; Viola, M.; Schuiringa, G.; Ilochonwu, B.C.; Masereeuw, R.; Vonk, L.; Malda, J.; Ito, K.; et al. Viscoelastic Chondroitin Sulfate and Hyaluronic Acid Double-Network Hydrogels with Reversible Cross-Links. *Biomacromolecules* **2022**, *23*, 1350–1365. [CrossRef] [PubMed]
109. Piola, B.; Sabbatini, M.; Gino, S.; Invernizzi, M.; Renò, F. 3D Bioprinting of Gelatin–Xanthan Gum Composite Hydrogels for Growth of Human Skin Cells. *Int. J. Mol. Sci.* **2022**, *23*, 539. [CrossRef] [PubMed]
110. Pan, H.; Zheng, B.; Shen, H.; Qi, M.; Shang, Y.; Wu, C.; Zhu, R.; Cheng, L.; Wang, Q. Strength-Tunable Printing of Xanthan Gum Hydrogel via Enzymatic Polymerization and Amide Bioconjugation. *Chem. Commun.* **2020**, *56*, 3457–3460. [CrossRef] [PubMed]
111. Kumari, S.; Mondal, P.; Chatterjee, K. Digital Light Processing-Based 3D Bioprinting of κ-Carrageenan Hydrogels for Engineering Cell-Loaded Tissue Scaffolds. *Carbohydr. Polym.* **2022**, *290*, 119508. [CrossRef] [PubMed]
112. Zhou, A.; Hu, Y.; Chen, C.; Mao, H.; Wang, L.; Zhang, S.; Huang, X. 3D Bioprintable Methacrylated Carrageenan/Sodium Alginate Dual Network Hydrogel for Vascular Tissue Engineering Scaffolding. *Int. J. Polym. Mater. Polym. Biomater.* **2023**, *72*, 550–560. [CrossRef]
113. Tavakoli, S.; Kharaziha, M.; Kermanpur, A.; Mokhtari, H. Sprayable and Injectable Visible-Light Kappa-Carrageenan Hydrogel for in-Situ Soft Tissue Engineering. *Int. J. Biol. Macromol.* **2019**, *138*, 590–601. [CrossRef]

114. Hu, S.; Martinez-Garcia, F.D.; Moeun, B.N.; Burgess, J.K.; Harmsen, M.C.; Hoesli, C.; de Vos, P. An Immune Regulatory 3D-Printed Alginate-Pectin Construct for Immunoisolation of Insulin Producing β-Cells. *Mater. Sci. Eng. C* **2021**, *123*, 112009. [CrossRef]
115. Johnson, D.L.; Ziemba, R.M.; Shebesta, J.H.; Lipscomb, J.C.; Wang, Y.; Wu, Y.; O'Connell, K.D.; Kaltchev, M.G.; van Groningen, A.; Chen, J.; et al. Design of Pectin-Based Bioink Containing Bioactive Agent-Loaded Microspheres for Bioprinting. *Biomed. Phys. Eng. Express* **2019**, *5*, 067004. [CrossRef]
116. Koch, L.; Kuhn, S.; Sorg, H.; Gruene, M.; Schlie, S.; Gaebel, R.; Polchow, B.; Reimers, K.; Stoelting, S.; Ma, N.; et al. Laser Printing of Skin Cells and Human Stem Cells. *Tissue Eng. Part C Methods* **2010**, *16*, 847–854. [CrossRef]
117. Farokhi, M.; Jonidi Shariatzadeh, F.; Solouk, A.; Mirzadeh, H. Alginate Based Scaffolds for Cartilage Tissue Engineering: A Review. *Int. J. Polym. Mater. Polym. Biomater.* **2020**, *69*, 230–247. [CrossRef]
118. Gonzalez-Fernandez, T.; Tenorio, A.J.; Campbell, K.T.; Silva, E.A.; Leach, J.K. Alginate-Based Bioinks for 3D Bioprinting and Fabrication of Anatomically Accurate Bone Grafts. *Tissue Eng. Part A* **2021**, *27*, 1168–1181. [CrossRef] [PubMed]
119. Lee, K.Y.; Mooney, D.J. Alginate: Properties and Biomedical Applications. *Prog. Polym. Sci.* **2012**, *37*, 106–126. [CrossRef] [PubMed]
120. Steiner, D.; Lingens, L.; Fischer, L.; Köhn, K.; Detsch, R.; Boccaccini, A.R.; Fey, T.; Greil, P.; Weis, C.; Beier, J.P.; et al. Encapsulation of Mesenchymal Stem Cells Improves Vascularization of Alginate-Based Scaffolds. *Tissue Eng. Part A* **2018**, *24*, 1320–1331. [CrossRef] [PubMed]
121. Gonzalez-Fernandez, T.; Sikorski, P.; Leach, J.K. Bio-Instructive Materials for Musculoskeletal Regeneration. *Acta Biomater.* **2019**, *96*, 20–34. [CrossRef]
122. Axpe, E.; Oyen, M. Applications of Alginate-Based Bioinks in 3D Bioprinting. *Int. J. Mol. Sci.* **2016**, *17*, 1976. [CrossRef]
123. Leonardo, M.; Prajatelistia, E.; Judawisastra, H. Alginate-Based Bioink for Organoid 3D Bioprinting: A Review. *Bioprinting* **2022**, *28*, e00246. [CrossRef]
124. Singh, R.; Shitiz, K.; Singh, A. Chitin and Chitosan: Biopolymers for Wound Management. *Int. Wound J.* **2017**, *14*, 1276–1289. [CrossRef]
125. Sahranavard, M.; Zamanian, A.; Ghorbani, F.; Shahrezaee, M.H. A Critical Review on Three Dimensional-Printed Chitosan Hydrogels for Development of Tissue Engineering. *Bioprinting* **2020**, *17*, e00063. [CrossRef]
126. Sun, Y.; You, Y.; Jiang, W.; Wu, Q.; Wang, B.; Dai, K. Generating Ready-to-Implant Anisotropic Menisci by 3D-Bioprinting Protein-Releasing Cell-Laden Hydrogel-Polymer Composite Scaffold. *Appl. Mater. Today* **2020**, *18*, 100469. [CrossRef]
127. Cruz, G.J.F.; Rimaycuna, J.; Alfaro, R.; Tripul, V.S.; Solis, R.L.; Gómez, M.M.; Paraguay-Delgado, F.; Solis, J.L. Antimicrobial Activity of Chitosan Composites against Bacterial Strains Isolated from Goat Meat and Cheese. *J. Phys. Conf. Ser.* **2019**, *1173*, 012005. [CrossRef]
128. Manouchehri, S.; Bagheri, B.; Rad, S.H.; Nezhad, M.N.; Kim, Y.C.; Park, O.O.; Farokhi, M.; Jouyandeh, M.; Ganjali, M.R.; Yazdi, M.K.; et al. Electroactive Bio-Epoxy Incorporated Chitosan-Oligoaniline as an Advanced Hydrogel Coating for Neural Interfaces. *Prog. Org. Coat.* **2019**, *131*, 389–396. [CrossRef]
129. Mohebbi, S.; Nezhad, M.N.; Zarrintaj, P.; Jafari, S.H.; Gholizadeh, S.S.; Saeb, M.R.; Mozafari, M. Chitosan in Biomedical Engineering: A Critical Review. *Curr. Stem Cell Res. Ther.* **2019**, *14*, 93–116. [CrossRef] [PubMed]
130. Bagheri, B.; Zarrintaj, P.; Samadi, A.; Zarrintaj, R.; Ganjali, M.R.; Saeb, M.R.; Mozafari, M.; Park, O.O.; Kim, Y.C. Tissue Engineering with Electrospun Electro-Responsive Chitosan-Aniline Oligomer/Polyvinyl Alcohol. *Int. J. Biol. Macromol.* **2020**, *147*, 160–169. [CrossRef] [PubMed]
131. Mahmodi, G.; Zarrintaj, P.; Taghizadeh, A.; Taghizadeh, M.; Manouchehri, S.; Dangwal, S.; Ronte, A.; Ganjali, M.R.; Ramsey, J.D.; Kim, S.-J.; et al. From Microporous to Mesoporous Mineral Frameworks: An Alliance between Zeolite and Chitosan. *Carbohydr. Res.* **2020**, *489*, 107930. [CrossRef]
132. Mora Boza, A.; Wlodarczyk-Biegun, M.K.; Del Campo, A.; Vázquez-Lasal, B.; San Román, J. Chitosan-Based Inks: 3D Printing and Bioprinting Strategies to Improve Shape Fidelity, Mechanical Properties, and Biocompatibility of 3D Scaffolds. *Biomecánica* **2019**, *27*, 7–16. [CrossRef]
133. Tarassoli, S.P.; Jessop, Z.M.; Kyle, S.; Whitaker, I.S. Candidate Bioinks for 3D Bioprinting Soft Tissue. In *3D Bioprinting for Reconstructive Surgery*; Elsevier: Amsterdam, The Netherlands, 2018; pp. 145–172.
134. Agarwal, T.; Chiesa, I.; Costantini, M.; Lopamarda, A.; Tirelli, M.C.; Borra, O.P.; Varshapally, S.V.S.; Kumar, Y.A.V.; Koteswara Reddy, G.; De Maria, C.; et al. Chitosan and Its Derivatives in 3D/4D (Bio) Printing for Tissue Engineering and Drug Delivery Applications. *Int. J. Biol. Macromol.* **2023**, *246*, 125669. [CrossRef]
135. Lazaridou, M.; Bikiaris, D.N.; Lamprou, D.A. 3D Bioprinted Chitosan-Based Hydrogel Scaffolds in Tissue Engineering and Localised Drug Delivery. *Pharmaceutics* **2022**, *14*, 1978. [CrossRef]
136. Xu, J.; Zhang, M.; Du, W.; Zhao, J.; Ling, G.; Zhang, P. Chitosan-Based High-Strength Supramolecular Hydrogels for 3D Bioprinting. *Int. J. Biol. Macromol.* **2022**, *219*, 545–557. [CrossRef]
137. Garg, H.G.; Hales, C.A. *Chemistry and Biology of Hyaluronan*; Elsevier: Amsterdam, The Netherlands, 2004; ISBN 9780080443829.
138. Shim, J.-H.; Kim, J.Y.; Park, M.; Park, J.; Cho, D.-W. Development of a Hybrid Scaffold with Synthetic Biomaterials and Hydrogel Using Solid Freeform Fabrication Technology. *Biofabrication* **2011**, *3*, 034102. [CrossRef]
139. Zhai, X.; Ruan, C.; Ma, Y.; Cheng, D.; Wu, M.; Liu, W.; Zhao, X.; Pan, H.; Lu, W.W. 3D-Bioprinted Osteoblast-Laden Nanocomposite Hydrogel Constructs with Induced Microenvironments Promote Cell Viability, Differentiation, and Osteogenesis Both In Vitro and In Vivo. *Adv. Sci.* **2018**, *5*, 1700550. [CrossRef] [PubMed]

140. Ehsanipour, A.; Nguyen, T.; Aboufadel, T.; Sathialingam, M.; Cox, P.; Xiao, W.; Walthers, C.M.; Seidlits, S.K. Injectable, Hyaluronic Acid-Based Scaffolds with Macroporous Architecture for Gene Delivery. *Cell Mol. Bioeng.* **2019**, *12*, 399–413. [CrossRef] [PubMed]
141. Nusgens, B.-V. Hyaluronic Acid and Extracellular Matrix: A Primitive Molecule? *Ann. Dermatol. Venereol.* **2010**, *137*, S3–S8. [CrossRef] [PubMed]
142. Petta, D.; D'Amora, U.; Ambrosio, L.; Grijpma, D.W.; Eglin, D.; D'Este, M. Hyaluronic Acid as a Bioink for Extrusion-Based 3D Printing. *Biofabrication* **2020**, *12*, 032001. [CrossRef]
143. Sekar, M.P.; Suresh, S.; Zennifer, A.; Sethuraman, S.; Sundaramurthi, D. Hyaluronic Acid as Bioink and Hydrogel Scaffolds for Tissue Engineering Applications. *ACS Biomater. Sci. Eng.* **2023**, *9*, 3134–3159. [CrossRef]
144. Bringmann, M.; Landrein, B.; Schudoma, C.; Hamant, O.; Hauser, M.-T.; Persson, S. Cracking the Elusive Alignment Hypothesis: The Microtubule–Cellulose Synthase Nexus Unraveled. *Trends Plant Sci.* **2012**, *17*, 666–674. [CrossRef]
145. Novotna, K.; Havelka, P.; Sopuch, T.; Kolarova, K.; Vosmanska, V.; Lisa, V.; Svorcik, V.; Bacakova, L. Cellulose-Based Materials as Scaffolds for Tissue Engineering. *Cellulose* **2013**, *20*, 2263–2278. [CrossRef]
146. Salmoria, G.V.; Klauss, P.; Paggi, R.A.; Kanis, L.A.; Lago, A. Structure and Mechanical Properties of Cellulose Based Scaffolds Fabricated by Selective Laser Sintering. *Polym. Test.* **2009**, *28*, 648–652. [CrossRef]
147. Curvello, R.; Raghuwanshi, V.S.; Garnier, G. Engineering Nanocellulose Hydrogels for Biomedical Applications. *Adv. Colloid. Interface Sci.* **2019**, *267*, 47–61. [CrossRef]
148. Piras, C.C.; Fernández-Prieto, S.; De Borggraeve, W.M. Nanocellulosic Materials as Bioinks for 3D Bioprinting. *Biomater. Sci.* **2017**, *5*, 1988–1992. [CrossRef]
149. Wan Jusoh, W.N.L.; Sajab, M.S.; Mohamed Abdul, P.; Kaco, H. Recent Advances in 3D Bioprinting: A Review of Cellulose-Based Biomaterials Ink. *Polymers* **2022**, *14*, 2260. [CrossRef]
150. Jovic, T.H.; Nicholson, T.; Arora, H.; Nelson, K.; Doak, S.H.; Whitaker, I.S. A Comparative Analysis of Pulp-Derived Nanocelluloses for 3D Bioprinting Facial Cartilages. *Carbohydr. Polym.* **2023**, *321*, 121261. [CrossRef]
151. Jovic, T.H.; Kungwengwe, G.; Mills, A.C.; Whitaker, I.S. Plant-Derived Biomaterials: A Review of 3D Bioprinting and Biomedical Applications. *Front. Mech. Eng.* **2019**, *5*, 19. [CrossRef]
152. Pitton, M.; Fiorati, A.; Buscemi, S.; Melone, L.; Farè, S.; Contessi Negrini, N. 3D Bioprinting of Pectin-Cellulose Nanofibers Multicomponent Bioinks. *Front. Bioeng. Biotechnol.* **2021**, *9*, 1186. [CrossRef]
153. Tortorella, S.; Inzalaco, G.; Dapporto, F.; Maturi, M.; Sambri, L.; Vetri Buratti, V.; Chiariello, M.; Comes Franchini, M.; Locatelli, E. Biocompatible Pectin-Based Hybrid Hydrogels for Tissue Engineering Applications. *New J. Chem.* **2021**, *45*, 22386–22395. [CrossRef]
154. Muthukrishnan, L. Imminent Antimicrobial Bioink Deploying Cellulose, Alginate, EPS and Synthetic Polymers for 3D Bioprinting of Tissue Constructs. *Carbohydr. Polym.* **2021**, *260*, 117774. [CrossRef] [PubMed]
155. Albrecht, F.B.; Dolderer, V.; Nellinger, S.; Schmidt, F.F.; Kluger, P.J. Gellan Gum Is a Suitable Biomaterial for Manual and Bioprinted Setup of Long-Term Stable, Functional 3D-Adipose Tissue Models. *Gels* **2022**, *8*, 420. [CrossRef] [PubMed]
156. Milivojevic, M.; Pajic-Lijakovic, I.; Bugarski, B.; Nayak, A.K.; Hasnain, M.S. Gellan Gum in Drug Delivery Applications. In *Natural Polysaccharides in Drug Delivery and Biomedical Applications*; Elsevier: Amsterdam, The Netherlands, 2019; pp. 145–186.
157. Morris, E.R.; Nishinari, K.; Rinaudo, M. Gelation of Gellan—A Review. *Food Hydrocoll.* **2012**, *28*, 373–411. [CrossRef]
158. Bonifacio, M.A.; Cometa, S.; Cochis, A.; Scalzone, A.; Gentile, P.; Scalia, A.C.; Rimondini, L.; Mastrorilli, P.; De Giglio, E. A Bioprintable Gellan Gum/Lignin Hydrogel: A Smart and Sustainable Route for Cartilage Regeneration. *Int. J. Biol. Macromol.* **2022**, *216*, 336–346. [CrossRef]
159. Li, Z.; Li, S.; Yang, J.; Ha, Y.; Zhang, Q.; Zhou, X.; He, C. 3D Bioprinted Gelatin/Gellan Gum-Based Scaffold with Double-Crosslinking Network for Vascularized Bone Regeneration. *Carbohydr. Polym.* **2022**, *290*, 119469. [CrossRef]
160. Cernencu, A.I.; Ioniță, M. The Current State of the Art in Gellan-Based Printing Inks in Tissue Engineering. *Carbohydr. Polym.* **2023**, *309*, 120676. [CrossRef] [PubMed]
161. Kumar, A.; Rao, K.M.; Han, S.S. Application of Xanthan Gum as Polysaccharide in Tissue Engineering: A Review. *Carbohydr. Polym.* **2018**, *180*, 128–144. [CrossRef] [PubMed]
162. Taniguchi Nagahara, M.H.; Caiado Decarli, M.; Inforçatti Neto, P.; Lopes da Silva, J.V.; Moraes, Â.M. Crosslinked Alginate-xanthan Gum Blends as Effective Hydrogels for 3D Bioprinting of Biological Tissues. *J. Appl. Polym. Sci.* **2022**, *139*, 52612. [CrossRef]
163. Garcia-Cruz, M.R.; Postma, A.; Frith, J.E.; Meagher, L. Printability and Bio-Functionality of a Shear Thinning Methacrylated Xanthan–Gelatin Composite Bioink. *Biofabrication* **2021**, *13*, 035023. [CrossRef]
164. Sun, G.; Mao, J.J. Engineering Dextran-Based Scaffolds for Drug Delivery and Tissue Repair. *Nanomedicine* **2012**, *7*, 1771–1784. [CrossRef]
165. Turner, P.R.; Murray, E.; McAdam, C.J.; McConnell, M.A.; Cabral, J.D. Peptide Chitosan/Dextran Core/Shell Vascularized 3D Constructs for Wound Healing. *ACS Appl. Mater. Interfaces* **2020**, *12*, 32328–32339. [CrossRef]
166. Musilová, L.; Achbergerová, E.; Vítková, L.; Kolařík, R.; Martínková, M.; Minařík, A.; Mráček, A.; Humpolíček, P.; Pecha, J. Cross-Linked Gelatine by Modified Dextran as a Potential Bioink Prepared by a Simple and Non-Toxic Process. *Polymers* **2022**, *14*, 391. [CrossRef]
167. Du, Z.; Li, N.; Hua, Y.; Shi, Y.; Bao, C.; Zhang, H.; Yang, Y.; Lin, Q.; Zhu, L. Physiological PH-Dependent Gelation for 3D Printing Based on the Phase Separation of Gelatin and Oxidized Dextran. *Chem. Commun.* **2017**, *53*, 13023–13026. [CrossRef]

168. Bean, S.R.; Zhu, L.; Smith, B.M.; Wilson, J.D.; Ioerger, B.P.; Tilley, M. Starch and Protein Chemistry and Functional Properties. In *Sorghum and Millets*; Elsevier: Amsterdam, The Netherlands, 2019; pp. 131–170.
169. Gopinath, V.; Saravanan, S.; Al-Maleki, A.R.; Ramesh, M.; Vadivelu, J. A Review of Natural Polysaccharides for Drug Delivery Applications: Special Focus on Cellulose, Starch and Glycogen. *Biomed. Pharmacother.* **2018**, *107*, 96–108. [CrossRef]
170. Zhuang, P.; Greenberg, Z.; He, M. Biologically Enhanced Starch Bio-Ink for Promoting 3D Cell Growth. *Adv. Mater. Technol.* **2021**, *6*, 2100551. [CrossRef]
171. Butler, H.M.; Naseri, E.; MacDonald, D.S.; Andrew Tasker, R.; Ahmadi, A. Optimization of Starch- and Chitosan-Based Bio-Inks for 3D Bioprinting of Scaffolds for Neural Cell Growth. *Materialia* **2020**, *12*, 100737. [CrossRef]
172. Marques, D.M.C.; Silva, J.C.; Serro, A.P.; Cabral, J.M.S.; Sanjuan-Alberte, P.; Ferreira, F.C. 3D Bioprinting of Novel κ-Carrageenan Bioinks: An Algae-Derived Polysaccharide. *Bioengineering* **2022**, *9*, 109. [CrossRef] [PubMed]
173. Li, H.; Tan, Y.J.; Li, L. A Strategy for Strong Interface Bonding by 3D Bioprinting of Oppositely Charged κ-Carrageenan and Gelatin Hydrogels. *Carbohydr. Polym.* **2018**, *198*, 261–269. [CrossRef] [PubMed]
174. Kim, M.H.; Lee, Y.W.; Jung, W.-K.; Oh, J.; Nam, S.Y. Enhanced Rheological Behaviors of Alginate Hydrogels with Carrageenan for Extrusion-Based Bioprinting. *J. Mech. Behav. Biomed. Mater.* **2019**, *98*, 187–194. [CrossRef]
175. Boonlai, W.; Tantishaiyakul, V.; Hirun, N. Characterization of K-carrageenan/Methylcellulose/Cellulose Nanocrystal Hydrogels for 3D Bioprinting. *Polym. Int.* **2022**, *71*, 181–191. [CrossRef]
176. Lim, W.; Kim, G.J.; Kim, H.W.; Lee, J.; Zhang, X.; Kang, M.G.; Seo, J.W.; Cha, J.M.; Park, H.J.; Lee, M.-Y.; et al. Kappa-Carrageenan-Based Dual Crosslinkable Bioink for Extrusion Type Bioprinting. *Polymers* **2020**, *12*, 2377. [CrossRef] [PubMed]
177. Mugnaini, G.; Resta, C.; Poggi, G.; Bonini, M. Photopolymerizable Pullulan: Synthesis, Self-Assembly and Inkjet Printing. *J. Colloid. Interface Sci.* **2021**, *592*, 430–439. [CrossRef]
178. Koffi, K.L.; Yapo, B.M.; Besson, V. Extraction and Characterization of Gelling Pectin from the Peel of Poncirus Trifoliata Fruit. *Agric. Sci.* **2013**, *4*, 614–619. [CrossRef]
179. Cui, X.; Li, J.; Hartanto, Y.; Durham, M.; Tang, J.; Zhang, H.; Hooper, G.; Lim, K.; Woodfield, T. Advances in Extrusion 3D Bioprinting: A Focus on Multicomponent Hydrogel-Based Bioinks. *Adv. Healthc. Mater.* **2020**, *9*, 1901648. [CrossRef]
180. Methacanon, P.; Krongsin, J.; Gamonpilas, C. Pomelo (Citrus Maxima) Pectin: Effects of Extraction Parameters and Its Properties. *Food Hydrocoll.* **2014**, *35*, 383–391. [CrossRef]
181. Stealey, S.; Guo, X.; Ren, L.; Bryant, E.; Kaltchev, M.; Chen, J.; Kumpaty, S.; Hua, X.; Zhang, W. Stability Improvement and Characterization of Bioprinted Pectin-Based Scaffold. *J. Appl. Biomater. Funct. Mater.* **2019**, *17*, 228080001880710. [CrossRef] [PubMed]
182. Banks, A.; Guo, X.; Chen, J.; Kumpaty, S.; Zhang, W. Novel Bioprinting Method Using a Pectin Based Bioink. *Technol. Health Care* **2017**, *25*, 651–655. [CrossRef] [PubMed]
183. Andriotis, E.G.; Eleftheriadis, G.K.; Karavasili, C.; Fatouros, D.G. Development of Bio-Active Patches Based on Pectin for the Treatment of Ulcers and Wounds Using 3D-Bioprinting Technology. *Pharmaceutics* **2020**, *12*, 56. [CrossRef]
184. Lapomarda, A.; Cerqueni, G.; Geven, M.A.; Chiesa, I.; De Acutis, A.; De Blasi, M.; Montemurro, F.; De Maria, C.; Mattioli-Belmonte, M.; Vozzi, G. Physicochemical Characterization of Pectin-Gelatin Biomaterial Formulations for 3D Bioprinting. *Macromol. Biosci.* **2021**, *21*, 2100168. [CrossRef] [PubMed]
185. Lapomarda, A.; Pulidori, E.; Cerqueni, G.; Chiesa, I.; De Blasi, M.; Geven, M.A.; Montemurro, F.; Duce, C.; Mattioli-Belmonte, M.; Tiné, M.R.; et al. Pectin as Rheology Modifier of a Gelatin-Based Biomaterial Ink. *Materials* **2021**, *14*, 3109. [CrossRef] [PubMed]
186. Lee, C.; Abelseth, E.; de la Vega, L.; Willerth, S.M. Bioprinting a Novel Glioblastoma Tumor Model Using a Fibrin-Based Bioink for Drug Screening. *Mater. Today Chem.* **2019**, *12*, 78–84. [CrossRef]
187. Kim, S.D.; Jin, S.; Kim, S.; Son, D.; Shin, M. Tyramine-Functionalized Alginate-Collagen Hybrid Hydrogel Inks for 3D-Bioprinting. *Polymers* **2022**, *14*, 3173. [CrossRef]
188. Stratesteffen, H.; Köpf, M.; Kreimendahl, F.; Blaeser, A.; Jockenhoevel, S.; Fischer, H. GelMA-Collagen Blends Enable Drop-on-Demand 3D Printablility and Promote Angiogenesis. *Biofabrication* **2017**, *9*, 045002. [CrossRef]
189. Trucco, D.; Sharma, A.; Manferdini, C.; Gabusi, E.; Petretta, M.; Desando, G.; Ricotti, L.; Chakraborty, J.; Ghosh, S.; Lisignoli, G. Modeling and Fabrication of Silk Fibroin–Gelatin-Based Constructs Using Extrusion-Based Three-Dimensional Bioprinting. *ACS Biomater. Sci. Eng.* **2021**, *7*, 3306–3320. [CrossRef]
190. Restan Perez, M.; Sharma, R.; Masri, N.Z.; Willerth, S.M. 3D Bioprinting Mesenchymal Stem Cell-Derived Neural Tissues Using a Fibrin-Based Bioink. *Biomolecules* **2021**, *11*, 1250. [CrossRef]
191. Abelseth, E.; Abelseth, L.; De la Vega, L.; Beyer, S.T.; Wadsworth, S.J.; Willerth, S.M. 3D Printing of Neural Tissues Derived from Human Induced Pluripotent Stem Cells Using a Fibrin-Based Bioink. *ACS Biomater. Sci. Eng.* **2019**, *5*, 234–243. [CrossRef] [PubMed]
192. Ramakrishnan, R.; Kasoju, N.; Raju, R.; Geevarghese, R.; Gauthaman, A.; Bhatt, A. Exploring the Potential of Alginate-Gelatin-Diethylaminoethyl Cellulose-Fibrinogen Based Bioink for 3D Bioprinting of Skin Tissue Constructs. *Carbohydr. Polym. Technol. Appl.* **2022**, *3*, 100184. [CrossRef]
193. Im, S.; Choe, G.; Seok, J.M.; Yeo, S.J.; Lee, J.H.; Kim, W.D.; Lee, J.Y.; Park, S.A. An Osteogenic Bioink Composed of Alginate, Cellulose Nanofibrils, and Polydopamine Nanoparticles for 3D Bioprinting and Bone Tissue Engineering. *Int. J. Biol. Macromol.* **2022**, *205*, 520–529. [CrossRef] [PubMed]

194. Liu, Y.; Hu, Q.; Dong, W.; Liu, S.; Zhang, H.; Gu, Y. Alginate/Gelatin-Based Hydrogel with Soy Protein/Peptide Powder for 3D Printing Tissue-Engineering Scaffolds to Promote Angiogenesis. *Macromol. Biosci.* **2022**, *22*, 2100413. [CrossRef] [PubMed]
195. Thanh, T.N.; Laowattanatham, N.; Ratanavaraporn, J.; Sereemaspun, A.; Yodmuang, S. Hyaluronic Acid Crosslinked with Alginate Hydrogel: A Versatile and Biocompatible Bioink Platform for Tissue Engineering. *Eur. Polym. J.* **2022**, *166*, 111027. [CrossRef]
196. Maturavongsadit, P.; Narayanan, L.K.; Chansoria, P.; Shirwaiker, R.; Benhabbour, S.R. Cell-Laden Nanocellulose/Chitosan-Based Bioinks for 3D Bioprinting and Enhanced Osteogenic Cell Differentiation. *ACS Appl. Bio Mater.* **2021**, *4*, 2342–2353. [CrossRef]
197. Ngo, T.B.; Spearman, B.S.; Hlavac, N.; Schmidt, C.E. Three-Dimensional Bioprinted Hyaluronic Acid Hydrogel Test Beds for Assessing Neural Cell Responses to Competitive Growth Stimuli. *ACS Biomater. Sci. Eng.* **2020**, *6*, 6819–6830. [CrossRef]
198. Janarthanan, G.; Shin, H.S.; Kim, I.-G.; Ji, P.; Chung, E.-J.; Lee, C.; Noh, I. Self-Crosslinking Hyaluronic Acid–Carboxymethylcellulose Hydrogel Enhances Multilayered 3D-Printed Construct Shape Integrity and Mechanical Stability for Soft Tissue Engineering. *Biofabrication* **2020**, *12*, 045026. [CrossRef]
199. Shopperly, L.K.; Spinnen, J.; Krüger, J.; Endres, M.; Sittinger, M.; Lam, T.; Kloke, L.; Dehne, T. Blends of Gelatin and Hyaluronic Acid Stratified by Stereolithographic Bioprinting Approximate Cartilaginous Matrix Gradients. *J. Biomed. Mater. Res. B Appl. Biomater.* **2022**, *110*, 2310–2322. [CrossRef]
200. Ma, N.; Cheung, D.Y.; Butcher, J.T. Incorporating Nanocrystalline Cellulose into a Multifunctional Hydrogel for Heart Valve Tissue Engineering Applications. *J. Biomed. Mater. Res. A* **2022**, *110*, 76–91. [CrossRef]
201. Lan, X.; Ma, Z.; Szojka, A.R.A.; Kunze, M.; Mulet-Sierra, A.; Vyhlidal, M.J.; Boluk, Y.; Adesida, A.B. TEMPO-Oxidized Cellulose Nanofiber-Alginate Hydrogel as a Bioink for Human Meniscus Tissue Engineering. *Front. Bioeng. Biotechnol.* **2021**, *9*, 766399. [CrossRef]
202. Kapr, J.; Petersilie, L.; Distler, T.; Lauria, I.; Bendt, F.; Sauter, C.M.; Boccaccini, A.R.; Rose, C.R.; Fritsche, E. Human Induced Pluripotent Stem Cell-Derived Neural Progenitor Cells Produce Distinct Neural 3D In Vitro Models Depending on Alginate/Gellan Gum/Laminin Hydrogel Blend Properties. *Adv. Healthc. Mater.* **2021**, *10*, 2100131. [CrossRef]
203. Gulrez, S.K.; Al-Assaf, S.; Phillips, G.O. Hydrogels: Methods of Preparation, Characterisation and Applications. In *Progress in Molecular and Environmental Bioengineering—From Analysis and Modeling to Technology Applications*; InTech: Lahore, Pakistan, 2011.
204. van Kampen, K.A.; Scheuring, R.G.; Terpstra, M.L.; Levato, R.; Groll, J.; Malda, J.; Mota, C.; Moroni, L. Biofabrication: From Additive Manufacturing to Bioprinting. In *Reference Module in Biomedical Sciences*; Elsevier: Amsterdam, The Netherlands, 2019.
205. Varghese, S.A.; Rangappa, S.M.; Siengchin, S.; Parameswaranpillai, J. Natural Polymers and the Hydrogels Prepared from Them. In *Hydrogels Based on Natural Polymers*; Elsevier: Amsterdam, The Netherlands, 2020; pp. 17–47. ISBN 9780128164211.
206. Akhtar, M.F.; Hanif, M.; Ranjha, N.M. Methods of Synthesis of Hydrogels . . . A Review. *Saudi Pharm. J.* **2016**, *24*, 554–559. [CrossRef]
207. GhavamiNejad, A.; Ashammakhi, N.; Wu, X.Y.; Khademhosseini, A. Crosslinking Strategies for 3D Bioprinting of Polymeric Hydrogels. *Small* **2020**, *16*, 2002931. [CrossRef]
208. Hennink, W.E.; van Nostrum, C.F. Novel Crosslinking Methods to Design Hydrogels. *Adv. Drug Deliv. Rev.* **2012**, *64*, 223–236. [CrossRef]
209. Ullah, F.; Othman, M.B.H.; Javed, F.; Ahmad, Z.; Akil, H.M. Classification, Processing and Application of Hydrogels: A Review. *Mater. Sci. Eng. C* **2015**, *57*, 414–433. [CrossRef]
210. Slager, J.; Domb, A.J. Biopolymer Stereocomplexes. *Adv. Drug Deliv. Rev.* **2003**, *55*, 549–583. [CrossRef]
211. Varaprasad, K.; Raghavendra, G.M.; Jayaramudu, T.; Yallapu, M.M.; Sadiku, R. A Mini Review on Hydrogels Classification and Recent Developments in Miscellaneous Applications. *Mater. Sci. Eng. C* **2017**, *79*, 958–971. [CrossRef]
212. Lee, M.; Rizzo, R.; Surman, F.; Zenobi-Wong, M. Guiding Lights: Tissue Bioprinting Using Photoactivated Materials. *Chem. Rev.* **2020**, *120*, 10950–11027. [CrossRef]
213. Elomaa, L.; Yang, Y.P. Additive Manufacturing of Vascular Grafts and Vascularized Tissue Constructs. *Tissue Eng. Part B Rev.* **2017**, *23*, 436–450. [CrossRef]
214. Gómez-Blanco, J.C.; Galván-Chacón, V.; Patrocinio, D.; Matamoros, M.; Sánchez-Ortega, Á.J.; Marcos, A.C.; Duarte-León, M.; Marinaro, F.; Pagador, J.B.; Sánchez-Margallo, F.M. Improving Cell Viability and Velocity in μ-Extrusion Bioprinting with a Novel Pre-Incubator Bioprinter and a Standard FDM 3D Printing Nozzle. *Materials* **2021**, *14*, 3100. [CrossRef] [PubMed]
215. Barron, J.A.; Krizman, D.B.; Ringeisen, B.R. Laser Printing of Single Cells: Statistical Analysis, Cell Viability, and Stress. *Ann. Biomed. Eng.* **2005**, *33*, 121–130. [CrossRef] [PubMed]
216. Dou, C.; Perez, V.; Qu, J.; Tsin, A.; Xu, B.; Li, J. A State-of-the-Art Review of Laser-Assisted Bioprinting and Its Future Research Trends. *ChemBioEng Rev.* **2021**, *8*, 517–534. [CrossRef]
217. Li, J.; Chen, M.; Fan, X.; Zhou, H. Recent Advances in Bioprinting Techniques: Approaches, Applications and Future Prospects. *J. Transl. Med.* **2016**, *14*, 271. [CrossRef]
218. Fu, Z.; Angeline, V.; Sun, W. Evaluation of Printing Parameters on 3D Extrusion Printing of Pluronic Hydrogels and Machine Learning Guided Parameter Recommendation. *Int. J. Bioprinting* **1970**, *7*, 434. [CrossRef]
219. Malekpour, A.; Chen, X. Printability and Cell Viability in Extrusion-Based Bioprinting from Experimental, Computational, and Machine Learning Views. *J. Funct. Biomater.* **2022**, *13*, 40. [CrossRef]

220. Erben, A.; Hörning, M.; Hartmann, B.; Becke, T.; Eisler, S.A.; Southan, A.; Cranz, S.; Hayden, O.; Kneidinger, N.; Königshoff, M.; et al. Precision 3D-Printed Cell Scaffolds Mimicking Native Tissue Composition and Mechanics. *Adv. Healthc. Mater.* **2020**, *9*, 2000918. [CrossRef]
221. Ji, Y.; Yang, Q.; Huang, G.; Shen, M.; Jian, Z.; Thoraval, M.-J.; Lian, Q.; Zhang, X.; Xu, F. Improved Resolution and Fidelity of Droplet-Based Bioprinting by Upward Ejection. *ACS Biomater. Sci. Eng.* **2019**, *5*, 4112–4121. [CrossRef]
222. Nguyen, A.K.; Goering, P.L.; Elespuru, R.K.; Sarkar Das, S.; Narayan, R.J. The Photoinitiator Lithium Phenyl (2,4,6-Trimethylbenzoyl) Phosphinate with Exposure to 405 Nm Light Is Cytotoxic to Mammalian Cells but Not Mutagenic in Bacterial Reverse Mutation Assays. *Polymers* **2020**, *12*, 1489. [CrossRef]
223. Boularaoui, S.; Al Hussein, G.; Khan, K.A.; Christoforou, N.; Stefanini, C. An Overview of Extrusion-Based Bioprinting with a Focus on Induced Shear Stress and Its Effect on Cell Viability. *Bioprinting* **2020**, *20*, e00093. [CrossRef]
224. Yu, C.; Ma, X.; Zhu, W.; Wang, P.; Miller, K.L.; Stupin, J.; Koroleva-Maharajh, A.; Hairabedian, A.; Chen, S. Scanningless and Continuous 3D Bioprinting of Human Tissues with Decellularized Extracellular Matrix. *Biomaterials* **2019**, *194*, 1–13. [CrossRef] [PubMed]
225. Beheshtizadeh, N.; Lotfibakhshaiesh, N.; Pazhouhnia, Z.; Hoseinpour, M.; Nafari, M. A Review of 3D Bio-Printing for Bone and Skin Tissue Engineering: A Commercial Approach. *J. Mater. Sci.* **2020**, *55*, 3729–3749. [CrossRef]
226. Galván-Chacón, V.P.; Zampouka, A.; Hesse, B.; Bohner, M.; Habibovic, P.; Barata, D. Bone-on-a-Chip: A Microscale 3D Biomimetic Model to Study Bone Regeneration. *Adv. Eng. Mater.* **2022**, *24*, 2101467. [CrossRef]
227. Marino, A.; Filippeschi, C.; Genchi, G.G.; Mattoli, V.; Mazzolai, B.; Ciofani, G. The Osteoprint: A Bioinspired Two-Photon Polymerized 3-D Structure for the Enhancement of Bone-like Cell Differentiation. *Acta Biomater.* **2014**, *10*, 4304–4313. [CrossRef] [PubMed]
228. Ma, Z.; Koo, S.; Finnegan, M.A.; Loskill, P.; Huebsch, N.; Marks, N.C.; Conklin, B.R.; Grigoropoulos, C.P.; Healy, K.E. Three-Dimensional Filamentous Human Diseased Cardiac Tissue Model. *Biomaterials* **2014**, *35*, 1367–1377. [CrossRef]
229. Marino, A.; Tricinci, O.; Battaglini, M.; Filippeschi, C.; Mattoli, V.; Sinibaldi, E.; Ciofani, G. A 3D Real-Scale, Biomimetic, and Biohybrid Model of the Blood-Brain Barrier Fabricated through Two-Photon Lithography. *Small* **2018**, *14*, 1702959. [CrossRef]
230. Skylar-Scott, M.A.; Liu, M.-C.; Wu, Y.; Dixit, A.; Yanik, M.F. Guided Homing of Cells in Multi-Photon Microfabricated Bioscaffolds. *Adv. Healthc. Mater.* **2016**, *5*, 1233–1243. [CrossRef]
231. Rayner, S.G.; Howard, C.C.; Mandrycky, C.J.; Stamenkovic, S.; Himmelfarb, J.; Shih, A.Y.; Zheng, Y. Multiphoton-Guided Creation of Complex Organ-Specific Microvasculature. *Adv. Healthc. Mater.* **2021**, *10*, 2100031. [CrossRef]
232. Tromayer, M.; Gruber, P.; Markovic, M.; Rosspeintner, A.; Vauthey, E.; Redl, H.; Ovsianikov, A.; Liska, R. A Biocompatible Macromolecular Two-Photon Initiator Based on Hyaluronan. *Polym. Chem.* **2017**, *8*, 451–460. [CrossRef]
233. Fu, Z.; Naghieh, S.; Xu, C.; Wang, C.; Sun, W.; Chen, X. Printability in Extrusion Bioprinting. *Biofabrication* **2021**, *13*, 033001. [CrossRef]
234. Michael, S.; Sorg, H.; Peck, C.-T.; Koch, L.; Deiwick, A.; Chichkov, B.; Vogt, P.M.; Reimers, K. Tissue Engineered Skin Substitutes Created by Laser-Assisted Bioprinting Form Skin-Like Structures in the Dorsal Skin Fold Chamber in Mice. *PLoS ONE* **2013**, *8*, e57741. [CrossRef] [PubMed]
235. Domingos, M.; Intranuovo, F.; Russo, T.; De Santis, R.; Gloria, A.; Ambrosio, L.; Ciurana, J.; Bartolo, P. The First Systematic Analysis of 3D Rapid Prototyped Poly(ε-Caprolactone) Scaffolds Manufactured through BioCell Printing: The Effect of Pore Size and Geometry on Compressive Mechanical Behaviour and in Vitro HMSC Viability. *Biofabrication* **2013**, *5*, 045004. [CrossRef] [PubMed]
236. Łabowska, M.B.; Cierluk, K.; Jankowska, A.M.; Kulbacka, J.; Detyna, J.; Michalak, I. A Review on the Adaption of Alginate-Gelatin Hydrogels for 3D Cultures and Bioprinting. *Materials* **2021**, *14*, 858. [CrossRef] [PubMed]
237. Roche, C.D.; Sharma, P.; Ashton, A.W.; Jackson, C.; Xue, M.; Gentile, C. Printability, Durability, Contractility and Vascular Network Formation in 3D Bioprinted Cardiac Endothelial Cells Using Alginate-Gelatin Hydrogels. *Front. Bioeng. Biotechnol.* **2021**, *9*, 636257. [CrossRef]
238. Singh, N.K.; Han, W.; Nam, S.A.; Kim, J.W.; Kim, J.Y.; Kim, Y.K.; Cho, D.-W. Three-Dimensional Cell-Printing of Advanced Renal Tubular Tissue Analogue. *Biomaterials* **2020**, *232*, 119734. [CrossRef]
239. Gudapati, H.; Dey, M.; Ozbolat, I. A Comprehensive Review on Droplet-Based Bioprinting: Past, Present and Future. *Biomaterials* **2016**, *102*, 20–42. [CrossRef]
240. Yanez, M.; Rincon, J.; Dones, A.; De Maria, C.; Gonzales, R.; Boland, T. In Vivo Assessment of Printed Microvasculature in a Bilayer Skin Graft to Treat Full-Thickness Wounds. *Tissue Eng. Part A* **2015**, *21*, 224–233. [CrossRef]
241. Cui, X.; Boland, T. Human Microvasculature Fabrication Using Thermal Inkjet Printing Technology. *Biomaterials* **2009**, *30*, 6221–6227. [CrossRef]

Disclaimer/Publisher's Note: The statements, opinions and data contained in all publications are solely those of the individual author(s) and contributor(s) and not of MDPI and/or the editor(s). MDPI and/or the editor(s) disclaim responsibility for any injury to people or property resulting from any ideas, methods, instructions or products referred to in the content.

Article

Microbial Polysaccharide-Based Formulation with Silica Nanoparticles; A New Hydrogel Nanocomposite for 3D Printing

Maria Minodora Marin [1,2], Ioana Catalina Gifu [3,*], Gratiela Gradisteanu Pircalabioru [4,5,6], Madalina Albu Kaya [2], Rodica Roxana Constantinescu [2], Rebeca Leu Alexa [2], Bogdan Trica [3], Elvira Alexandrescu [3], Cristina Lavinia Nistor [3], Cristian Petcu [3] and Raluca Ianchis [3,*]

1. Advanced Polymer Materials Group, Faculty of Applied Chemistry and Materials Science, Politehnica University of Bucharest, 1–7 Polizu Street, 01106 Bucharest, Romania; maria_minodora.marin@upb.ro
2. Department of Collagen, National Research and Development Institute for Textile and Leather, Division Leather and Footwear Research Institute, 93 Ion Minulescu Str., 031215 Bucharest, Romania; madalina.albu@icpi.ro (M.A.K.)
3. National Research and Development Institute for Chemistry and Petrochemistry ICECHIM—Spl. Independentei 202, 6th District, 0600021 Bucharest, Romania
4. eBio-Hub Research Center, University Politehnica of Bucharest—CAMPUS, 6 Iuliu Maniu Boulevard, 061344 Bucharest, Romania; gratiela87@gmail.com
5. Research Institute of University of Bucharest (ICUB), University of Bucharest, 030018 Bucharest, Romania
6. Academy of Romanian Scientists, 010719 Bucharest, Romania
* Correspondence: catalina.gifu@icechim-pd.ro (I.C.G.); raluca.ianchis@icechim-pd.ro (R.I.)

Abstract: Natural polysaccharides are highly attractive biopolymers recommended for medical applications due to their low cytotoxicity and hydrophilicity. Polysaccharides and their derivatives are also suitable for additive manufacturing, a process in which various customized geometries of 3D structures/scaffolds can be achieved. Polysaccharide-based hydrogel materials are widely used in 3D hydrogel printing of tissue substitutes. In this context, our goal was to obtain printable hydrogel nanocomposites by adding silica nanoparticles to a microbial polysaccharide's polymer network. Several amounts of silica nanoparticles were added to the biopolymer, and their effects on the morpho-structural characteristics of the resulting nanocomposite hydrogel inks and subsequent 3D printed constructs were studied. FTIR, TGA, and microscopy analysis were used to investigate the resulting crosslinked structures. Assessment of the swelling characteristics and mechanical stability of the nanocomposite materials in a wet state was also conducted. The salecan-based hydrogels displayed excellent biocompatibility and could be employed for biomedical purposes, according to the results of the MTT, LDH, and Live/Dead tests. The innovative, crosslinked, nanocomposite materials are recommended for use in regenerative medicine.

Keywords: salecan; silica nanoparticles; 3D printing

1. Introduction

For obtaining three-dimensional (3D) objects from a computer design, 3D printing was created more than 30 years ago. A quick and affordable design cycle is made possible by this layer-by-layer method for creating customized biomaterials [1,2]. Many polymers have been employed to create materials used in pharmaceutical and medical fields using three-dimensional printing technology [3]. Moreover, multiple materials can be combined in a single structure using multi-material 3D printing [4]. With the development of 3D printing technology, it is now possible to reconstruct tissues using active components and a growing variety of materials that have strong and stable mechanical properties, and good biocompatibility [5,6]. It is interesting how the development of functional hydrogels opens up several opportunities for incorporating hydrophilic networks into 3D-printed scaffolds that are similar to the extracellular matrix [7].

Hydrogels are three-dimensional polymeric structures that can absorb large amounts of water without disintegrating or losing their physical integrity [8,9]. Hydrogels have drawn a lot of interest as prospective biomaterials for medical applications such as drug delivery, tissue engineering, cell culture scaffolds, wound dressings, and filtration/separation techniques due to their high water content and inherent mechanical strength [10,11]. Hydrogels can be made from synthetic or natural polymers using techniques that include physical crosslinking, chemical gelation, or self-assembly [8,12–14]. Natural polymer-based hydrogels, particularly those made of polysaccharides, have seen increased use in recent years because of their outstanding biocompatibility, bioactivity, biodegradability, hydrophilicity, and low toxicity [9,15,16]. In previous research works, polysaccharides such as starch, dextran, cellulose, and their derivatives have been used to develop polysaccharide-based hydrogels [17,18]. Moreover, polysaccharides feature active groups that can be functionalized by utilizing a variety of methods in order to improve their mechanical properties and stability; furthermore, they can be produced from renewable sources [19]. Alginate, chitosan, cellulose, k-carrageenan, pectin, and other polysaccharides are frequently used in the manufacture of composite hydrogels [20–24].

Salecan is a new water-soluble glucan (natural polysaccharide), produced by the strain *Agrobacterium* sp. ZX09, which has recently been identified and commercialized [25–27]. Its backbone features a distinctive design "→ 3)-β-D-Glcp-(1 → 3-[β-D-Glcp-(1 → 3)-β-D-Glcp-(1 → 3)]3-α-D-Glcp-(1 → 3)-α-D-Glcp-(1→", groups that are all connected by α-(1–3) and β-(1–3) glycosidic connections [28]. Salecan, which differs significantly from the other β-glucans, offers a variety of benefits including biocompatibility, great immune boosting, high solubility, antioxidation properties, and biodegradability. The medical, pharmaceutical, cosmetic, and food sectors all greatly value these qualities [29–31]. One of the most important characteristics of salecan is the rheological property, which is incredibly valuable in our current inquiry. For drug delivery purposes, our research team has developed polymer nanocomposites containing salecan [32]. We recently investigated the possibility of synthesizing exclusively salecan green crosslinked materials, as well as the first use of salecan in additive manufacturing [25].

Organic-inorganic nanocomposite particles have attracted a lot of attention recently in research areas and industry. The combined and, in some cases, synergistic effects of the organic and inorganic components in these hybrid materials make them useful for a variety of applications, including catalysis, coatings, photonics, and biomedicine [33,34]. To create organic-inorganic nanocomposite materials, several inorganic particles, including silica, Laponite clay, magnetite, zinc oxide, titanium oxide, and graphene oxide, have been used [33,35]

The silicate minerals that make up the majority of the silica in the Earth's crust are also present in plants and grains [36,37]. Due to the presence of silanol groups (Si-OH), silica nanoparticles (SiO_2) have good biocompatibility, homogeneous pore size, adjustable particle size, a large surface area, and are easily modifiable [38]. Furthermore, the proliferation and differentiation of cell cultures can all be aided by the addition of silica nanoparticles [39]. Based on these features, the inorganic silica skeleton is more stable than conventional degradable biomaterials in the presence of temperature changes, organic solvents, and acidic environments [40]. The production of silica nanoparticles has increased, making them the second most common nanomaterial produced globally [41]. Functionalized silica nanoparticles, present free OH groups on their surface, and these groups are highly affine to the COO–groups found in biopolymers such as sodium alginate, gelatin, agar, etc., forming hydrogen bonds [39]. For example, Roopavath et al. prepared a 3D printable alginate/gelatin/SiO_2 ink with increased viscosity that presented a potential for bone regeneration and nanomedicine [39]. Moreover, using a salecan graft copolymer and Fe_3O_4@SiO_2 nanoparticles as the drug carrier, Hu et al. created a pH-sensitive magnetic composite hydrogel [17].

In the present work, a novel 3D-printed nanocomposite hydrogel was synthesized and characterized. The main objective of this study was to introduce silicon dioxide nanoparticles in the polymer network of a microbial polysaccharide (salecan) in order to obtain new 3D printable nanocomposite hydrogels. Thus, we have followed enhanced printing fidelity by varying the amount of nanofiller in the polysaccharide formulation. As far as we know, our study is the first report investigating an ink based on salecan with entrapped silica nanoparticles developed for additive manufacturing with potential for biomedical applications.

2. Results and Discussion

2.1. Rheological Behavior

All of the samples displayed normal filaments and kept the 3D-printed structure they were originally created with depending on the composition. As was seen both during and after the 3D printing process, the 3D printed shape demonstrated good stability and maintained its design. Even though they can be printed up to 10 layers thick, the neat hydrogel sample S0 and S1 showed collapsed layers in contrast to the nanocomposite structures made from the printing inks S2 and S3.

To evaluate the printability of the newly formulated inks and the influence of silica nanoparticles amount over the salecan, the shear viscosity was investigated. Figure 1 shows the variation of viscosity with a shear rate for S0, S1, S2, and S3. The bicomponent samples forming the printing inks offer specific rheological characteristics. The shear viscosity decreases with the applied shear rate for all samples, which suggests a non-Newtonian pseudoplastic fluid with a shear-thinning character. This rheological characteristic is typical of polysaccharide materials and is linked to their vast hydrodynamic size, which comes from the grouping of linear and stiff macromolecules, which results in high viscosities and pseudoplasticity [25,42]. Furthermore, the salecan hydrogels' viscosity is reduced from 10^5 Pa·s to 10^0 Pa·s with an increased shear rate, displaying shear-thinning behavior and suggesting the feasibility of injecting salecan hydrogel-based inks [25].

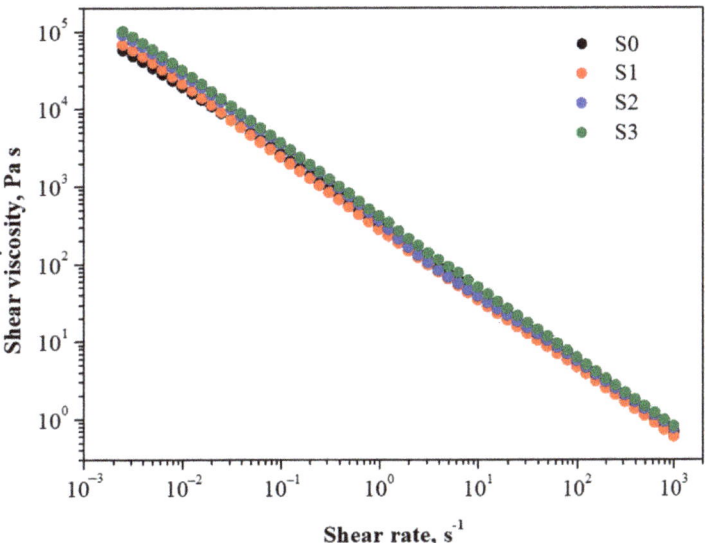

Figure 1. Viscosity as a function of shear rate for S0, S1, S2, and S3 hydrogel-based inks.

As intended, the viscosity of the silica—salecan composites slightly increased with the increasing silica concentration at a constant polysaccharide concentration. The difference between the studied samples is the viscosity value, which was the highest for S3 and the lowest for S0, while S1 and S2 presented intermediate values. The slightly increased viscosity conferred by the inclusion of silica nanoparticles is beneficial for the extrusion of hydrogel and for producing 3D structures with precise geometry [43].

2.2. FT-IR Analyses

To understand the influence of silica nanoparticle content on the structural properties of the newly obtained salecan/silica composite, we investigated the compositional behaviors of designed hydrogels by using the FT-IR technique. The FT-IR spectra of the newly obtained biomaterials based on salecan and silica nanoparticles are given in Figure 2.

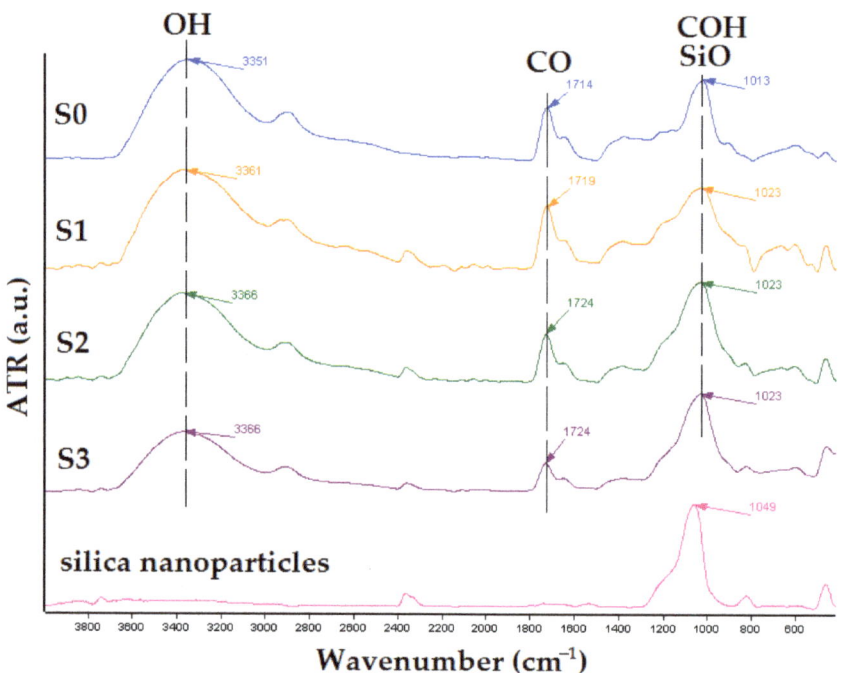

Figure 2. FTIR Spectra of the obtained samples.

FT-IR spectroscopy showed the specific peak of salecan at around 3300 cm^{-1} due to its abundant OH groups [17]. Other peaks appear at 2800 and 1013 cm^{-1}, which are attributed to CH_2 and C–OH stretching frequencies in the glucopyranose ring from the salecan structure, respectively [25,30]. Moreover, stretching vibration of the C=O peak from the citric acid structure appears on the FT-IR curve at around 1700 cm^{-1} [44,45]. The FTIR curve of silica nanoparticles exhibits a peak at 1049 cm^{-1}, which is ascribed to the Si-O-Si bond from the silicon dioxide inorganic network, with no OH-specific peaks in the 3300 cm^{-1} area.

In the case of the produced nanocomposites, we discovered that the aforementioned peaks were slightly changed and diminished as the silica content in the polysaccharide matrix increased. The peak assigned to Si–O at 1049 cm^{-1} overlaps with the peak attributed to C–OH from salecan at 1013 cm^{-1}. The resulting peaks of nanocomposites are more of a combination of the two peaks with a gradual shift to higher wavenumbers (1023 cm^{-1}). Unlike other investigations which used functionalized silica nanoparticles [17,39], the present study used amorphous SiO_2 nanoparticles which are only physically confined in biopolymeric matrices and do not participate in the crosslinking processes.

2.3. Swelling Behavior

The hydrogel's ability to swell in contact with fluids is a crucial characteristic for scaffolds used in tissue engineering applications since it has a significant impact on the regeneration of the new tissue and can present an important influence on the mechanical properties and the degradation rate [14,46]. The water uptake of the new 3D-printed composite hydrogels was thoroughly assessed for that aim. The swelling behavior for the studied samples was evaluated for 48 h using a simulated fluid with pH 5.5, 7.4, and 11. Figure 3 presents the values of swelling degree (SD) and illustrates how salecan and SiO_2 content affect the water absorption capacity. The maximum amount of water was absorbed by the S0 sample (sample without silica nanoparticles) which presented the best hydrophilicity at pH 7.4.

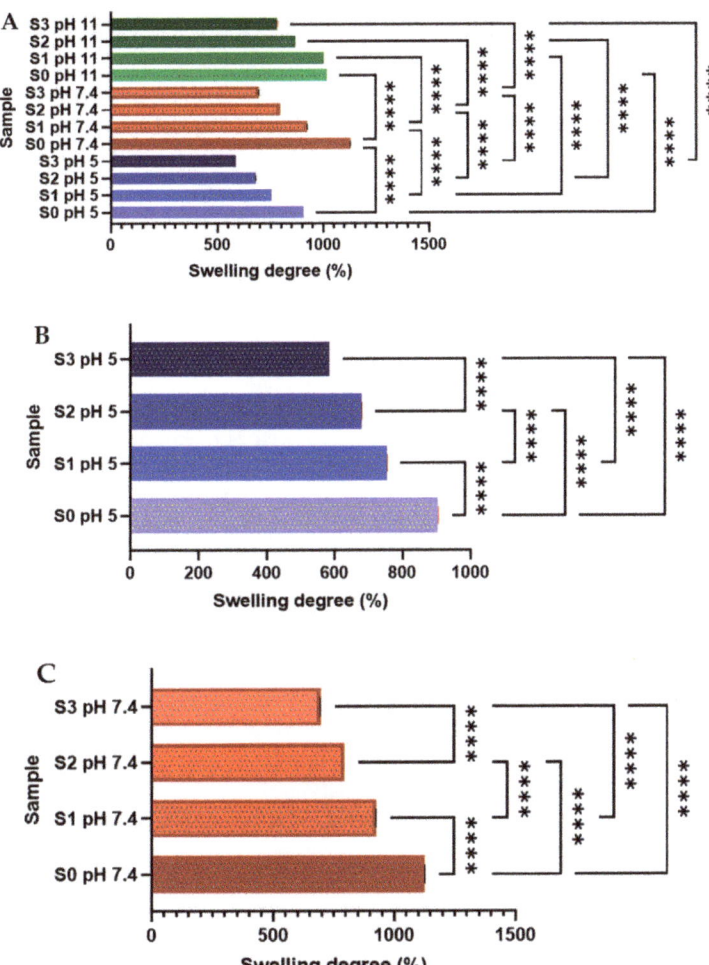

Figure 3. *Cont.*

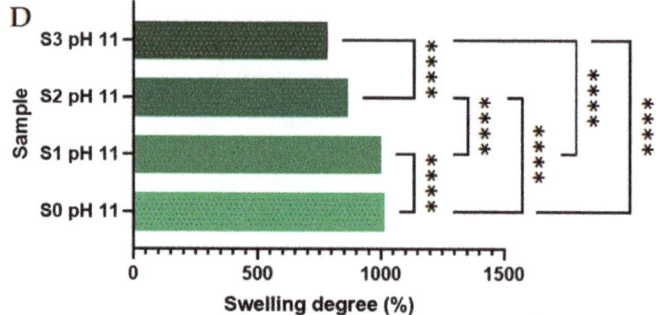

Figure 3. (**A**) Equilibrium swelling behavior in different pH media of the 3D printed samples. Statistical analysis (**A**): The influence of different pH media over the samples swelling degree; (**B**–**D**): The influence of composition for the samples kept in the same pH media ((**B**)-pH 5, (**C**)-pH 7.4, (**D**)-pH 11: The resulted values consist of average values with additional standard errors. Statistical significance: **** $p < 0.0001$. ONE WAY ANOVA TEST.

The inclusion of silica nanoparticles in the polymeric matrix leads to the limitation of the hydrogel swelling behavior in different pH media as a consequence of the replacement of biopolymer with silica nanoparticles. Moreover, with the exception of S0, whose SD value is maximum at pH 7.4, the swelling degree was higher with the increase of the pH for all the samples because more COO– groups are forming inside the hydrogels (from the citric acid structure) [25,47].

All of the 3D structures that were printed have the potential to quickly absorb fluids in various pH media in about one hour. The produced 3D printed composites consist of durable biopolymer networks, evidenced by the fact that all the swelled samples retained their stability in wet settings after 24 h, the crosslinking procedure being successful in the presence of the nanofiller. Moreover, after one month, minimal degradation of less than 3% was found for samples preserved in acidic media, of ~10% for neutral pH, while the degradation reached its peak for samples preserved in the basic medium after only 30 h, regardless of sample type. This behavior is governed by the polysaccharide matrix, with a distinct behavior observed in cases where salecan crosslinked with citric acid was obtained at various concentrations of reactants [25].

Because hydrogel behavior is pH-dependent, this novel family of materials can therefore be used in any application that requires higher stability in an acidic environment and gradual disintegration over time in environments with greater pH's.

2.4. Thermal Properties

To confirm that the inclusion of the silica nanoparticles can improve the thermal stability of new 3D printed composites, further TGA investigation was performed. Figure 4 presents the TGA curves of new 3D printed composites tested in nitrogen conditions with a heating rate of 10 °C/min.

Thermogravimetric parameters of the analyzed samples are presented in Table 1.

Table 1. Thermogravimetric of the samples.

Samples	Step 1		Step 2		Step 3	
	T, °C	Weight Loss, %	T, °C	Weight Loss, %	T, °C	Weight Loss, %
S0	80–200	16.81	200–320	57.49	320–700	39.86
S1		11.95		22.07		34.35
S2		10.84		22.31		26.72
S3		9.29		18.48		26.10

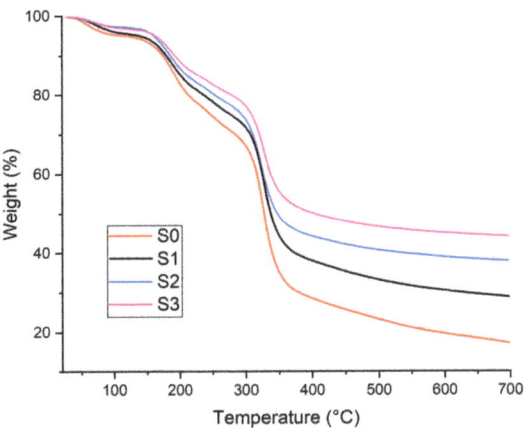

Figure 4. Thermal properties of the obtained sample, TGA.

All the samples exhibited three different steps of decomposition. The first step ranging between 80 to 200 °C with a weight loss of around 10–17% was attributed to the removal of adsorbed water and slightly volatile substances [48]. The second step ranged in temperature from 200 to 320 °C and presented a weight loss of about 19–58% representing the cleavage of side chains [30]. The third decomposition stage correlated with dehydration of carbohydrate chains, which was presented between 320 and 700 °C, respectively, showed a weight loss between 26–40% [17,30]. As can be observed in Table 2, the weight loss for all the steps decreases with the addition of the silica nanoparticles, so, the sample with the highest residual mass was S3.

Table 2. Elastic modulus calculated from stress-strain curves.

Sample	Elastic Modulus (KPa)								
	1–6 KPa	STDV	R^2	20–25 KPa	STDV	R^2	25–35 KPa	STDV	R^2
S0	6.57	±0.66	0.9906	12.74	±0.44	0.9990	13.46	±0.10	0.9971
S1	15.35	±2.10	0.9979	24.37	±0.98	0.9988	25.90	±0.18	0.9977
S2	7.32	±1.29	0.9913	27.38	±1.45	0.9998	31.09	±1.36	0.9993
S3	9.42	±2.36	0.9988	27.47	±1.50	0.9977	32.14	±1.88	0.9984

2.5. Mechanical Properties of the Crosslinked Materials in Wet Conditions

The mechanical properties can be observed in Figure 5.

The sweeps were carried out to determine the storage modulus and loss modulus. It was observed that the storage modulus (G′) of all samples was greater than the loss modulus (G″), indicating a crosslinked state for all of the samples analyzed. With a frequency range of 0.1 to 10 Hz, the change in G′ was practically constant and still greater than that of G″, indicating that hydrogels have elastic solid characteristics and excellent stability. Generally, the storage modulus increased as SiO_2 loadings increased. This suggested that the inclusion of silica nanofiller increased the stiffness of the salecan matrix and effectively boosted the elastic characteristics of the hydrogel nanocomposites due to the rigid inorganic nanoparticles impeding the mobility and deformability of the biopolymeric chains. This phenomenon was observed in other research studies that investigated the influence of silica nanoparticles on the mechanical properties of hydrogels [49–51].

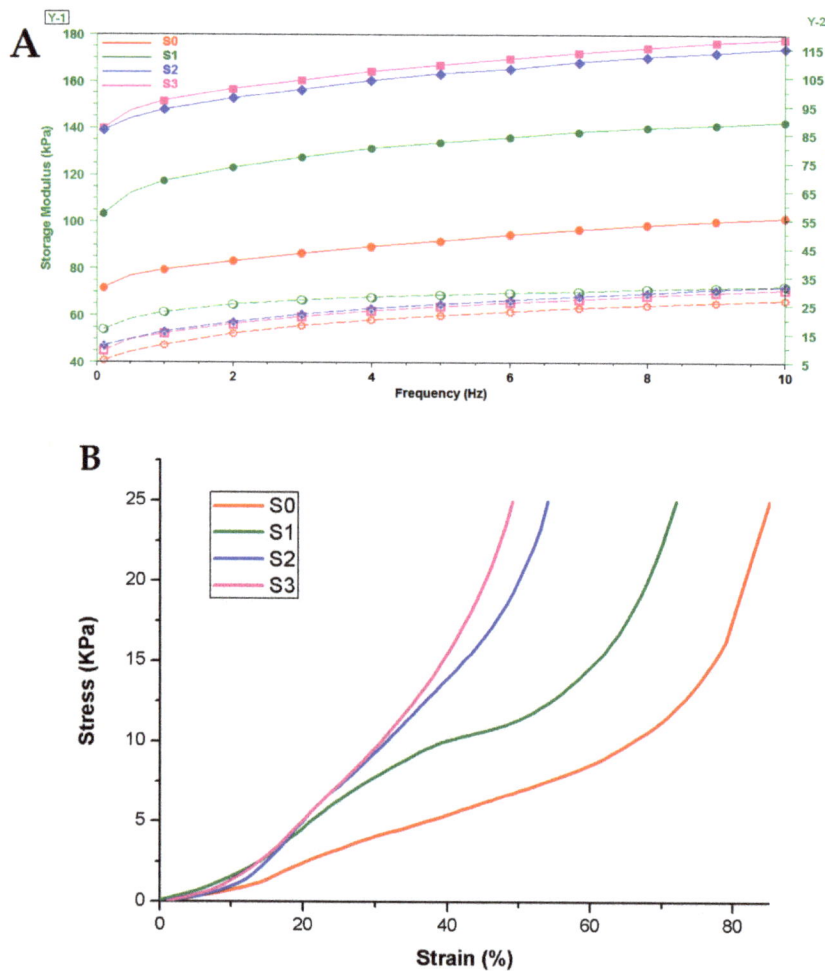

Figure 5. DMA analyses of the obtained hydrogels swelled at equilibrium; (**A**). Frequency sweep tests (filled symbols represent G′ and open symbols represent G″); (**B**). Stress-strain analyses.

A stress-strain test is an effective instrument for measuring how material changes in response to different loading situations. Stresses of ~26 KPa (according to a static force of ~6.5 N) have been applied for all samples, thus, the obtained hydrogels may satisfy the requirements of soft tissues [52,53]. The swollen cylindrical hydrogels exhibited a linear increase of stress reaching ~40–85% strain till achieving a constant compression. The crosslinked hydrogels with silica loading had a stiffer behavior that was better able to withstand stress till maximum compression. The addition of high loads of silica nanoparticles resulted in a considerable improvement in the compressive strength of the composite wet samples compared to the salecan neat sample. The origin of this phenomenon could be the decreased mobility of the biopolymer macromolecules in the presence of the inorganic filler. Another probable reason is that the absorption capacity of crosslinked hydrogels changes depending on the sample composition, with the presence of additional water molecules in the biopolymeric network impairing mechanical behavior. As depicted in Figure 5, the pure salecan sample retained more water, while the samples loaded with silica nanofiller had a lower swelling degree.

It is important to note that the samples recovered quickly following the imposed mechanical stress showing an elastic behavior, the water ejected under stress circumstances being quickly reabsorbed. The elastic modulus of hydrogels was determined using stress-strain curves, and the three most linear stress-strain regimes were considered [54,55]. The results are presented in Table 2.

The elastic modulus of the produced hydrogels ranged between 6.5 and 32 KPa, revealing their elastic property suited for soft tissue engineering [55]. The elastic modulus rose in the presence of silica nanoparticles across the ranges investigated, showing an increase in the stiffness of the nanocomposites. Moreover, the elastic modulus generally increased as the concentration of inorganic filler increased, with higher stress being more pronounced. The fact that the first interval deviates from the norm is intriguing, and this is very probably related to the distribution of silica particles within the hydrogel sample. Under great stress, the increased amount of silica nanoparticles are likely to restrict the movement of the biopolymer networks, resulting in the expected reliance on the amount of silica added to the system, which is consistent with other studies [56–58].

Thus, the inclusion of silicon dioxide nanoparticles into salecan crosslinked hydrogels can improve the mechanical behavior of the resulting hydrogel nanocomposites due to possible physical interaction between inorganic nanoparticles and salecan biopolymer chains. The salecan hydrogel's mechanical stability in wet conditions can therefore be easily modified by adding silica nanoparticles to suit different applications such as soft tissue engineering or wound dressings.

2.6. Morphological Evaluation

SEM analysis was used to examine the hydrogel's porous microarchitecture. In Figure 6 the microstructure aspect of new 3D printed scaffolds can be observed.

The SEM images of the novel 3D printed composite reveal porous scaffolds, with an architecture strongly dependent on the composition of the hydrogels as visible in Figure 6. Furthermore, all the samples presented interconnected pores without any phase separation between the polymer matrix and silica nanoparticles. The neat salecan sample presented the spongiest structure with larger pores than salecan/SiO_2 composite materials. The addition of silica nanoparticles to the polymer matrix causes shrinkage of the pores, creating a structure with more pores of smaller sizes. The network structure of the pores is crucial when aiming for regenerative medicine applications since it helps to guide and promote the creation of new tissue [59]. The same effect was obtained by Hu et al. [17] in their research with an increase in $Fe_3O_4@SiO_2$ content; the average pore size was diminished. It is possible that the hydroxyl groups on nanoparticle surfaces can establish hydrogen bonds with the carboxyl and hydroxyl groups from the biopolymer/polymer chains, leading to the development of more crosslinking points as nanoparticle content increased, with a reduction in the average pore size [17,60]. Furthermore, the differences presented in SEM images suggest that in the presents of a higher content of silica nanoparticles, the 3D printed composites maintained their profile better. The roundness of open pores could be used to quantify the printing accuracy of 3D forms. Using SEM images and the H. Wadell equation, the average roundness of the open pores of the scaffolds was estimated [61]. R = 1 is obtained for a perfectly round object, but irregular shapes have values less than 1. Thus, when the macropores of 3D printed form approach a rectangular shape, R values tend to 0 while R values approaching 1 indicate that the macropores are nearly circular. The computed value of the S0 sample was ~0.58, whilst the nanocomposite samples loaded with silica nanoparticles roundness value ranged between 0.5–0.3 (Figure 6B). These differences imply that in the presence of silica nanoparticles, 3D-printed objects kept their shape better when printed. S3 had the lowest R values, indicating that the macropores are close to a rectangular shape, with the 3D printed shape matching the designed 3D model the best. This fact reinforces the observations made throughout the 3D-printing process, as well as the evaluation of the stability of the printed constructions produced when nanoparticles were used. These observations correlated well with the rheological behavior of nanocomposite inks, which showed that the silica-salecan composites' viscosity slightly increased commensurate with their silica content.

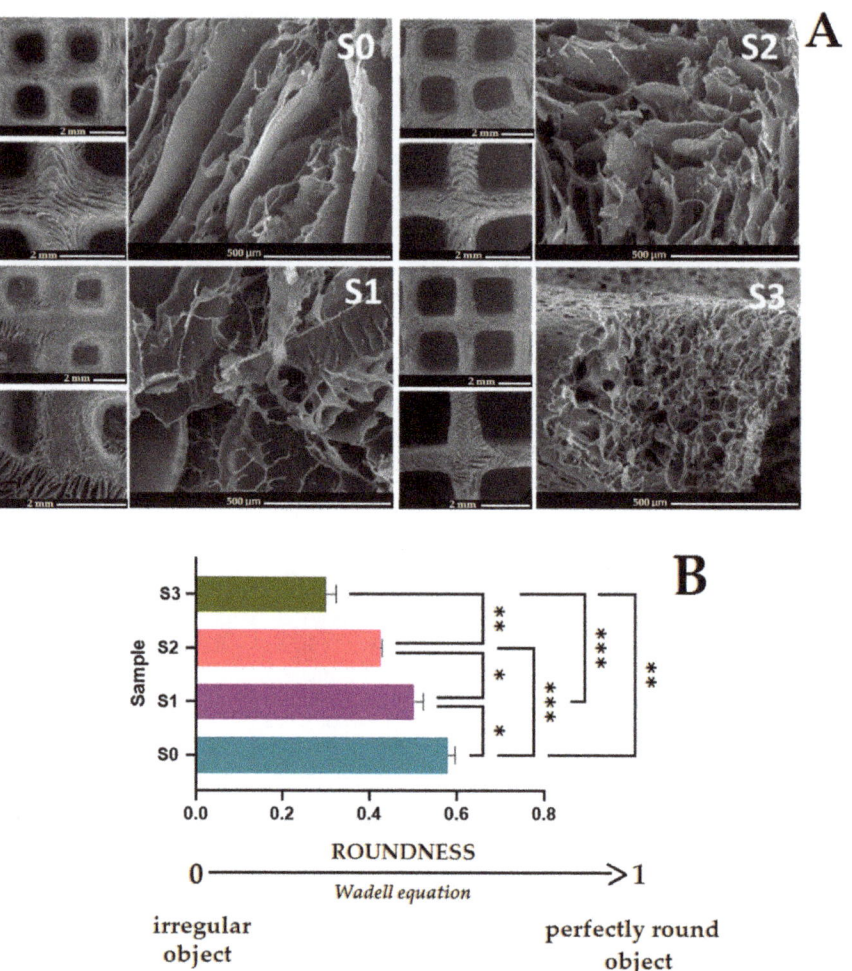

Figure 6. (**A**). SEM images showing the microstructure aspect of 3D printed scaffolds (30×, 50×, 250×); (**B**). The roundness of open pores is calculated using the Wadell equation. The resulting values consist of average values with additional standard errors. Statistical significance: * $p < 0.05$; ** $p < 0.005$; *** $p < 0.0005$. ONE WAY ANOVA TEST.

Figure 7 presents the TEM images of the nanocomposite sample S3, which have a higher concentration of silica nanoparticles. From the TEM images for sample S3, nanometric agglomerates can be observed due to the presence of spherical silica particles.

Furthermore, the TEM images revealed that the silica nanoparticles used for the development of the new 3D printed composites presented nanometer dimensions with an apparent average particle size between 10–30 nm.

2.7. Antimicrobial Activity

A key characteristic of many hydrogels utilized as scaffolds for regenerative medicine is their antibacterial activity [62]. All 3D printed scaffolds were tested against two gram-positive bacteria, Staphylococcus aureus and Escherichia coli (gram-negative bacteria). The antibacterial activity of every one of the obtained 3D printed salecan-based hydrogels was assessed (Figure 8).

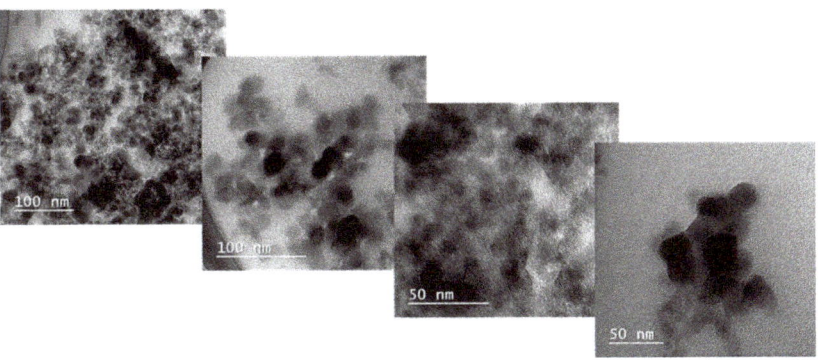

Figure 7. TEM images of the nanocomposite sample (S3).

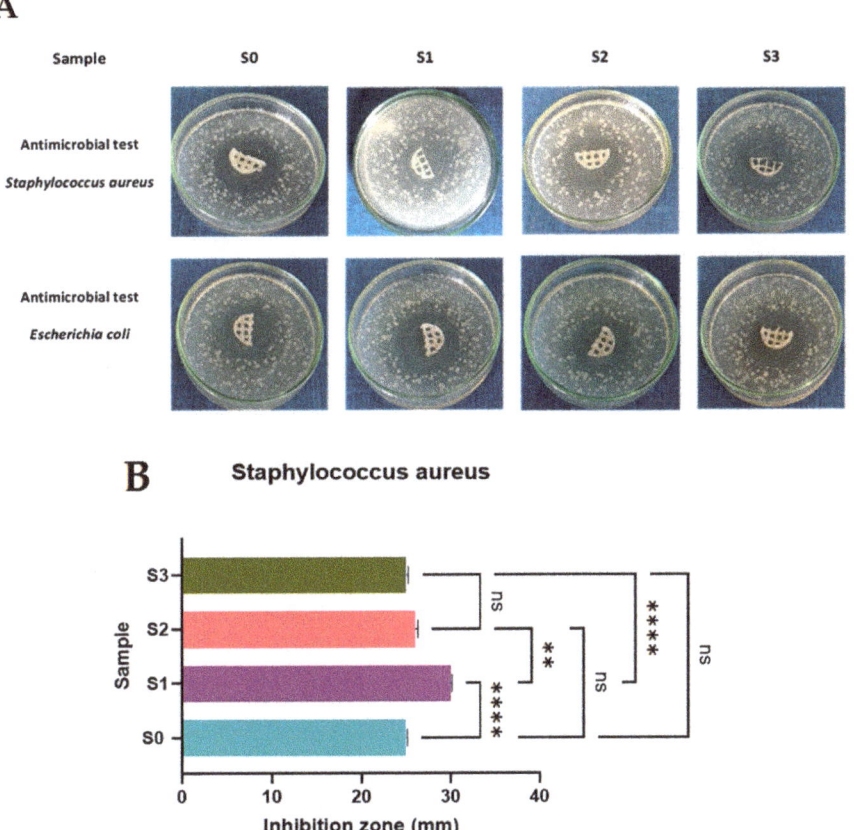

Figure 8. *Cont.*

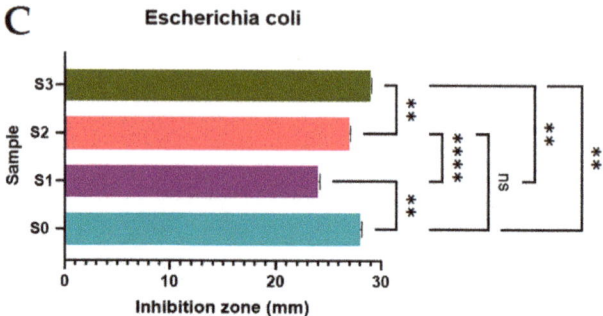

Figure 8. Antimicrobial activity (*Staphylococcus aureus* and *Escherichia coli*) of the 3D printed samples. (**A**). Images with the seeded samples; (**B**). Inhibition zone diameters for the samples seeded with *Staphylococcus aureus*; (**C**). Inhibition zone diameters for the samples seeded with *Escherichia coli*. Statistical significance: ns $p < 0.5$; ** $p < 0.005$, **** $p < 0.0001$. ONE WAY ANOVA TEST.

It can be observed from Figure 8 that every sample showed antibacterial activity against the two strains that were put to the test. Due to the samples' citric acid content, which is known to have an antibacterial effect, *Staphylococcus aureus* and *Escherichia coli* were inhibited in a detectable zone [62,63]. According to their chemical makeup, citric acid-derived polymers were shown to suppress bacterial proliferation to various degrees. Because of the leakage of unreacted citric acid molecules present in the tested samples, citrate-based polymers' inherent antibacterial characteristics allow them to impede bacterial development [64–66].

All of the samples had bacterial inhibition diameters for any of the microorganisms examined, according to the results presented in Figure 8.

Thus, all the samples presented good antimicrobial activity against *Staphylococcus aureus* and *Escherichia coli* and the presence of silica nanoparticles does not alter the antimicrobial behavior of the green crosslinked polysaccharide.

2.8. Biological Assessment of the Crosslinked 3D Printed Polysaccharide Constructs

One of the most fundamental things to take into account when choosing materials for biomedical purposes is cytotoxicity. The influence of salecan-based hydrogels on cells' viability was established by in vitro methods ((Live/Dead, MTT (3-[4,5-dimethylthiazol-2-yl]-2,5 diphenyl tetrazolium bromide) and LDH (lactate dehydrogenase) tests) assessed on HeLa cells.

Using Live/Dead staining, the cell viability of HeLa cells, cultivated in the salecan-based hydrogels, was evaluated. After 24 h of incubation, the green fluorescence, presented in Figure 9, showed that the majority of the cells embedded in the polysaccharide hydrogels were still alive. Furthermore, the MTT test confirmed the good biocompatibility of the new 3D printed polysaccharide constructs. Additionally, no significant difference in the cytotoxicity results between the salecan hydrogels and the negative control sample was observed from the LDH test, demonstrating the non-toxicity of these 3D printed hydrogels to HeLa cells.

The results of the MTT, LDH, and Live/Dead tests showed that the salecan-based hydrogels had great biocompatibility and can be used for biomedical applications. The outcomes are in line with those of prior research in which salecan-containing hydrogels showed promising biological characteristics [31,67,68].

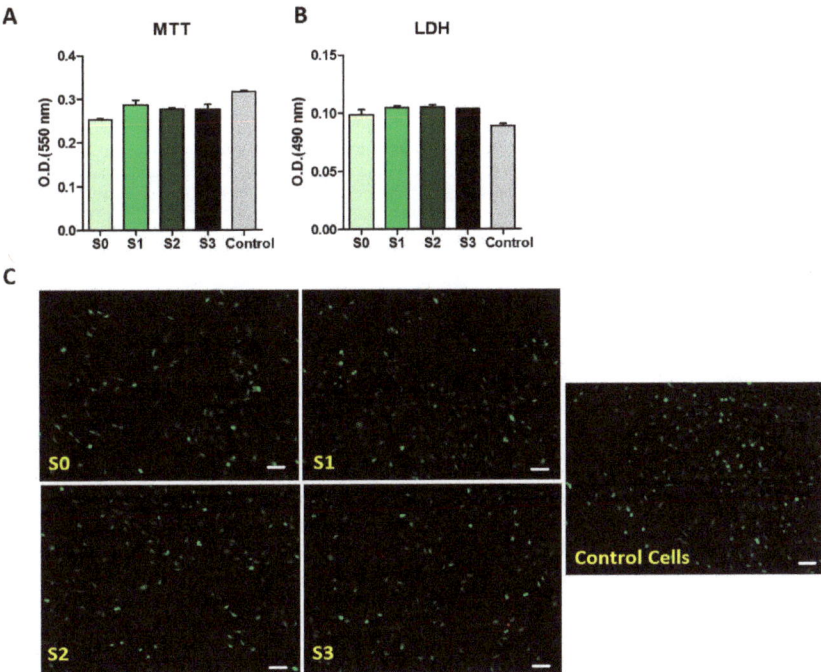

Figure 9. Biological assessment of the 3D printed constructs. Results were obtained from: (**A**). MTT assay, (**B**). LDH tests and (**C**). Live/Dead tests (scale bar-200 μm).

3. Conclusions

The synthesis of nanocomposite hydrogels with appropriate rheology for the production of precise-shaped 3D constructions was accomplished. Green-crosslinked biopolymer nanosilica materials were created by adding varying quantities of silica nanoparticles to a polysaccharide hydrogel.

According to our findings, the hydrogel's pore size, swelling, and mechanical properties under wet circumstances can all be easily modified by varying the quantity of silica nanoparticles in the composition used for hydrogel 3D printing. Moreover, increased printing ink viscosity was obtained by adding silica nanoparticles to the salecan biopolymer matrix, which led to the creation of 3D constructions with improved shape fidelity. MTT, LDH, and Live/Dead test outcomes demonstrated the salecan-based hydrogels' good biocompatibility and potential for usage in biomedical applications.

Due to the adaptability of the internal and external architecture, unique multi-functional hydrogels containing bioactive ingredients may be developed with particular demands and a regulated degradation profile that corresponds to the targeted tissue functionality. These investigations are presently ongoing.

Therefore, our hydrogel might be used as an easy-to-use and adaptable hydrogel for the creation of customized 3D printed constructs with a range of uses.

4. Materials and Methods

4.1. Materials and Synthesis

Suzhou Chemicals (Suzhou, China) provided the microbial polysaccharide known as Salecan (>90% purity), and SC Remed Prodimpex SRL (Pantelimon, Romania) provided the citric acid (>99.5% purity). Sigma-Aldrich (Steinheim, Germany) supplied the silica nanoparticles (amorphous silicon dioxide, 99.8%, 12 nm average size). Using HCl 37% and NaOH (>99.3% purity) from SC Chimreactiv SRL, Romania, we created simulated biological fluids with various pH values (5.5, 7.4, and 11) in our laboratory and Na_2HPO_4 (>99.9% purity) from Reactivul, Voinesti, Romania.

The new formulations were prepared according to the composition from Table 3 using an adapted preparation process [62].

Table 3. Composition of the obtained composites.

Sample	Salecan [g]	Citric Acid Solution 5% [mL]	Silica Nanoparticles [g]
S0	1.125	15	-
S1	1.125	15	0.281
S2	1.125	15	0.562
S3	1.125	15	0.843

Briefly, in order to obtain the new nanocomposite based on salecan, different concentrations of silica nanoparticles were dispersed in citric acid solution using magnetic stirring (600 rpm, 90 min, 22–25 °C using Arex heating magnetic stirrer, Velp Scientifica, Usmate Velate, Italy) and sonication treatment (5 min, 20% amplitude, 20 khz frequency, ice bath, using High Intensity Ultrasonic Processor, CPX 750, CV33, COLE PARMER, Cambridgeshire, UK); salecan was then added in the silica dispersion under mechanical stirring thus obtaining the hydrogel ink. A part of the composition was used for molding and the other part was used for 3D printing. All obtained samples were freeze-dried and then exposed to thermal treatment. The schematic synthesis steps can be observed in Figure 10.

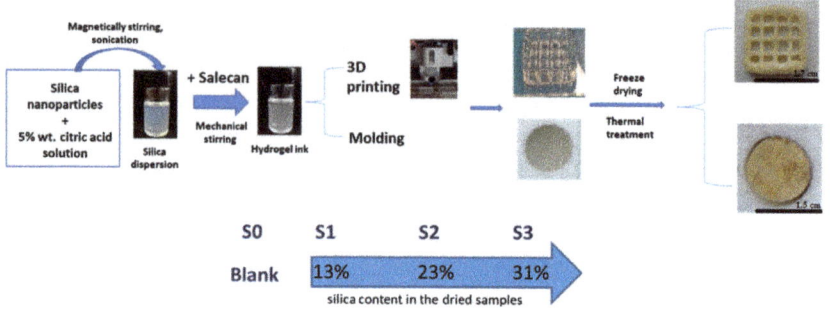

Figure 10. Schematic representation of the formation and characterization of the new polysaccharide-based nanocomposites.

Four inks were developed in order to construct porous biocompatible scaffolds that could support cell attachment and proliferation because the main objective of this research study was to create a new nanocomposite printing ink that could be used in tissue engineering. A 3D bioprinter from RegenHU Ltd., Villaz-St-Pierre, Switzerland, was used to print the nanocomposite ink samples. Room temperature (25–30 °C) was used for the printing procedure. The experiments were conducted using a direct dispensing printhead and a 5 mL syringe with an attached cylindrical nozzle that has a 0.41 mm diameter at various pressures between 190 and 350 kPa. Up to 10 layers of each mixture were printed. The 3D printed hydrogel structures underwent the previously indicated processes of lyophilization and heat treatment after the additive manufacturing process. The constructions were washed to remove all of the unreacted polysaccharides, and a previously known technique was used to quantify the degree of crosslinking between the polysaccharide and citric acid [8,25,27]. The crosslinking percentage was found to be ~95% for all samples regardless of composition. This demonstrated that the biopolymer crosslinking process was not impacted by the presence of silica nanoparticles, a highly crosslinked structure being synthesized.

4.2. Methods

4.2.1. Rheological Behavior

The shear viscosity vs. shear rate behavior was estimated using a Kinexus Pro Rheometer (Malvern, UK). The temperature of 25 °C was controlled with a Julabo CF41 cryo-compact circulator. In parallel plate geometry, the samples were put (10 mm diameter) with a gap of 0.5 mm. The shear viscosity was measured at an applied shear rate between 0.001 and 1000 s^{-1}. The results were represented logarithmically.

4.2.2. Fourier Transform Infrared Spectrometry (FT-IR)

The samples were structurally evaluated using Vertex 70 Bruker FTIR spectrometer (Billerica, MA, USA). All the FT-IR analyses were achieved in the 4000–600 cm^{-1} wavelength scale. The grounded samples were used for qualitative FTIR analysis in ATR mode.

4.2.3. Swelling Behavior

The swelling degree of the obtained 3D printed composites was analyzed by incubation in simulated biological fluids having different pH values (pH = 5.5; 7.4; 11) and a temperature of 37 °C. After a predicted time, the samples were removed from fluids, and filter paper was used to remove any surface-adsorbed water. The increase in weight (w − w0) from the initial weight (w0) represented the degree of swelling.

4.2.4. Thermo Gravimetric Analysis (TGA)

With a NETZSCH TG 209 F1 Libra instrument (Selb, Germany) (controlled atmosphere utilizing a nitrogen flow rate of around 20 mL/min, scanning from 25 to 700 °C and a heating degree of 10 °C/min), the thermogravimetric analysis of 3D printed nanocomposite was evaluated in triplicate. The samples utilized for all measurements ranged in mass from 3.5 to 5 mg.

4.2.5. Scanning Electron Microscopy (SEM)

The internal structure and morphology of the 3D printed nanocomposites were examined using the scanning electron microscope ESEM-FEI Quanta 200 (Eindhoven, The Netherlands). All the scaffolds were studied without any coating.

Wadell's Equation (1) was used to calculate the roundness of open pores (R). R is therefore equal to the average radius of curvature of the four corners of the 3D construct open pores to the radius of the greatest inscribed circle of the analyzed pore, as specified in Equation (1):

$$R = \left(\frac{1}{n}\sum_{i=1}^{n} ri\right) / rmax \quad (1)$$

where *ri* = radius of the corner curvature, *rmax* = radius of the greatest inscribed circle, *n*—number of corners (*n* = 4). The R-value for every 3D construct is the estimated average of the open pores shown in the SEM picture. R is greater than 0 for all other objects and equals 1 for perfectly spherical objects [61,69].

4.2.6. Transmission Electron Microscopy (TEM)

The freeze-dried 3D printed composites were examined using transmission electron microscopy (TEM). On a TECNAI F20 G2 TWIN Cryo-TEM (FEI, Hillsboro, OR, USA), all of the samples were examined in BF-TEM (Bright Field Transmission Electron Microscopy) mode at an accelerating voltage of 200 kV.

4.2.7. Mechanical Tests

The resulting hydrogel samples' mechanical behavior was assessed using a DMA Q800 TA Instruments (New Castle, DE, USA). The analyses were conducted at 37 °C in compression mode using cylindrical samples with a 15 mm diameter and ~8 mm thickness. All of the equilibrium swollen composites were compressed, with a ramp force of 0.2 N/min to 6.5 N/min till a constant strain plateau was registered.

Elastic modulus was calculated from stress-strain curves using the following equation $E = \sigma/\varepsilon$, where compressive stress is denoted as σ (N/m^2) and ε represents the related strain. Three linear regions of stress-strain curves were considered to calculate the elastic modulus, namely: 1–6, 20–25, and 25–35% strain compressions.

Dynamic frequency sweeps were carried out over a frequency range of 0.1–10 Hz with a continuous strain of 0.1% (in the linear viscoelastic area) at 25 °C in order to record the storage (G′) and loss (G″) moduli of the water-swollen samples at equilibrium.

4.2.8. Antimicrobial Activity

Two standard strains from the ICPI Institute's Microbiology Department collection—*Escherichia coli* ATCC 11229, a Gram-negative bacteria strain, and *Staphylococcus aureus* ATCC 25923, a Gram-positive bacterium strain—were used to determine the antibacterial activity for the 3D printed nanocomposites. Using a modified spot diffusion approach, the antibacterial properties were qualitatively screened. Our prior research study included a full description of this method's procedure [70]. Following 24–48 h of microbial cultures developed on Muller Hinton agar (MHA), bacterial and yeast suspensions of $(1-5) \times 10^8$ µcf/mL (corresponding with 0.5 McFarland standard density) were obtained. These suspensions were put in Petri dishes, left at room temperature to ensure equal diffusion of the compound in the medium, and then incubated with 3D printed composites at 37 °C for 24–48 h. The samples' inhibitory zone diameters were evaluated in triplicate after 24 h.

4.2.9. Biological Assessment of the 3D Printed Salecan-Based Hydrogels

The biocompatibility of the materials was tested on HeLa cells. Cells were cultivated at a density of 10^5 cells/well in Dulbecco's Modified Eagle Medium with 10% fetal bovine serum. The materials were UV sterilized and added on top of the cells for 24 h. Cytotoxicity was measured using the LDH Cytotoxicity kit (Sigma) following the manufacturer's instructions. Absorbance was read at $\lambda = 490$ nm using a NanoQuant Infinite M200 Pro instrument. The viability of the cells was analyzed using the Live/Dead assay (cat. No. L3224). Imaging was performed at $\lambda = 494/517$ (live cells) and at $\lambda = 517/617$ (dead cells) using a Zeiss fluorescence microscope.

Cell proliferation was quantified using the CyQUANT™ MTT Cell Viability Assay (Thermos Scientific, Zeiss Oberkochen, Baden-Wurttenberg, Germany) following the manufacturer's instructions.

4.2.10. Statistical Analyses

The data are presented as mean standard deviation. A one-way ANOVA was used to assess the significance of differences. If the *p*-value was less than 0.05, the significance was evaluated.

Author Contributions: Conceptualization, M.M.M. and R.I.; methodology, R.L.A. and M.M.M.; formal analysis, R.R.C., C.L.N., G.G.P. and I.C.G.; investigation, B.T., E.A., C.P., M.M.M., G.G.P. and R.I.; resources, M.M.M., M.A.K. and R.I.; writing—original draft preparation, M.M.M. and R.I.; writing—review and editing, M.M.M., M.A.K., I.C.G. and R.I.; visualization, M.M.M. and R.I.; supervision, R.I.; funding acquisition, M.A.K. and R.I. All authors have read and agreed to the published version of the manuscript.

Funding: This work was supported by the Ministry of Research, Innovation and Digitization through Program 1—Development of the national research and development system, Subprogram 1.2—Institutional performance—Projects to finance excellence in RDI, Contract no. 15PFE/2021 and Nucleu Program 2023-2026, project code PN 23 26 01 01.

Institutional Review Board Statement: Not applicable.

Informed Consent Statement: Not applicable.

Data Availability Statement: Not applicable.

Acknowledgments: We greatly acknowledge the scientific and technical support of Ludmila Aricov and Dorian Radu.

Conflicts of Interest: The authors declare no conflict of interest.

References

1. Bácskay, I.; Ujhelyi, Z.; Fehér, P.; Arany, P. The Evolution of the 3D-Printed Drug Delivery Systems: A Review. *Pharmaceutics* **2022**, *14*, 1312. [CrossRef] [PubMed]
2. Norman, J.; Madurawe, R.D.; Moore, C.M.V.; Khan, M.A.; Khairuzzaman, A. A New Chapter in Pharmaceutical Manufacturing: 3D-Printed Drug Products. *Adv. Drug Deliv. Rev.* **2017**, *108*, 39–50. [CrossRef]
3. Araújo, M.R.P.; Sa-Barreto, L.L.; Gratieri, T.; Gelfuso, G.M.; Cunha-Filho, M. The Digital Pharmacies Era: How 3D Printing Technology Using Fused Deposition Modeling Can Become a Reality. *Pharmaceutics* **2019**, *11*, 128. [CrossRef] [PubMed]
4. Guarch-Pérez, C.; Shaqour, B.; Riool, M.; Verleije, B.; Beyers, K.; Vervaet, C.; Cos, P.; Zaat, S.A.J. 3D-Printed Gentamicin-Releasing Poly-ε-Caprolactone Composite Prevents Fracture-Related Staphylococcus Aureus Infection in Mice. *Pharmaceutics* **2022**, *14*, 1363. [CrossRef]
5. Zhang, Q.; Zhou, J.; Zhi, P.; Liu, L.; Liu, C.; Fang, A.; Zhang, Q. 3D Printing Method for Bone Tissue Engineering Scaffold. *Med. Nov. Technol. Devices* **2023**, *17*, None. [CrossRef] [PubMed]
6. Vikram Singh, A.; Hasan Dad Ansari, M.; Wang, S.; Laux, P.; Luch, A.; Kumar, A.; Patil, R.; Nussberger, S. The Adoption of Three-Dimensional Additive Manufacturing from Biomedical Material Design to 3D Organ Printing. *Appl. Sci.* **2019**, *9*, 811. [CrossRef]
7. Miao, Y.; Chen, Y.; Luo, J.; Liu, X.; Yang, Q.; Shi, X.; Wang, Y. Black Phosphorus Nanosheets-Enabled DNA Hydrogel Integrating 3D-Printed Scaffold for Promoting Vascularized Bone Regeneration. *Bioact. Mater.* **2023**, *21*, 97–109. [CrossRef]
8. Qi, X.; Hu, X.; Wei, W.; Yu, H.; Li, J.; Zhang, J.; Dong, W. Investigation of Salecan/Poly(Vinyl Alcohol) Hydrogels Prepared by Freeze/Thaw Method. *Carbohydr. Polym.* **2015**, *118*, 60–69. [CrossRef]
9. Ianchis, R.; Ninciuleanu, C.; Gifu, I.C.; Alexandrescu, E.; Nistor, C.L.; Nitu, S.G.; Petcu, C. Hydrogel-Clay Nanocomposites as Carriers for Controlled Release. *CMC* **2018**, *25*. [CrossRef]
10. El-Sherbiny, I.M. Enhanced PH-Responsive Carrier System Based on Alginate and Chemically Modified Carboxymethyl Chitosan for Oral Delivery of Protein Drugs: Preparation and in-Vitro Assessment. *Carbohydr. Polym.* **2010**, *80*, 1125–1136. [CrossRef]
11. El-Sherbiny, I.M.; Yacoub, M.H. Hydrogel Scaffolds for Tissue Engineering: Progress and Challenges. *Glob. Cardiol. Sci. Pract.* **2013**, *2013*, 316–342. [CrossRef]
12. Dragan, E.S.; Apopei, D.F. Multiresponsive Macroporous Semi-IPN Composite Hydrogels Based on Native or Anionically Modified Potato Starch. *Carbohydr. Polym.* **2013**, *92*, 23–32. [CrossRef]
13. Almany, L.; Seliktar, D. Biosynthetic Hydrogel Scaffolds Made from Fibrinogen and Polyethylene Glycol for 3D Cell Cultures. *Biomaterials* **2005**, *26*, 2467–2477. [CrossRef]
14. Dinu, M.V.; Přádný, M.; Drăgan, E.S.; Michálek, J. Morphogical and Swelling Properties of Porous Hydrogels Based on Poly(Hydroxyethyl Methacrylate) and Chitosan Modulated by Ice-Templating Process and Porogen Leaching. *J. Polym. Res.* **2013**, *20*, 285. [CrossRef]
15. Aouada, F.A.; Moura, M.R.d.; Lopes da Silva, W.T.; Muniz, E.C.; Mattoso, L.H.C. Preparation and Characterization of Hydrophilic, Spectroscopic, and Kinetic Properties of Hydrogels Based on Polyacrylamide and Methylcellulose Polysaccharide. *J. Appl. Polym. Sci.* **2011**, *120*, 3004–3013. [CrossRef]
16. Mandal, S.; Nagi, G.K.; Corcoran, A.A.; Agrawal, R.; Dubey, M.; Hunt, R.W. Algal Polysaccharides for 3D Printing: A Review. *Carbohydr. Polym.* **2023**, *300*, 120267. [CrossRef]
17. Hu, X.; Wang, Y.; Zhang, L.; Xu, M.; Zhang, J.; Dong, W. Design of a PH-Sensitive Magnetic Composite Hydrogel Based on Salecan Graft Copolymer and Fe3O4@SiO2 Nanoparticles as Drug Carrier. *Int. J. Biol. Macromol.* **2018**, *107*, 1811–1820. [CrossRef] [PubMed]
18. Li, F.; Wu, H.; Fan, L.; Zhang, H.; Zhang, H.; Gu, C. Study of Dual Responsive Poly[(Maleilated Dextran)-Graft-(N-Isopropylacrylamide)] Hydrogel Nanoparticles: Preparation, Characterization and Biological Evaluation. *Polym. Int.* **2009**, *58*, 1023–1033. [CrossRef]
19. Thambi, T.; Phan, V.H.G.; Lee, D.S. Stimuli-Sensitive Injectable Hydrogels Based on Polysaccharides and Their Biomedical Applications-Thambi-2016-Macromolecular Rapid Communications-Wiley Online Library. Available online: https://onlinelibrary.wiley.com/doi/full/10.1002/marc.201600371 (accessed on 4 April 2023).
20. Aderibigbe, B.; Buyana, B. Alginate in Wound Dressings. *Pharmaceutics* **2018**, *10*, 42. [CrossRef] [PubMed]
21. Beltran-Vargas, N.E.; Peña-Mercado, E.; Sánchez-Gómez, C.; Garcia-Lorenzana, M.; Ruiz, J.-C.; Arroyo-Maya, I.; Huerta-Yepez, S.; Campos-Terán, J. Sodium Alginate/Chitosan Scaffolds for Cardiac Tissue Engineering: The Influence of Its Three-Dimensional Material Preparation and the Use of Gold Nanoparticles. *Polymers* **2022**, *14*, 3233. [CrossRef]
22. Dai, L.; Cheng, T.; Duan, C.; Zhao, W.; Zhang, W.; Zou, X.; Aspler, J.; Ni, Y. 3D Printing Using Plant-Derived Cellulose and Its Derivatives: A Review. *Carbohydr. Polym.* **2019**, *203*, 71–86. [CrossRef] [PubMed]
23. Tavakoli, S.; Kharaziha, M.; Kermanpur, A.; Mokhtari, H. Sprayable and Injectable Visible-Light Kappa-Carrageenan Hydrogel for In-Situ Soft Tissue Engineering. *Int. J. Biol. Macromol.* **2019**, *138*, 590–601. [CrossRef]

24. Mohammed, A.S.A.; Naveed, M.; Jost, N. Polysaccharides; Classification, Chemical Properties, and Future Perspective Applications in Fields of Pharmacology and Biological Medicine (A Review of Current Applications and Upcoming Potentialities). *J. Polym. Environ.* **2021**, *29*, 2359–2371. [CrossRef]
25. Ianchis, R.; Alexa, R.L.; Gifu, I.C.; Marin, M.M.; Alexandrescu, E.; Constantinescu, R.; Serafim, A.; Nistor, C.L.; Petcu, C. Novel Green Crosslinked Salecan Hydrogels and Preliminary Investigation of Their Use in 3D Printing. *Pharmaceutics* **2023**, *15*, 373. [CrossRef]
26. Qi, X.; Wei, W.; Shen, J.; Dong, W. Salecan Polysaccharide-Based Hydrogels and Their Applications: A Review. *J. Mater. Chem. B* **2019**, *7*, 2577–2587. [CrossRef] [PubMed]
27. Fan, Z.; Cheng, P.; Gao, Y.; Wang, D.; Jia, G.; Zhang, P.; Prakash, S.; Wang, Z.; Han, J. Understanding the Rheological Properties of a Novel Composite Salecan/Gellan Hydrogels. *Food Hydrocoll.* **2022**, *123*, 107162. [CrossRef]
28. Fan, Z.; Cheng, P.; Prakash, S.; Zhang, P.; Mei, L.; Ji, S.; Wang, Z.; Han, J. Rheological Investigation of a Versatile Salecan/Curdlan Gel Matrix. *Int. J. Biol. Macromol.* **2021**, *193*, 2202–2209. [CrossRef] [PubMed]
29. Fan, Z.; Cheng, P.; Chu, L.; Han, J. Exploring the Rheological and Structural Characteristics of Novel Pectin-Salecan Gels. *Polymers* **2022**, *14*, 4619. [CrossRef] [PubMed]
30. Qi, X.; Su, T.; Tong, X.; Xiong, W.; Zeng, Q.; Qian, Y.; Zhou, Z.; Wu, X.; Li, Z.; Shen, L.; et al. Facile Formation of Salecan/Agarose Hydrogels with Tunable Structural Properties for Cell Culture. *Carbohydr. Polym.* **2019**, *224*, 115208. [CrossRef]
31. Fan, Z.; Cheng, P.; Yin, G.; Wang, Z.; Han, J. In Situ Forming Oxidized Salecan/Gelatin Injectable Hydrogels for Vancomycin Delivery and 3D Cell Culture. *J. Biomater. Sci. Polym. Ed.* **2020**, *31*, 762–780. [CrossRef]
32. Florian, P.E.; Icriverzi, M.; Ninciuleanu, C.M.; Alexandrescu, E.; Trica, B.; Preda, S.; Ianchis, R.; Roseanu, A. Salecan-Clay Based Polymer Nanocomposites for Chemotherapeutic Drug Delivery Systems; Characterization and In Vitro Biocompatibility Studies. *Materials* **2020**, *13*, 5389. [CrossRef] [PubMed]
33. Ip, H.T.; Liu, L.; Hong, L.; Ngai, T. Synthesis of Polystyrene/Silica and Poly(Styrene-Co-Butyl Acrylate)/Silica Nanocomposite Particles by Pickering Emulsion Polymerization with Non-Functionalized Silica Nanoparticles. *Colloids Surf. A Physicochem. Eng. Asp.* **2022**, *654*, 130104. [CrossRef]
34. Zhao, N.; Yan, L.; Zhao, X.; Chen, X.; Li, A.; Zheng, D.; Zhou, X.; Dai, X.; Xu, F.-J. Versatile Types of Organic/Inorganic Nanohybrids: From Strategic Design to Biomedical Applications. *Chem. Rev.* **2019**, *119*, 1666–1762. [CrossRef]
35. Marin, M.M.; Ianchis, R.; Leu Alexa, R.; Gifu, I.C.; Kaya, M.G.A.; Savu, D.I.; Popescu, R.C.; Alexandrescu, E.; Ninciuleanu, C.M.; Preda, S.; et al. Development of New Collagen/Clay Composite Biomaterials. *Int. J. Mol. Sci.* **2022**, *23*, 401. [CrossRef] [PubMed]
36. Huang, Y.; Li, P.; Zhao, R.; Zhao, L.; Liu, J.; Peng, S.; Fu, X.; Wang, X.; Luo, R.; Wang, R.; et al. Silica Nanoparticles: Biomedical Applications and Toxicity. *Biomed. Pharmacother.* **2022**, *151*, 113053. [CrossRef]
37. Diab, R.; Canilho, N.; Pavel, I.A.; Haffner, F.B.; Girardon, M.; Pasc, A. Silica-Based Systems for Oral Delivery of Drugs, Macromolecules and Cells. *Adv. Colloid Interface Sci.* **2017**, *249*, 346–362. [CrossRef]
38. Choi, Y.; Kim, J.; Yu, S.; Hong, S. PH- and Temperature-Responsive Radially Porous Silica Nanoparticles with High-Capacity Drug Loading for Controlled Drug Delivery. *Nanotechnology* **2020**, *31*, 335103. [CrossRef]
39. Kiran Roopavath, U.; Soni, R.; Mahanta, U.; Suresh Deshpande, A.; Narayan Rath, S. 3D Printable SiO_2 Nanoparticle Ink for Patient Specific Bone Regeneration. *RSC Adv.* **2019**, *9*, 23832–23842. [CrossRef]
40. Florek, J.; Caillard, R.; Kleitz, F. Evaluation of Mesoporous Silica Nanoparticles for Oral Drug Delivery–Current Status and Perspective of MSNs Drug Carriers. *Nanoscale* **2017**, *9*, 15252–15277. [CrossRef]
41. Hsu, S.-Y.; Morris, R.; Cheng, F. Signaling Pathways Regulated by Silica Nanoparticles. *Molecules* **2021**, *26*, 1398. [CrossRef]
42. Shao, H.; Zhang, H.; Tian, Y.; Song, Z.; Lai, P.; Ai, L. Composition and Rheological Properties of Polysaccharide Extracted from Tamarind (*Tamarindus indica* L.) Seed. *Molecules* **2019**, *24*, 1218. [CrossRef] [PubMed]
43. Unagolla, J.M.; Jayasuriya, A.C. Hydrogel-Based 3D Bioprinting: A Comprehensive Review on Cell-Laden Hydrogels, Bioink Formulations, and Future Perspectives. *Appl. Mater. Today* **2020**, *18*, 100479. [CrossRef] [PubMed]
44. Gerezgiher, A.G.; Szabó, T. Crosslinking of Starch Using Citric Acid. *J. Phys. Conf. Ser.* **2022**, *2315*, 012036. [CrossRef]
45. Caprarescu, S.; Radu, A.-L.; Purcar, V.; Sarbu, A.; Vaireanu, D.-I.; Ianchis, R.; Ghiurea, M. Removal of Copper Ions from Simulated Wastewaters Using Different Bicomponent Polymer Membranes. *Water Air Soil Pollut* **2014**, *225*, 2079. [CrossRef]
46. Ninciuleanu, C.M.; Ianchiş, R.; Alexandrescu, E.; Mihăescu, C.I.; Scomoroşcenco, C.; Nistor, C.L.; Preda, S.; Petcu, C.; Teodorescu, M. The Effects of Monomer, Crosslinking Agent, and Filler Concentrations on the Viscoelastic and Swelling Properties of Poly(Methacrylic Acid) Hydrogels: A Comparison. *Materials* **2021**, *14*, 2305. [CrossRef]
47. Ninciuleanu, C.M.; Ianchiş, R.; Alexandrescu, E.; Mihăescu, C.I.; Burlacu, S.; Trică, B.; Nistor, C.L.; Preda, S.; Scomoroscenco, C.; Gîfu, C.; et al. Adjusting Some Properties of Poly(Methacrylic Acid) (Nano)Composite Hydrogels by Means of Silicon-Containing Inorganic Fillers. *Int. J. Mol. Sci.* **2022**, *23*, 10320. [CrossRef]
48. Asabuwa Ngwabebhoh, F.; Saha, N.; Nguyen, H.T.; Brodnjak, U.V.; Saha, T.; Lengalova, A.; Saha, P. Preparation and Characterization of Nonwoven Fibrous Biocomposites for Footwear Components. *Polymers* **2020**, *12*, 3016. [CrossRef]
49. Boonmahitthisud, A.; Nakajima, L.; Nguyen, K.D.; Kobayashi, T. Composite Effect of Silica Nanoparticle on the Mechanical Properties of Cellulose-Based Hydrogels Derived from Cottonseed Hulls. *J. Appl. Polym. Sci.* **2017**, *134*. [CrossRef]
50. Li, X.; Rombouts, W.; Van Der Gucht, J.; De Vries, R.; Dijksman, J.A. Mechanics of Composite Hydrogels Approaching Phase Separation. *PLoS ONE* **2019**, *14*, e0211059. [CrossRef] [PubMed]

51. Xing, W.; Tang, Y. On Mechanical Properties of Nanocomposite Hydrogels: Searching for Superior Properties. *Nano Mater. Sci.* **2022**, *4*, 83–96. [CrossRef]
52. Liu, J.; Zheng, H.; Poh, P.S.P.; Machens, H.-G.; Schilling, A.F. Hydrogels for Engineering of Perfusable Vascular Networks. *Int. J. Mol. Sci.* **2015**, *16*, 15997–16016. [CrossRef] [PubMed]
53. Advanced BioMatrix-Stiffness Range of Living Tissue. Available online: https://advancedbiomatrix.com/stiffness-range-of-living-tissue.html (accessed on 4 April 2023).
54. Low, Z.W.; Chee, P.L.; Kai, D.; Loh, X.J. The Role of Hydrogen Bonding in Alginate/Poly(Acrylamide-Co-Dimethylacrylamide) and Alginate/Poly(Ethylene Glycol) Methyl Ether Methacrylate-Based Tough Hybrid Hydrogels. *RSC Adv.* **2015**, *5*, 57678–57685. [CrossRef]
55. Suflet, D.M.; Popescu, I.; Pelin, I.M.; Ichim, D.L.; Daraba, O.M.; Constantin, M.; Fundueanu, G. Dual Cross-Linked Chitosan/PVA Hydrogels Containing Silver Nanoparticles with Antimicrobial Properties. *Pharmaceutics* **2021**, *13*, 1461. [CrossRef]
56. Hocken, A.; Beyer, F.L.; Lee, J.S.; Grim, B.J.; Mithaiwala, H.; Green, M.D. Covalently Integrated Silica Nanoparticles in Poly(Ethylene Glycol)-Based Acrylate Resins: Thermomechanical, Swelling, and Morphological Behavior. *Soft Matter* **2022**, *18*, 1019–1033. [CrossRef] [PubMed]
57. Kato, K.; Matsui, D.; Mayumi, K.; Ito, K. Synthesis, Structure, and Mechanical Properties of Silica Nanocomposite Polyrotaxane Gels. *Beilstein J. Org. Chem.* **2015**, *11*, 2194–2201. [CrossRef]
58. Ahn, S.; Kim, S.; Kim, B.C.; Shim, K.; Cho, B. Mechanical Properties of Silica Nanoparticle Reinforced Poly(Ethylene 2, 6-Naphthalate). *Macromol. Res.* **2004**, *12*, 293–302. [CrossRef]
59. Loh, Q.L.; Choong, C. Three-Dimensional Scaffolds for Tissue Engineering Applications: Role of Porosity and Pore Size. *Tissue Eng Part B Rev* **2013**, *19*, 485–502. [CrossRef]
60. Liu, W.; Li, Y.; Feng, S.; Ning, J.; Wang, J.; Gou, M.; Chen, H.; Xu, F.; Du, Y. Magnetically Controllable 3D Microtissues Based on Magnetic Microcryogels. *Lab Chip* **2014**, *14*, 2614–2625. [CrossRef]
61. Wadell, H. Volume, Shape, and Roundness of Rock Particles. *J. Geol.* **1932**, *40*, 443–451. [CrossRef]
62. Marin, M.M.; Albu Kaya, M.; Kaya, D.A.; Constantinescu, R.; Trica, B.; Gifu, I.C.; Alexandrescu, E.; Nistor, C.L.; Alexa, R.L.; Ianchis, R. Novel Nanocomposite Hydrogels Based on Crosslinked Microbial Polysaccharide as Potential Bioactive Wound Dressings. *Materials* **2023**, *16*, 982. [CrossRef]
63. Botet-Carreras, A.; Marimon, M.B.; Millan-Solsona, R.; Aubets, E.; Ciudad, C.J.; Noé, V.; Montero, M.T.; Domènech, Ò.; Borrell, J.H. On the Uptake of Cationic Liposomes by Cells: From Changes in Elasticity to Internalization. *Colloids Surf. B Biointerfaces* **2023**, *221*, 112968. [CrossRef]
64. Su, L.-C.; Xie, Z.; Zhang, Y.; Nguyen, K.T.; Yang, J. Study on the Antimicrobial Properties of Citrate-Based Biodegradable Polymers. *Front. Bioeng. Biotechnol.* **2014**, *2*, 23. [CrossRef] [PubMed]
65. Wang, M.; Xu, P.; Lei, B. Engineering Multifunctional Bioactive Citrate-Based Biomaterials for Tissue Engineering. *Bioact. Mater.* **2023**, *19*, 511–537. [CrossRef]
66. Xi, Y.; Guo, Y.; Wang, M.; Ge, J.; Liu, Y.; Niu, W.; Chen, M.; Xue, Y.; Winston, D.D.; Dai, W.; et al. Biomimetic Bioactive Multifunctional Poly(Citrate-Siloxane)-Based Nanofibrous Scaffolds Enable Efficient Multidrug-Resistant Bacterial Treatment/Non-Invasive Tracking In Vitro/In Vivo. *Chem. Eng. J.* **2020**, *383*, 123078. [CrossRef]
67. Qi, X.; Su, T.; Zhang, M.; Tong, X.; Pan, W.; Zeng, Q.; Zhou, Z.; Shen, L.; He, X.; Shen, J. Macroporous Hydrogel Scaffolds with Tunable Physicochemical Properties for Tissue Engineering Constructed Using Renewable Polysaccharides. *ACS Appl. Mater. Interfaces* **2020**, *12*, 13256–13264. [CrossRef] [PubMed]
68. Fan, Z.; Cheng, P.; Ling, L.; Han, J. Dynamic Bond Crosslinked Poly(γ-Glutamic Acid)/Salecan Derived Hydrogel as a Platform for 3D Cell Culture. *Mater. Lett.* **2020**, *273*, 127936. [CrossRef]
69. Cruz-Matías, I.; Ayala, D.; Hiller, D.; Gutsch, S.; Zacharias, M.; Estradé, S.; Peiró, F. Sphericity and Roundness Computation for Particles Using the Extreme Vertices Model. *J. Comput. Sci.* **2019**, *30*, 28–40. [CrossRef]
70. Marin, M.M.; Albu Kaya, M.G.; Iovu, H.; Stavarache, C.E.; Chelaru, C.; Constantinescu, R.R.; Dinu-Pîrvu, C.-E.; Ghica, M.V. Obtaining, Evaluation, and Optimization of Doxycycline-Loaded Microparticles Intended for the Local Treatment of Infectious Arthritis. *Coatings* **2020**, *10*, 990. [CrossRef]

Disclaimer/Publisher's Note: The statements, opinions and data contained in all publications are solely those of the individual author(s) and contributor(s) and not of MDPI and/or the editor(s). MDPI and/or the editor(s) disclaim responsibility for any injury to people or property resulting from any ideas, methods, instructions or products referred to in the content.

Article

Three-Dimensional Printing of Red Algae Biopolymers: Effect of Locust Bean Gum on Rheology and Processability

Sónia Oliveira [1], Isabel Sousa [1], Anabela Raymundo [1] and Carlos Bengoechea [2,*]

1. LEAF—Linking Landscape, Environment, Agriculture and Food—Research Center, Associated Laboratory TERRA, Instituto Superior de Agronomia, Universidade de Lisboa, Tapada da Ajuda, 1349-017 Lisboa, Portugal; soliveira@isa.ulisboa.pt (S.O.); isabelsousa@isa.ulisboa.pt (I.S.); anabraymundo@isa.ulisboa.pt (A.R.)
2. Departamento de Ingeniería Química, Escuela Politécnica Superior, Universidad de Sevilla, 41012 Sevilla, Spain
* Correspondence: cbengoechea@us.es

Citation: Oliveira, S.; Sousa, I.; Raymundo, A.; Bengoechea, C. Three-Dimensional Printing of Red Algae Biopolymers: Effect of Locust Bean Gum on Rheology and Processability. *Gels* 2024, 10, 166. https://doi.org/10.3390/gels10030166

Academic Editors: Enrique Aguilar and Helena Herrada-Manchón

Received: 31 January 2024
Revised: 19 February 2024
Accepted: 21 February 2024
Published: 23 February 2024

Copyright: © 2024 by the authors. Licensee MDPI, Basel, Switzerland. This article is an open access article distributed under the terms and conditions of the Creative Commons Attribution (CC BY) license (https://creativecommons.org/licenses/by/4.0/).

Abstract: Seaweeds, rich in high-value polysaccharides with thickening/gelling properties (e.g., agar, carrageenan, and alginate), are extensively used in the food industry for texture customization and enhancement. However, conventional extraction methods for these hydrocolloids often involve potentially hazardous chemicals and long extraction times. In this study, three red seaweed species (*Chondrus crispus*, *Gelidium Corneum*, and *Gracilaria gracilis*) commercialized as food ingredients by local companies were chosen for their native gelling biopolymers, which were extracted using water-based methodologies (i.e., (1) hydration at room temperature; (2) stirring at 90 °C; and (3) centrifugation at 40 °C) for production of sustainable food gels. The potential use of these extracts as bioinks was assessed employing an extrusion-based 3D printer. The present work aimed to study the gelation process, taken place during printing, and assess the effectiveness of the selected green extraction method in producing gels. To improve the definition of the printed gel, two critical printing parameters were investigated: the addition of locust bean gum (LBG) at different concentrations (0, 0.5, 1, 1.5, 2, and 2.5%) and printing temperature (30, 40, 60, and 80 °C). Rheological results from a controlled-stress rheometer indicated that gels derived from *G. corneum* and *G. gracilis* exhibited a lower gel strength (lower G' and G'') and excessive material spreading during deposition (lower viscosity) than *C. crispus*. Thus, G' was around 5 and 70 times higher for *C. crispus* gels than for *G. corneum* and *G. gracilis*, respectively. When increasing LBG concentration (0.5 to 2.5% w/w) and lowering the printing temperature (80 to 30 °C), an enhanced gel matrix definition for *G. corneum* and *G. gracilis* gels was found. In contrast, gels from *C. crispus* demonstrated greater stability and were less influenced by these parameters, showcasing the potential of the seaweed to develop sustainable clean label food gels. Eventually, these results highlight the feasibility of using algal-based extracts obtained through a green procedure as bioinks where LBG was employed as a synergic ingredient.

Keywords: seaweeds; gels; locust bean gum; hydrocolloids; green extraction; 3D food printing; rheology

1. Introduction

Seaweeds, also known as macroalgae, are a group of macroscopic multicellular species, traditionally classified into Phaeophyta (brown), Rhodophyta (red), and Chlorophyta (green), based on their pigmentation. Carbohydrates represent the major component of the biomass of seaweeds, varying between 25 and 50% (green), 30 and 60% (red), and 30 and 50% (brown) dry weight (DW) [1]. Phycocolloids are large molecules present in intercellular spaces in algae, conferring consistency/flexibility to their cell walls [2]. Phycocolloids in red algae (Rhodophyta) are floridean starch and floridoside, which are similar to general starch. The major polysaccharide constituent of red algae are galactans, mostly represented by carrageenan (up to 75% DW) and agar (up to 52% DW) [3]. Industrially, commercial carrageenans are mainly extracted from *Chondrus* spp., *Eucheuma* spp., and *Kappaphycus* spp. [3,4]. However, seaweeds produce hybrid structures of the kappa, iota, and lambda forms of carrageenans, which are composed D-galactopyranose residues bonded by regularly alternating α-(1 → 3) and β-(1 → 4) bonds [5]. Kappa (κ)-carrageenan has one sulfate ester, while iota (ι)- and lambda (λ)- carrageenan contain two and three sulfates per dimer, respectively. Their ability to act as a gelling or thickening agent is dependent on the presence of the anhydro-galactose bridge of the 4-linked galactose residue, crucial for the formation of the helical structure and, consequently, the ability to form a gel. While both the κ-carrageenan and ι-carrageenan share the ability to form gels, they distinguish themselves by producing strong and brittle gels for the κ-type and soft gels for the ι-type. The λ type, usually only described as a thickening agent, has been proved to form gels based on a trivalent cation complexation [6]. Commercial carrageenans find extensive application as additives in the pharmaceutical, cosmetic, and food industries for enhancing and customizing the texture of functional food products, showcasing its phycocolloids relevance from a boasting market size of 872 million USD in 2022 [7]. Commercial agar is mainly extracted from *Gracilaria* spp. (53%) and *Gelidium* spp. (44%) [8,9]. Agar presents unique physicochemical properties, acting as a gelling, thickening, and stabilizing agent, also exhibiting notable biodegradability and a substantial water-holding capacity. Its applications include the food, pharmaceutical, cosmetic, biotechnological, and biomedical sectors, presenting a notable market size valued at 264 million USD in 2022 [10]. Agar is mainly composed of agarose and agaropectin, the former being the gel-forming component, with a linear chain of repeating units of (1,3)-linked-β-D-galactose and (1,4)-linked-3,6-anhydro-α-L-galactose [11]. The gelling ability of agar is hindered by the substitution of the hydroxyl groups of l-galactose by sulphate esters, methyl ethers, or pyruvate acid ketals.

The quality of phycocolloids is industrially valued by its yield, gel strength, and purity. Gel properties of phycocolloids are usually species-dependent and dictated by the environmental conditions and extraction and isolation methods of the polysaccharides. Conventional extraction methods generally include a washing step followed by pre-treatments (employing alkali (e.g., NaOH) or acidified water) and a final hot-water extraction [3]. Alkali treatment has commonly been carried out due to the reportedly weak gel properties of the gels extracted with hot water [12]. However, these multi-step methods are time-, energy-, water-, and solvent-consuming, generating large amounts of wastewater. Therefore, the search for alternative polysaccharide extraction methods that are eco-friendlier, due to the absence of a strong alkali, and predictably cheaper, as fewer steps are included in the extraction process, has boosted over increasing ecological concerns [11,13]. These include microwave-, ultrasound-, and enzyme-assisted extraction, which are relatively environment-friendly technologies but still rather limited to the laboratory scale [14]. Finding adequate operational conditions to adjust the mechanical properties of the formed gels is crucial for the industrial scale-up process.

Polymers extracted from algae have proven to be valid bioinks with the properties required for 3D printing. As those properties are largely influenced by the extraction technique employed, when mechanical features of algal polysaccharides are poor, either mixing with other polymers or chemical modification can be considered to make them suitable for 3D printing [15,16]. Also, the examination of the rheological properties of the bio-inks developed is of paramount importance, as the extrudability, shape retention, and overall quality of the printed materials depend largely on them [17].

Locust bean gum (LBG) is a galactomannan naturally occurring in the endosperm of some *Leguminosae*, mainly extracted from carob tree seeds. It is composed of a linear mannose (M) backbone bearing side chains with a single galactose (G) unit. Galactomannans by themselves can form viscous solutions; however, they are also able to synergistically interact with other biopolymers. The synergistic effects between κ-carrageenan–galactomannan systems depend on the mannose/galactose ratio of the galactomannan backbone [3]. In these mixed systems, the mannose-free regions of the galactomannans are able to associate with carrageenan helices, leading to the formation of elastic and strong gels, offering the potential of improving food products and reducing production costs [18].

In this paper, a very simple and eco-friendly extraction method—water-based—was explored to develop 3D-printed biopolymer-based hydrogels. Three different red seaweed species (*Chondrus crispus*, *Gelidium Corneum*, and *Gracilaria gracilis*) were chosen for its native phycocolloids (carrageenan and agar). The aim of this work was as follows: first, to determine if the green extraction method selected was adequate to obtain bioinks, which form a gel during the printing process; and, second, to establish the best formulation and processing conditions (i.e., optimal thickening agent content and printing temperature) to improve the structure of the printed hydrogel. This evaluation is carried out by analyzing the rheological behavior of the biopolymer solutions in terms of mechanical properties and flow. The influence of the parameters on the gel structure was confirmed by scanning electron microscope (SEM) analysis.

2. Results and Discussion

2.1. Temperature and Time Sweep: Gels Rheological Behavior during Thermal Cycle

The effect of the distinct composition of each seaweed extract, with and without LBG additions, on the gelation mechanism of the phycocolloids-based extracts was rheologically characterized, and the results are presented in Figure 1. As observed, the samples generally showed a similar behavior for viscoelastic moduli as a function of cooling. At the beginning of the cooling cycle from 80 °C, the extracts were fluids with a predominantly viscous character ($G'' > G'$). Upon further decrease in temperature, the moduli values increased and a cross-over could be detected (i.e., $G'' = G'$), indicating the beginning of gel formation. This sol–gel transition was rheologically analyzed and is reported in Table 1. The crossover time was termed, in this work, as gelation time (t_{gel}) and crossover temperature as gelation temperature (T_{gel}). From that moment on, the elastic behavior was predominant ($G' > G''$), and eventually, plateau values were achieved for both viscoelastic moduli.

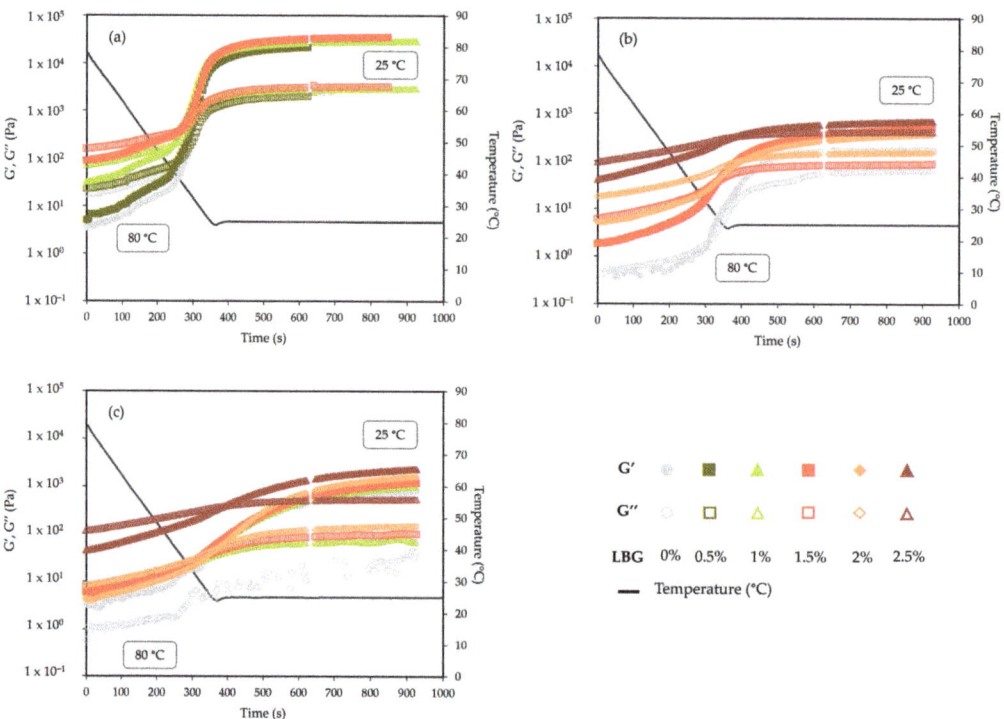

Figure 1. Temperature and time sweeps of *C. crispus* (**a**), *G. gracilis* (**b**), and *G. corneum* (**c**).

Table 1. Viscoelastic properties of seaweed (*C. crispus*, *G. gracilis*, and *G. corneum*) extracts, with different LBG concentrations, at the crosspoint $G' = G''$: gelation time (t_{gel}), gelation temperature, (T_{gel}), and gelation modulus (G_{gel}).

Seaweed Extract	LBG (%)	t_{gel} (s)	T_{gel} (°C)	G'_{gel} (Pa)
C. crispus	0.0	257.39 ± 23.88 [a]	35.45 ± 3.99 [a]	64.17 ± 19.36 [c]
	0.5	244.15 ± 12.05 [a]	37.19 ± 2.64 [a]	85.47 ± 8.65 [c]
	1.0	229.13 ± 16.10 [a]	39.31 ± 2.67 [a]	179.83 ± 8.72 [b]
	1.5	225.59 ± 46.39 [a]	39.90 ± 7.72 [a]	358.43 ± 41.90 [a]
G. gracilis	0.0	298.92 ± 31.93 [b]	27.88 ± 5.06 [a]	2.66 ± 0.50 [a]
	1.5	318.84 ± 2.11 [b]	25.73 ± 2.11 [a]	45.27 ± 10.30 [b]
	2.0	397.79 ± 0.02 [a]	24.99 ± 0.02 [a]	116.90 ± 13.29 [c]
	2.5	394.61 ± 0.01 [a]	25.00 ± 0.01 [a]	325.64 ± 15.52 [d]
G. corneum	0.0	No crosspoint		
	1.0	298.54 ± 12.27 [c]	29.89 ± 2.68 [a]	23.02 ± 3.19 [a]
	1.5	328.92 ± 8.17 [b]	25.36 ± 1.32 [b]	32.34 ± 5.03 [b]
	2.0	377.16 ± 3.73 [a]	25.03 ± 0.01 [b]	56.74 ± 11.73 [b]
	2.5	387.76 ± 12.25 [a]	25.01 ± 0.01 [b]	422.33 ± 52.79 [b]

Values are given as mean ± standard deviation (n = 3). Means (for the same sample) with different letters, within the same column, differ significantly ($p < 0.05$).

Concerning the effect of LBG, for *C. crispus* gels, gelation time tended to decrease with increasing additions of LBG (0–1.5% w/w), whereas gelation temperature seemed to increase. However, no significant influence was detected, so no major effect of the addition of LBG on the gelling behavior of *C. crispus* extracts could be implied. Despite this, the moduli significantly increased, as they increased almost six times their original values when LBG content was 1.5% w/w. Thus, the addition of LBG promoted an enhancement of their gel strength, influencing the gelation mechanism, which suggests that LBG is playing a significant role in the gel formation process. All *C. crispus* printed gels presented a good shape definition and a self-supporting structure yet maintaining a soft and flexible consistency. Previous studies have shown that LBG additions to κ-carrageenan gels lead to gels with enhanced texture properties [19]. This synergetic behavior between κ-carrageenan and LBG has been extensively reported in the literature [5,20–22]. Thus, their good miscibility has been associated with the intermolecular interactions formed between both polysaccharides, which, for example, has led to the formation of films with adequate water vapor barrier and mechanical properties [5]. More specifically, an interaction between the double-stranded helix of κ-carrageenan and the unbranched "smooth" segments of the d-mannose backbone of the galactomannan molecules has been proposed to explain the synergistic mechanism [23]. Previous studies have suggested that in mixed systems containing κ-carrageenan or agar, the gelling mechanism is primarily influenced by these phycocolloids, with other compounds functioning as active particle fillers within the gel network [24,25]. Consequently, analyzing the relevant phycocolloids extracted from seaweed biomass can provide valuable insights into the rheological behavior of the gels. A previous study reports for *C. crispus* (from the same biomass producer as present study—IMTA-cultivated *C. crispus*), a carrageenan yield (aqueous extraction) of 30.6–36.3%. This study also found that the native state of *C. crispus* is a source of copolymers of carrageenan κ/ι [26]. Several studies have shown that FTIR spectroscopy can be used to analyze algae-driven polysaccharides [27,28]. FTIR results (Figure 2 revealed typical bands associated with the presence of copolymers of carrageenan κ/ι. These results showed strong bands associated with κ-carrageenan at 845 cm^{-1}, which is assigned to D-galactose-4-sulphate (G4S), and at 930 cm^{-1}, indicative of the presence of 3,6-anhydro-D-galactose (DA). Moreover, a lower absorbance at 805 cm^{-1} (3,6-anhydro-D-galactose-2-sulphate) was also exhibited, indicating the presence of sulphate ester in the 2-position of the anhydro-D-galactose residues (DA2S), a characteristic band of ι-carrageenan [29]. Pereira et al. (2009) found similar bands for *C. crispus* collected from the central zone of the western coast of Portugal [27]. Moreover, the typical band (830–820 cm^{-1}) for λ-carrageenan was not present in the *C. crispus* gel. The presence of the κ/ι hybrid copolymers contributes to explaining how the obtained *C. crispus* extracts could form stable yet soft gels. While pure κ-carrageenan tends to form hard and brittle gels, in the presence of ι-carrageenan, soft and elastic gels take over [30].

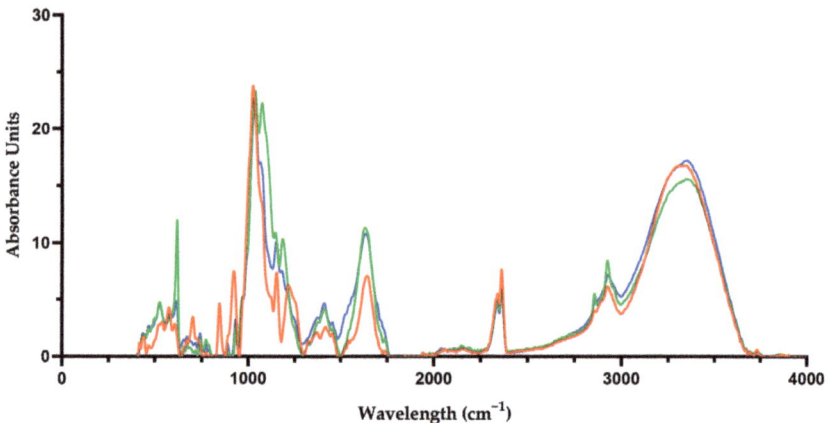

Figure 2. FTIR spectra of seaweed gels: *C. crispus* (━), *G. gracilis* (━), and *G. corneum* (━).

G. corneum and *G. gracilis* are the largest sources of agar worldwide, and the crossover between G″ and G′ can be attributed to the formation of a network by aggregation of the agarose helices into larger bundles [31]. No crossover point (G″ = G′) was detected for the *G. corneum* extract without LBG addition, as it already presented a solid-like behavior (G′ > G″ or tan δ < 1), even at 80 °C. In any case, similarly to *C. crispus* extracts, both G′ and G″ increased during the remaining cooling stage from 80 to 25 °C, which can be attributed to additional molecular interactions, leading to the formation of three-dimensional networks [32]. However, the moduli (G′ and G″) after gelation (i.e., at 25 °C) always presented lower values, independently of the LBG content, when compared to *C. crispus* gels. It has been demonstrated that in extraction methods where no pre-treatments (e.g., alkaline solutions) are applied, as is the case in our work, less purified agar-based extracts are produced. These agar-based extracts were reported to form softer gels [31]. Some authors have found that extracts resulting from hot-water extraction processes include other components that, even if they did not have an important effect on the nature of the interactions formed between the agarose chains, did lead to fewer or smaller agarose aggregates, which are responsible for holding the hydrogel network structure. This has been associated with the lower agar (and agarose) content found in hot-water extraction when compared to extraction processes where strong alkali are employed [33].

Conversely to *C. crispus* gels, the addition of LBG led to crossover points (G″ = G′) to happen later (i.e., at a lower temperature) in the cooling step. A similar behavior was reported by Sousa and Gonçalves (2015), where agar/LBG gels presented lower gelation temperatures and higher gelation times, compared to pure agar gels [25]. Previous studies reported gelling temperatures for *G. gracilis* and *G. corneum* biomasses (from the same producers as the present study—Algaplus and Iberagar) of 42.2 °C and 34.4 °C, respectively [34,35]. In our study, gelling temperatures from pure *G. gracilis* and *G. gelidium* extracts (i.e., without LBG) presented lower T_{gel} compared to the previous reported values. These findings could indicate an influence of other compounds present in the extract (e.g., proteins, lipids, ash, starch), and/or their interactions with agar, on the formation of the gels [36].

As mentioned before, no cross-over point (G″ = G′) was detected for the *G. corneum* extract without LBG addition, as that extract already displayed a predominantly elastic behavior at 80 °C. However, an increase in G′ with cooling reveals a gel structure reinforcement. Sousa and Gonçalves (2015) reported no cross-point for a mixed system of agar/LBG (25–75 ratio, respectively). However, at a 50–50 ratio, a crossover of the moduli was reported [25]. In our study, although increasing additions of LBG did not hinder the occurrence of the crossover between G′ and G″, the gelling properties of *G. gracilis* and *G. gelidium* gels seem to be greatly influenced by increasing additions of LBG. Particularly, *G. gelidium* finds a reversal from solid-like behavior to fluid-like behavior at 80 °C due to the addition of LBG. This should be related to the reported less coarse network found when LBG was added to agar solutions, leading to a decrease in the gelation point and a reduction in elastic moduli [26]. The reversal from solid-like to fluid-like behavior at 80 °C might be a good point for 3D printing, as those extracts might flow in an easier way through the nozzle and then gel once they are deposited on the printing bed at lower temperatures. This, together with the higher viscoelastic moduli due to the presence of the thickening agent, can contribute to a better definition for printed samples containing LBG.

Immediately after the temperature sweep, the gel maturation kinetics of the extracts were monitored through the evolution with time of G′ and G″ moduli during an isothermal stage at 25 °C (Figure 1). While *C. crispus* gels reached the steady state revealing constant values of both moduli, *G. gracilis* and *G. corneum* gels, particularly the latter, showed a slower equilibration step. The maturation pattern observed in the gels of *G. gracilis* and *G. corneum* follows the typical behavior of biopolymer gelation. Initially, G′ shows a rapid increase, followed by a slower progression. This is generally attributed to the ongoing reorganization of polymeric molecules within the gel network. [37].

Table 2 reports the viscoelastic properties (G′, G″, and tan δ) for the seaweed gels, at 80 °C (before gelation) and 25 °C (after gelation). Overall, increasing additions of LBG led to a decrease in elasticity (tan δ increase) of all seaweed gels. Specifically, the gels of *G. gracilis* displayed a significant increase of 244% in tan δ when transitioning from 0% LBG to 2.5% LBG gels. Sousa and Gonçalves (2015) also reported significant elasticity losses (tan δ increase) in agar/LBG binary systems [25]. In contrast, all *C. crispus* gels presented a comparable elastic response (tan δ circa 0.07–0.1). The same behavior was observed for *G. corneum* gels, up to 2% LBG addition.

For all seaweed extracts, G′ at 25 °C is significantly higher than at 80 °C, as expected after gelation. At 25 °C, *C. crispus* gels exhibited a decrease in G′ values (G′ $_{25\ °C}$), when 0.5% LBG was added to the gel. However, further additions led to an increase in the elastic modulus, reaching a similar G′ value of the gel without LBG. This could be explained by the interference effect of the low content of LBG on the self-association of κ-carrageenan chains [5]. Further LBG additions seem to have reinforced the gel's network structure (increased G′ values). A similar structuring effect can be observed for *G. corneum* and *G. gracilis* gels. Overall, LBG additions contributed to elevate gel strength of these seaweeds. The exception is for the *G. gracilis* gel at 2% LBG addition, where G′ exhibits a sudden decrease. A previous study also reported a synergy for agar extracted from a *Gelidium* sp. and LBG in 1:9 proportion, when a maximum hardness of the gel was achieved. When this ratio changed to 1:4, the gel returned to exhibit the gel strength of agar, i.e., the synergistic behavior was lost [38]. The same authors stated that such agar-LBG synergism is not possible for agars extracted from *Gracilaria* spp. Findings from another study confirm a lack of agar/LBG synergy for *Gracilaria vermiculophylla*. This study reported a gradual decrease in gel rigidity (G′ decrease) with LBG addition in mixed systems of agar/LBG, with higher concentrations of LBG [26]. In a more detailed analysis, Khoobbakht et al. (2024) observed that agar/LBG (commercial hydrocolloids) formed harder gels when mixed in a 4:1 ratio compared to a 1:1 ratio [39]. Thus, while previous studies suggest a limited synergy between agar and LBG, our study presents a contrasting perspective. Our findings indicate an enhancement of the viscoelastic properties of gels derived from *G. corneum* and *G. gracilis* when LBG is added.

Contrarily to the viscoelastic properties at the crossover point (Table 1), at 25 °C (Table 2), *G. corneum* gels exhibit higher viscoelastic properties (G′, G″) than *G. gracilis*. These findings reflect the gel strength of *G. corneum*, accordingly shown on the mechanical spectra (Figure 3). These results can generally be attributed to (i) the yield of extracted agar and to (ii) the agar quality of each of the seaweeds. In the context of agar yield, previous studies report considerable variability of agar yields, extracted from *G. corneum* (from the same producer as in our study). A previous study on the influence of seasonal harvesting on the biochemical characteristics of *G. corneum* revealed that its agar aqueous extraction yield ranged between 5.99% (spring) and 8.70% DW (winter) [40]. Martínez-Sans et al. (2021) reported a yield of 10–12% for *G. corneum* (as *G. sesquipedale*), while Gomes-Dias et al. (2022) obtained an extraction yield of 8.4% for native agar using a conventional hot-water bath extraction method (95 °C, 90 min) [35,41]. On the other hand, other studies have reported higher yields for agar extracted from *G. corneum* through aqueous extraction methods, ranging from 21% to 48% [42,43]. Native agar yields from *Gracilaria* typically range around 10–15% but can increase when using alkali pretreatment (15–33%) [44]. A previous study reported an agar yield of 14% for a similar *Gracilaria gracilis* biomass, obtained from the same producer as the present study [34]. Regarding the agar quality, *Gelidium* spp. is known for presenting a better agar quality than *Gracilaria* spp. This is due to the natural internal desulfation of *Gelidium*, modulated by enzymatic processes. On the other hand, *Gracilaria* presents polysaccharides which are typically more sulfated, and the conversion of these sulfated groups (i.e., L-galactose-6-sulphate to 3,6-anhydro-L-galactose) does not occur in sufficient quantity during the seaweed's lifespan [38]. As a result, the *Gracilaria* species typically requires alkali pre-treatments to improve the gelling capacity of its agar gels [45,46]. In summary, the findings suggest that native agars from *G. corneum* exhibit superior gelling properties compared to *G. gracilis*, explaining the significantly higher values of G′ at 25 °C for the former (Table 2). Nonetheless, it is noteworthy to note that despite the increase in $G'_{25\,°C}$ values (over $G'_{80\,°C}$), the values obtained for *G. cornem* and *G. gracilis* gels without LBG addition (737.8 and 446.9 Pa, respectively) are substantially lower than values reported in the literature. Gomes-Dias et al. (2024) reports 8838 Pa for G′ at 20 °C, for hot-water extracted agars (95 °C, 180 min) from *G. corneum*. The same study reports a contrastingly high G′ value (2567 Pa) at 20 °C for agar extracted from *Gracilaria vermiculophyla* [47]. One possible explanation for the lower values obtained in our study can be the purification step for agar that most studies perform when analyzing gelling properties of native seaweed-derived hydrocolloids. The lack of the purification step in the present study could have led to the presence of other components in the seaweed extracts, such as proteins, which have shown to give rise to the formation of softer hydrogels [31]. Nevertheless, non-purification is a deliberate strategy in this study, considering the positioning of the gels within the context of clean label foods. Moreover, even when comparing the extraction method employed in this study with similar extraction methods (hot-water extractions), it appears that most studies have used longer extraction times (>180 min) and higher temperatures (>95 °C) than ours (70 min, 90 °C).

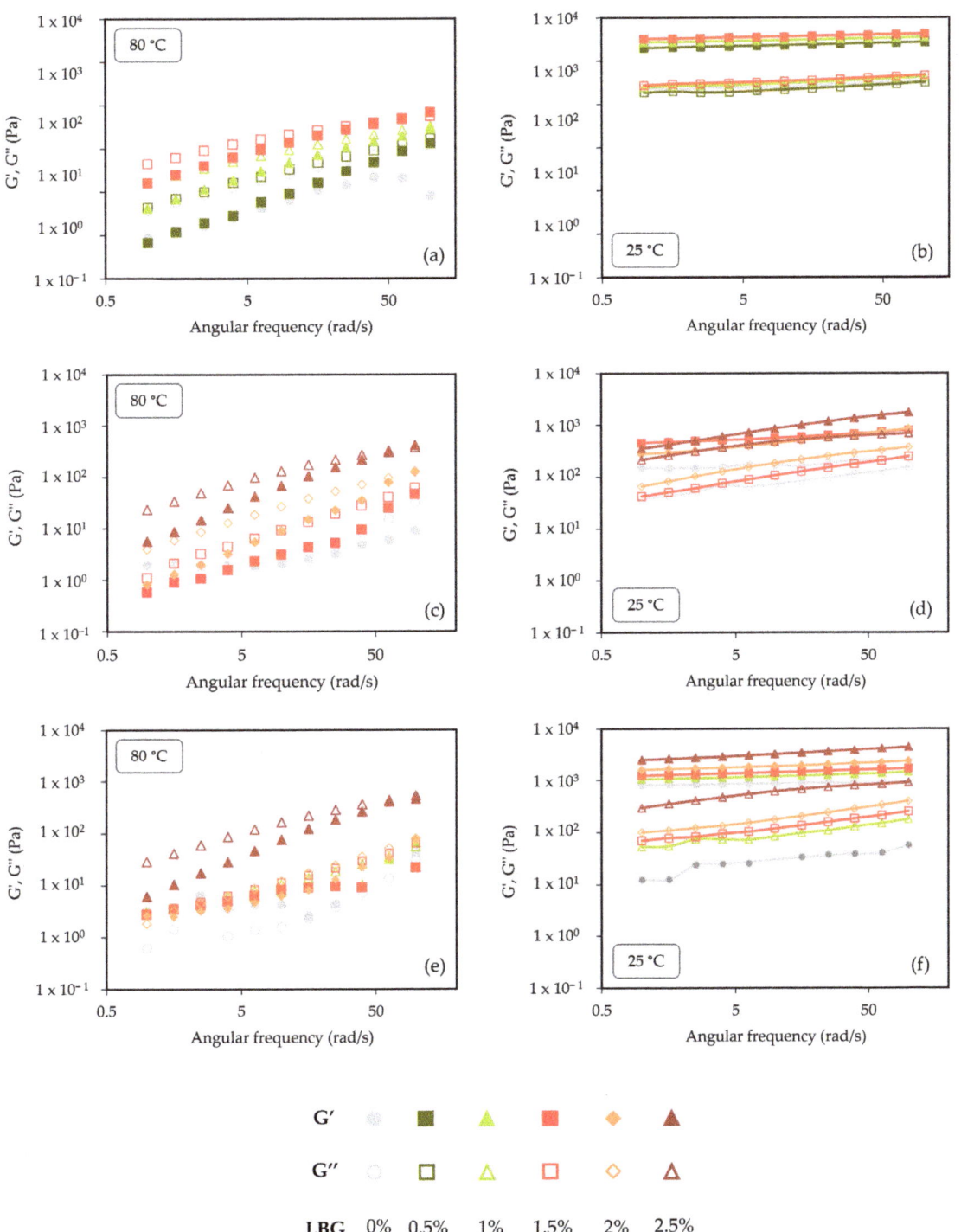

Figure 3. Frequency sweeps of *C. crispus* (**a**,**b**), *G. gracilis* (**c**,**d**), and *G. corneum* (**e**,**f**), at 80 and 25 °C, respectively.

Table 2. Viscoelastic properties of seaweed (*C. crispus*, *G. gracilis*, and *G. corneum*) extracts, with different LBG concentrations, before (80 °C) and after (25 °C) at the crosspoint: elastic modulus at 80 °C ($G'_{80\,°C}$) and at 25 °C ($G'_{25\,°C}$), tan δ at 80 °C (tan δ $_{80\,°C}$) and at 25 °C (tan δ $_{25\,°C}$).

Seaweed Extract	LBG (%)	$G'_{80\,°C}$ (Pa)	$G'_{25\,°C}$ (Pa)	tan δ $_{80\,°C}$	tan δ $_{25\,°C}$
C. crispus	0.0	3.95 ± 0.76 [b]	30,746.25 ± 5277.21 [a]	3.91 ± 1.55 [a]	0.07 ± 0.00 [b]
	0.5	5.73 ± 1.03 [b]	20,580.05 ± 1413.15 [b]	4.16 ± 0.71 [a]	0.09 ± 0.00 [a]
	1.0	36.54 ± 6.65 [b]	27,675.65 ± 23,10.27 [a,b]	2.12 ± 0.22 [b]	0.10 ± 0.01 [a]
	1.5	129.75 ± 60.01 [a]	30,137.03 ± 4766.15 [a]	1.53 ± 0.26 [b]	0.10 ± 0.01 [a]
G. gracilis	0.0	0.32 ± 1.14 [b]	446.99 ± 309.49 [a]	1.03 ± 1.79 [a]	0.18 ± 0.18 [b]
	1.5	5.47 ± 5.11 [b]	579.51 ± 110.26 [a]	2.60 ± 1.13 [a]	0.19 ± 0.02 [b]
	2.0	5.45 ± 0.13 [b]	429.84 ± 80.71 [a]	3.56 ± 0.25 [a]	0.40 ± 0.02 [a,b]
	2.5	39.14 ± 3.19 [a]	662.01 ± 44.31 [a]	2.35 ± 0.02 [a]	0.62 ± 0.01 [a]
G. corneum	0.0	3.17 ± 0.42 [b]	737.84 ± 65.50 [e]	0.29 ± 0.17 [d]	0.05 ± 0.01 [c]
	1.0	5.97 ± 0.31 [b]	943.61 ± 136.30 [d]	1.28 ± 0.00 [c]	0.07 ± 0.00 [b]
	1.5	6.44 ± 0.50 [b]	1259.55 ± 30.00 [c]	1.35 ± 0.06 [c]	0.09 ± 0.00 [b]
	2.0	5.88 ± 2.07 [b]	1718.15 ± 123.68 [b]	1.78 ± 0.2 [b]	0.09 ± 0.00 [b]
	2.5	66.50 ± 30.17 [a]	2338.98 ± 22.61 [a]	2.32 ± 0.41 [a]	0.24 ± 0.00 [a]

Values are given as mean ± standard deviation (n = 3). Means (for the same sample) with different letters, within the same column, differ significantly ($p < 0.05$).

2.2. Frequency Sweeps: Gels Characteristics

Frequency sweeps, shown in Figure 3, were performed both in the extract solutions at 80 °C and in the gels formed after cooling to 25 °C, and demonstrate the distinct behavior of the seaweed extracts at the two physical states for different LBG contents. At 80 °C (Figure 3a,c,e), seaweed extracts were presented in a liquid-like state. At this temperature, G' and G'' values were similar for all the seaweed extracts and increased with frequency. G'' was always above G', which is a behavior typical of entangled solutions [31]. While *C. crispus* extracts show an increase in elastic properties, manifested by a notable rise in G' and a reduction in tan δ (Table 2), as the content of LBG content increases; for *G. gracilis* and *G. corneum*, tan δ is always higher for the systems containing LBG, which would indicate a growing importance of their viscous behavior, which could be negative for attaining self-supporting gels in the 3D-printing process.

At 25 °C (Figure 3b,d,e), seaweed extracts acquire a solid-like texture and G' is always above G''. At 25 °C, *C. crispus* extracts showed a behavior typical of strong gels, where G' was at least one order of magnitude higher than G'', and both moduli showed little frequency dependence within the studied range (0.1–100 rad/s). This behavior should be beneficial for the 3D-printing process, as the spread of the sample on the printing bed would be hindered. The same could not be observed for *G. gracilis* and *G. corneum* gels. Although G' was always above G'', both these gels presented a greater frequency dependence, indicating a weak-gel-like behavior. Nonetheless, *G. corneum* gels presented enhanced gel strength (higher G' values) than *G. gracilis* gels (Table 2). Increasing additions of LBG to gels seemed to increase the slope of G' and G'' along the frequency range (Figure 3). Sousa and Gonçalves (2015) reported an increase in frequency dependence of the moduli at higher LBG concentrations of agar/LBG binary systems [25].

The results obtained from the mechanical spectra (Figure 3) can be further explained by the composition of seaweeds used in this study, as the polysaccharides' composition has a direct effect on the gelling mechanism and gel strength. According to Gomes-Dias et al. (2024), *Gelidium corneum* biomass (cultivated and harvested by the same manufacturer from the present study—Iberagar) presents galactans (37%) and glucans (14%) in its sugar composition, although 28% of these polysaccharides are of water extractives [47]. Silva-

Brito et al. (2021) studied the sugar composition of *G. gracilis* biomass (from the same producer as our biomass—Algaplus) and reported 24.4% of galactans, 3.5% of glucans, and 2.6% of arabians [35]. Moreover, Nova et al. (2023) reported for the same *G. gracilis* biomass (cultivated and harvested by the same producer), the presence of galactose as the main sugar residue (34 mol%), followed by 3,6-anhydrogalactose (25 mol%) and glucose (19 mol%) [48]. The presence of glucose is likely due to the presence of floridean starch, a storage polysaccharide [36].

The degree of sulfation and the content on 3,6-anhydro-L-galactose (3,6-AG) is also revealed to be an important factor to consider when evaluating the quality of agars [36]. The conversion of L-galactose-6-sulphate to 3,6-anhydro-L-galactose is associated with an increase in gel strength, as well as in gelling temperatures [49]. Consequently, the lower the sulphates/3,6-AG ratio, the higher the gelling properties [47]. Gomes-Dias et al. (2024) reported a sulfate content of 0.49% and a 3,6-AG content of 28.2% for agar extracted from *G. corneum* (hot-water extraction). The same study reports a sulfate content of 2.1% and a 3,6-AG content of 25.3% for agar extracted from *Gracilaria vermiculophyla* [47]. For the same seaweed, other studies have reported higher values for sulfate and 3.6- AG content. Sousa et al. (2010) reported a 3,6-AG content of 31.3% and a sulfate content of 1.78% for agar extracted from *G. vermiculophylla*, produced in an IMTA system and subjected to a traditional extraction method (alkali treatment and acid neutralization) [49]. For the same red seaweed (without alkali pre-treatments), Villanueva et al. (2009) reported a content of 3,6-AG of ~32–33% and a sulfate content of 2.2–2.4% [50]. Nonetheless, for agar extracted from *G. gracilis* (harvested from the southern Atlantic coast of Morocco) Belattmania et al. (2021) reported a 3,6-AG content of 5.7% and a sulfate content of 0.7% [51].

Previous studies have found a negative correlation between sulfate content and gel strength and gelling temperature, while 3,6-AG content has been found to have the opposite correlation with these gelation properties [49,50]. In the present study, all seaweed biomasses underwent a water-based extraction without alkaline pre-treatments. Consequently, the sulfate content was anticipated to be higher, while the 3,6-AG content was expected to be lower, in comparison to prior studies employing traditional or alternative extraction methods. Gomes-Dias et al. (2022) observed that conventionally extracted agar presented double the sulphate content of the autohydrolysis extracted agar, which led to lower viscoelastic moduli values [35]. Additionally, as previously mentioned, it is noteworthy that agars from *Gelidium* spp. typically present a low degree of substitution and, thus low sulfate content, resulting in agars with high gel strength, compared to *Gracilaria* spp. [52]. FTIR results (Figure 2) revealed that the band at 930 cm^{-1} in the spectra, corresponding to 3,6-anhydro-galactose, was more intense in the *G. gracilis* gels than in the *G. corneum*. However, a more intense broad band for *G. gracilis* in the 1240–1260 cm^{-1} spectral region, associated with the vibration of the sulphated groups, contributes to a higher sulphates/3,6-AG ratio. As sulfate groups are related to a lower gel strength of agar, agar in *G. gracilis* is considered to present lower agar quality, due to the higher presence of α-l-galactose 6-sulphate units detected in FTIR [53]. Sulphur content in the seaweed extracts can not only be attributed to the sulphate substitution in the agar component, but also to the presence of other sulphated components such as proteins [8]. *G. gracilis* biomass is reported to have a higher protein content (21%) than *G. corneum* biomass (10.4–19.4%).

This divergence in seaweed composition contributes to elucidating the heightened gelling properties observed in *G. corneum*, evidenced by increased gel strength (higher G' values) and a higher gelling temperature (T_{gel}) when lower LBG additions (<1% w/w) are applied (Table 1).

2.3. Flow Curves: Gel Viscosities

Figure 4 presents the steady shear flow behavior of seaweed extracts at 80 °C, over the selected range of shear rates (1–100 s^{-1}). Results obtained from viscosity curves showed a shear-thinning behavior, with higher viscosities at low shear rates and lower viscosities at high shear rates. Such flow behavior is necessary for a successful 3D extrusion. During printing, high shear rates are generated in the nozzle walls, and extrusion is thus facilitated if viscosity decreases. The process is reversed when a food material is layered on the printing platform, and viscosity increases due to the reduction in shear rates, contributing to forming a self-supporting material [54].

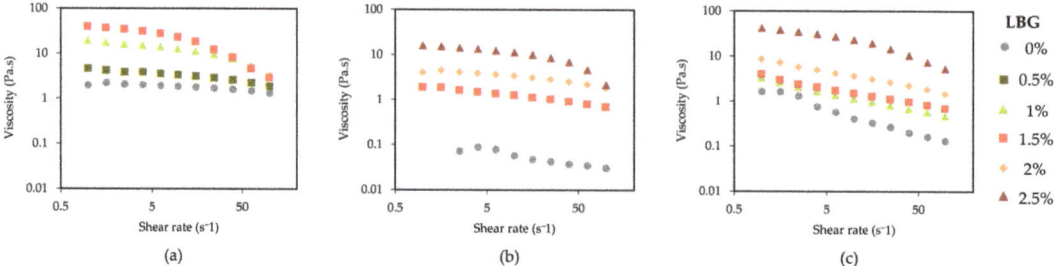

Figure 4. Flow curves of *C. crispus* (**a**), *G. gracilis* (**b**), and *G. corneum* (**c**), at 80 °C (1–100 s^{-1}).

Table 3 reports the viscosity values obtained at a selected shear rate (100 s^{-1}) for all seaweed extracts. This shear rate value was selected as it is the closest shear rate at the nozzle, based on estimated values obtained from the Weissenberg–Rabinowitsch–Mooney equation.

Table 3. Flow properties of seaweed (*C. crispus*, *G. gracilis*, and *G. corneum*) gelled extracts, with different LBG concentrations, at 80 °C: viscosity at (η_{100}).

Seaweed Extract	LBG (%)	η_{100} (Pa·s)
C. crispus	0.0	1.29 ± 0.05 [d]
	0.5	2.09 ± 0.19 [c]
	1.0	3.48 ± 0.04 [b]
	1.5	2.83 ± 0.15 [a]
G. gracilis	0.0	0.04 ± 0.03 [c]
	1.5	0.71 ± 0.04 [b]
	2.0	1.70 ± 0.48 [a]
	2.5	1.93 ± 0.65 [a]
G. corneum	0.0	0.11 ± 0.02 [c]
	1.0	0.49 ± 0.02 [b,c]
	1.5	0.65 ± 0.04 [b,c]
	2.0	1.37 ± 0.11 [b]
	2.5	4.32 ± 1.00 [a]

Values are given as mean ± standard deviation (n = 3). Means (for the same sample) with different letters, within the same column, differ significantly ($p < 0.05$).

Results show that flow properties were dependent on LBG concentration: higher viscosity with higher additions of LBG, demonstrating the synergetic effect: η algae$_{extract}$ + LBG > η algae$_{extract}$. Similarly to viscoelastic properties, *C. crispus* gels obtained the highest values, followed by *G. corneum* and *G. gracilis*.

Hernández et al. (2001) reported increased viscosity for a binary system composed of κ-carrageenan and λ-carrageenan with LBG (LBG + κ and LBG + λ), compared to single systems of any of the studied gums. These findings align with the results obtained in this study, reinforcing the presence of synergistic effects—specifically, an increase in viscosity—resulting from the combination of LBG with carrageenan [55].

As can be clearly observed in Supplementary Table S1, where shift factors for viscosity and shear rate were used to overlap every seaweed extract system to the corresponding system without LBG addition (LBG 0%) (master curve for every seaweed extract in Supplementary Figure S1), *G. corneum* is the seaweed most affected by the addition of LBG, as at 1.5% LBG, the viscosity enhancement was almost 4 or 5 times that found for *C. crispus* or *G. gracilis*, respectively.

2.4. Printed Gels

In Figure 5, the visual representation of printed gels from *C. crispus* (a), *G. gracilis* (b), and *G. corneum* (c) is shown. The figures illustrate the impact of cooling the printing temperature (from 80 to 30 °C) and increasing LBG content on the appearance of the seaweed gels.

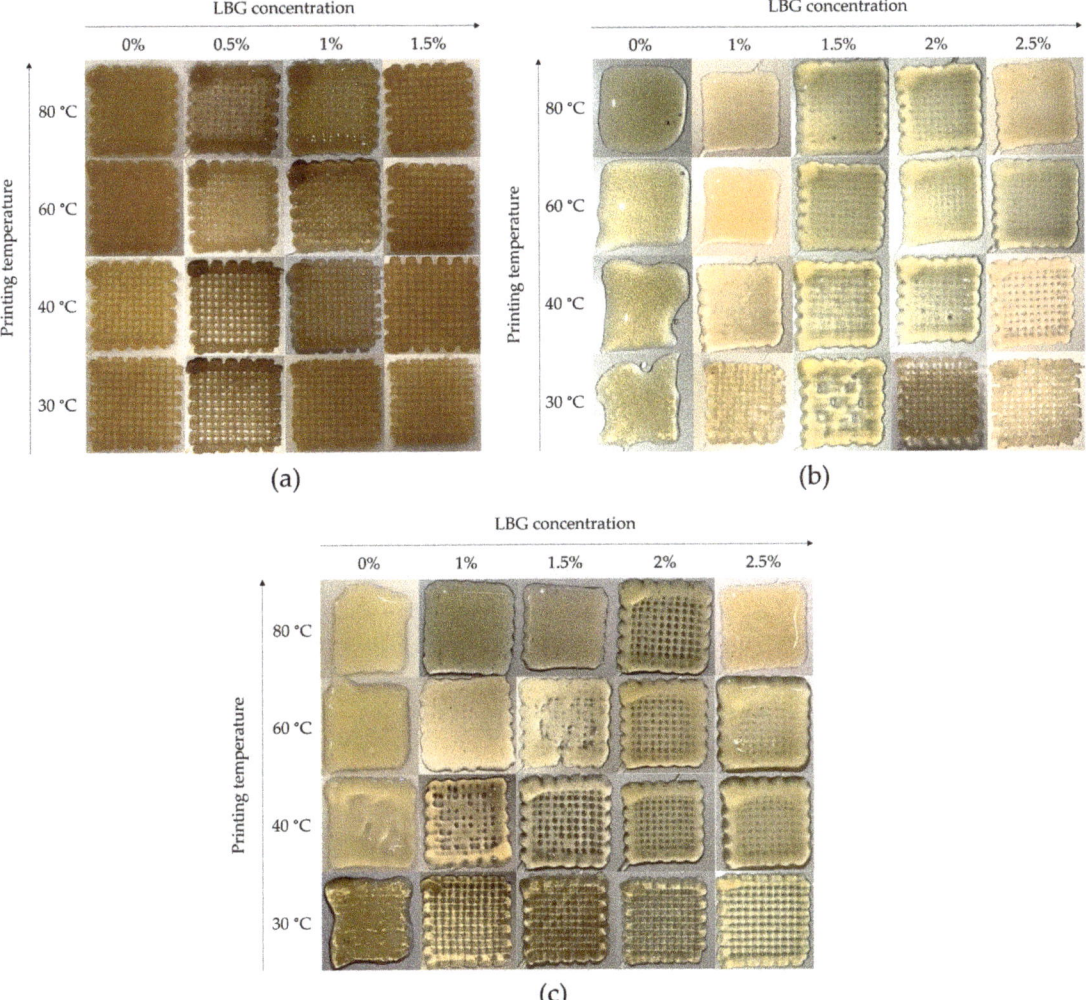

Figure 5. Visual effect of LBG additions (0–2.5% w/w) and printing temperature (80–30 °C) on gel's structure of *C. crispus* (**a**), *G. gracilis* (**b**), and *G. corneum* (**c**).

C. crispus exhibited the best gel structure, confirmed by its viscoelastic properties (G′, G″ and tan δ) (Tables 1 and 2). Moreover, all *C. crispus* printed gels (Figure 5a) presented a good shape definition and a self-supporting structure yet maintaining a soft and flexible consistency, which should be related to the abovementioned rheological properties, and to the fact that texture properties are affected by LBG additions to κ-carrageenan gels. During deposition of layers and cooling of the *C. crispus* extracts, a typical gel structure was obtained, and even for extracts without LBG addition, a well-defined 3D structure was produced. Contrastingly, printing definition was very poor for *G. corneum* and *G. gracilis* gels without LBG, as the material flowed when deposited (Figure 5b,c). With increasing LBG additions, both *G. corneum* and *G. gracilis* acquired an improved shape definition and increased gel strength (Tables 1 and 2). Nevertheless, it was observed that additional additions of LBG had a more pronounced impact on the gel structure of *G. gelidium* (Figure 5c) in comparison to *G. gracilis* gel (Figure 5b). This should be related to *G. corneum* gels exhibiting higher viscoelastic properties (G′, G″) than *G. gracilis*, which reflect the gel strength of *G. corneum*, accordingly shown on the mechanical spectra (Figure 3) and visibly noticeable in Figure 5c, where *G. corneum* gels acquire a better-defined shape with increasing additions of LBG. Moreover, a better shape definition of *G. corneum* gels can be further supported by the gels' higher values of viscosity, as evidenced in Table 3. In contrast, a reduction in the printing temperature seemed to produce a greater improvement of shape definition in *G. gracilis* gel, when compared to *G. gelidium* gel. Lower printing temperatures increase viscosity and may reduce the spreading effect of the filaments during deposition.

SEM was used to characterize the microstructure of the gels. Figure 6 shows the internal structures of the *C. crispus*, *G. gracilis*, and *G. corneum* gels, printed at 25 °C, with (1.5%) and without LBG addition. Under a 30× magnification, it was possible to observe the printed layer filaments of the gels at 1.5% LBG addition while, without LBG, the printed filaments could not be observed, except in the *C. crispus* gel. For this seaweed, the impact of LBG addition to the structure was lower, as can be observed in Figure 5a. These results agree with the mechanical spectra obtained from frequency sweeps (Figure 3), which show little frequency dependence, indicating the presence of a strong gel.

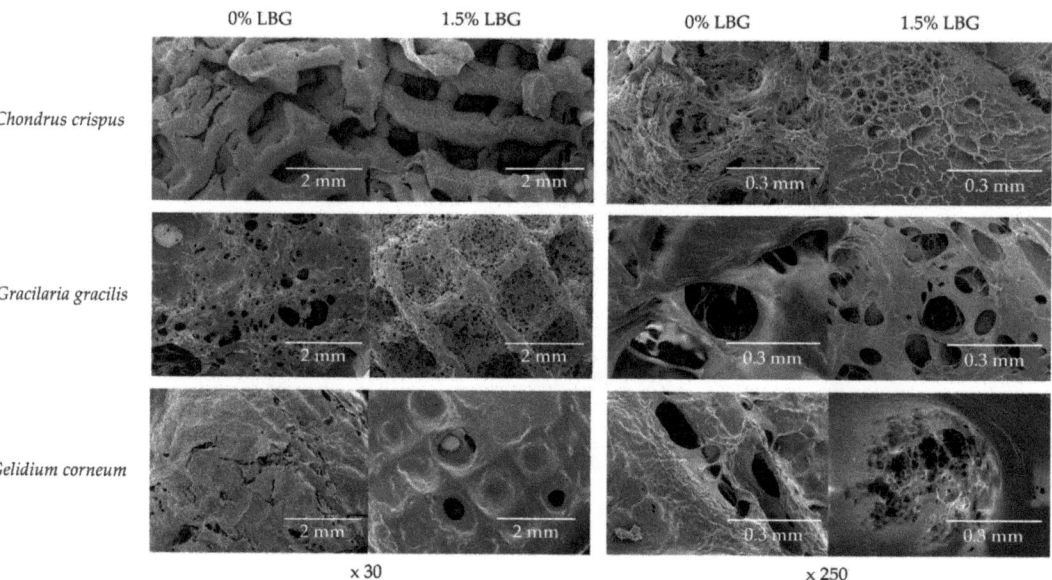

Figure 6. SEM images of *C. crispus*, *G. gracilis*, and *G. Corneum* 3D gels, with and without LBG addition, under 30× and 250× magnifications.

Under a 250× magnification, the morphology of the gel and the size and number of pores illustrate the molecular interactions established with and without LBG addition to seaweed gels. Overall, 1.5% LBG addition resulted in denser gels characterized by smaller-sized pores. This effect was particularly noticeable in *C. crispus* and *G. gelidium* gels, confirming the stronger synergistic interaction among the gelling components (κ/ι-carrageenan and agar, respectively) in these gels. In contrast, *G. gracilis* exhibited larger pores and a less dense microstructure. All the seaweed gels presented a heterogeneous structure. Previous studies have shown that gels of mixed systems generally present a more heterogenous structure than gels of individual gelling components [56].

3. Conclusions

The present study investigated the gel-forming kinetics and gel properties of algal polysaccharides obtained in a green extraction process from three red seaweed biomasses (*Chondrus crispus*, *Gracilaria gracilis*, and *Gelidium corneum*). Locust bean gum (LBG) was added to the final extracts, and gelation was monitored in situ. Results showed a similar gelation behavior for all seaweed extracts during the cooling cycle, transitioning from a liquid-like state to a final gel formation, at the crosspoint between G' and G''. *C. crispus* gels exhibited increased gel strength (higher G') and more stable gels (shorter maturation time to obtain steady gel structures). This structuring effect was additionally observed on the microstructure of the gels. *G. corneum* and *G. gracilis* exhibited lower viscoelastic properties, yet their gel strength noticeably improved with the addition of LBG, indicating a synergistic interaction between the seaweed extracts and LBG. This synergy proved to be necessary to obtain self-supporting structures from *G. gracilis* and *G. corneum* extracts, during the printing process. Similarly, viscosity of gels was likewise enhanced by the increasing additions to the seaweed extracts. Nevertheless, flow curves proved that gels presented a suitable flow behavior for extrusion: a shear-thinning behavior with increasing shear-rates. Additionally, printing at lower temperatures seemed to have improved the shape definition of the printed gels.

These findings confirm that a green-extraction can be applied to develop algal-based extracts with gelling capacity. Additionally, a synergetic interaction between extracts and LBG enhanced the gels' suitability for printing. This could be used in several applications, such as the design of novel food products. The effect of different thickening agents on the features of the bioinks and the resulting 3D-printed gels should be considered in future research.

4. Materials and Methods

4.1. Materials and Sample Preparation

Materials. Three different species of red algae (Rhodophyta) were selected for the preparation of the biopolymer solutions: *Chondrus crispus*, *Gelidium Corneum*, and *Gracilaria gracilis*. *Chondrus crispus* is typically rich in carrageenan (being κ-carrageenan and ι-carrageenan, able to form gels), while agar, a gelling, thickening, and stabilizing agent, is more abundant in *Gelidium Corneum* and *Gracilaria gracilis*. *C. crispus* and *G. gracilis* were purchased from AlgaPlus (Ilhavo, Portugal) and *G. corneum* from Iberagar-Sociedade Luso-Espanhola de Colóides Marinhos S.A. (Coina, Portugal). According to the datasheet of the producer, *C. crispus* and *G. gracilaria* were cultivated in an integrated multi-trophic aquaculture system (Aveiro, Portugal). The seaweeds were washed with seawater and dried in an air-tunnel set at 30 °C, until the desired moisture content (12%) was reached, followed by milling and sieving (<0.25 mm). Regarding *G. corneum*, according to the manufacturer, the algae undergo a drying process in the sun, are milled, and finally sieved (3 mm).

The proximate composition for *C. crispus*, provided by the manufacturer, is as follows: 12% moisture, 1% lipids, 17.7% carbohydrates (<0.01% sugars), 17.3% protein, and 38.1% dietary fiber. For *G. gracilaria*, it is as follows: 12% moisture, 1.2% lipids, 8.2% carbohydrates (0.2% sugars), 21% protein, and 38.8% dietary fiber. For *G. corneum*, a previous study using a similar biomass from the same manufacturer obtained the following: 9.2–19.8% ash, 0.7–4.7% lipids, 24.8–39.5% carbohydrates, and 10.4–19.4% crude protein [43,44,57]. These red algae were selected for their native phycocolloids and gelling ability, crucial for the 3D-printing process.

Sample preparation. Seaweed solutions were prepared following a 2-step alternative extraction method. First, seaweeds (in dry powder) were hydrated in water (1:10 w/w, seaweed/water), for 30 min, at room temperature. Then, the seaweed solutions were heated to 90 °C (70 min) with constant stirring. Finally, the solutions were centrifuged (40 °C, 10,000 rpm, 15 min), and the supernatants were collected to be printed. Once printed, gels from the different species of seaweeds did not all present the same definition of shape. To improve the shape quality of the printed gels, the addition of locust bean gum (LGB) to the supernatants was studied, as LBG presents a synergy with naturally present phycocolloids in Rhodophyta (carrageenan and agar). LBG was added to supernatants immediately following centrifugation, under strong stirring to ensure full dissolution. Based on preliminary tests, LBG additions to supernatants were set as follows: *Chondrus crispus* (0–1.5% w/w), *Gelidium Corneum* (0–2.5% w/w), and *Gracilaria gracilis* (0–2.5% w/w).

4.2. Three-Dimensional-Printing Process

The printing process was performed using the Genesis™ 3D printer (3D Biotechnology Solutions, Brazil), following the LBG addition to the supernatants.

Printing of the gels took place at a different range of temperatures (80–30 °C) to study the influence of printing temperature on the quality of shape of the printed gels. To do this, syringes (22 G nozzle) were loaded with the seaweed extracts and were pre-heated beforehand according to the selected printing temperature, while the printing platform was always kept at room temperature. After 15 min of equilibration time in the syringe, the extracts were printed at 40 mm/s, forming a total of 8 layers of a cube design (20 × 20 × 20 mm).

4.3. Rheology Characterization

Rheology characterization tests were performed to study the viscoelastic properties of the algae-based biopolymers solutions, in terms of storage (G′) and loss (G″) moduli, using a strain-controlled rheometer (Discovery HR 10, TA instruments, New Castle, DE, USA), equipped with a Peltier plate temperature system. A serrated parallel plates geometry (PP20) was used, with a gap of 1 mm.

Before starting measurements, hot biopolymer solutions were directly placed on the preheated parallel plates geometry and coated with paraffin oil to avoid evaporation.

Determining LVE. Firstly, the linear viscoelastic region (LVE) was determined via amplitude sweep tests, carried out at a fixed frequency (6.28 rad/s), at 80 °C. All measurements were performed within the LVE, as at this range, all dynamic rheological parameters are independent of applied shear strain.

First frequency sweep. Secondly, frequency sweep tests were carried out within a frequency range varying from 1 to 100 rad/s (ω), at a fixed temperature (80 °C), to study the viscoelastic properties.

Temperature sweep. Following frequency sweeps, temperature sweeps were run, applying a cooling temperature ramp from 80 to 25 °C (ω = 6.28 rad/s, 10 °C/min). During temperature sweeps, the transition from sol-gel was monitored, which corresponds to $\tan \delta = G''/G' = 1$. The temperature and time at which the crossover between G′ and G″ occurred, i.e., G″ > G′, was noted as an empirical indication of gelling temperature (T_{gel}) and time (t_{gel}), respectively [58].

Time sweep. Time sweeps were performed (30 min, ω = 6.28 rad/s, 25 °C) to obtain the gel maturation kinetics.

Final frequency sweep. A final frequency sweep was performed on the gels, from 1 to 100 rad/s (ω), at a fixed temperature (25 °C), to analyze the influence of the temperature ramp on the viscoelastic properties.

Flow curves. Flow sweeps were performed (80 °C) to study the viscosity (η) of the biopolymer solutions as a function of a selected range of shear rates (1–100 s^{-1}).

4.4. Fourier Transform Infrared (FTIR) Spectroscopy

Seaweed extracts at 80 °C were cast in Petri dishes and dried overnight. FTIR spectra of films formed from seaweed extracts were performed with an ATR objective in an Invenio X spectrometer (BRUKER, Billerica, MA, USA). Spectra were collected in the wavelength between 4000 and 400 cm^{-1}, with a resolution of 4 cm^{-1}.

4.5. Scanning Electron Microscope (SEM)

The microstructures of the gels were investigated via scanning electron microscopy (SEM) using a tabletop scanning electron microscope (SEM) (TM3030 PLUS, Hitachi, Japan). Samples were printed, and immediately, small cuts were performed on their surface using a scalpel to promote better observation of the inner microstructures. The samples were placed on the specimen holder, at −4 °C, in a vacuum chamber. Observations were performed under different magnifications (30× and 250×).

4.6. Statistical Analysis

Analysis of variance (ANOVA) was performed on Minitab (version 17, Pennsylvania, USA) to evaluate significant ($p < 0.05$) differences between samples. In case of significant differences, multiple pairwise comparisons were performed with Tukey's Honest Significant Difference test (HSD) at 95% confidence level.

Supplementary Materials: The following supporting information can be downloaded at: https://www.mdpi.com/article/10.3390/gels10030166/s1, Figure S1: Master curve for the different LBG contents for every seaweed extract, using every system without LBG as reference; Table S1: Shift factor for viscosity and shear rate to overlap every seaweed extract system to the corresponding system with-out LBG addition (LBG 0%).

Author Contributions: Conceptualization, A.R., I.S. and C.B.; methodology, S.O. and C.B.; software, S.O. and C.B.; writing—original draft preparation, S.O.; writing—review and editing, A.R., I.S. and C.B.; supervision, A.R., I.S. and C.B.; funding acquisition, C.B., A.R. and I.S. All authors have read and agreed to the published version of the manuscript.

Funding: This research work was funded by national funds through FCT—Fundação para a Ciência e a Tecnologia, I.P., under the project UIDB/04129/2020 of LEAF-Linking Landscape, Environment, Agriculture and Food, Research Unit, PhD Grant 2020.07207.BD and Project Vertical Algae, S2F—Agenda Mobilizadora PRR Pacto da Bioeconomia Azul. The authors also acknowledge the project PID2021-124294OB-C21 funded by MCIN/AEI/10.13039/501100011033/ and by "ERDF A way of making Europe" for supporting this study. Authors also acknowledge the financial support by Ministerio de Universidades (PRX22/00448) for the research stay of C.B. in the Instituto Superior de Agronomia.

Institutional Review Board Statement: Not applicable.

Informed Consent Statement: Not applicable.

Data Availability Statement: All data and materials are available on request from the corresponding author. The data are not publicly available due to ongoing research using a part of the data.

Acknowledgments: The authors would like to thank the CITIUS for granting access and assistance with the Microanalysis services.

Conflicts of Interest: The authors declare no conflicts of interest.

References

1. Synytsya, A.; Čopíková, J.; Kim, W.J.; Park, Y. Il Cell Wall Polysaccharides of Marine Algae. In *Springer Handbook of Marine Biotechnology*; Kim, S.-K., Ed.; Springer: Berlin/Heidelberg, Germany, 2015; pp. 543–590. ISBN 978-3-642-53970-1.
2. Pereira, L. Atlantic Algae as Food and Their Extracts. *Explor. Foods Foodomics* **2023**, *1*, 15–31. [CrossRef]
3. Alba, K.; Kontogiorgos, V. Seaweed Polysaccharides (Agar, Alginate Carrageenan). In *Encyclopedia of Food Chemistry*; Elsevier: Amsterdam, The Netherlands, 2019; pp. 240–250. [CrossRef]
4. Aswathi Mohan, A.; Robert Antony, A.; Greeshma, K.; Yun, J.H.; Ramanan, R.; Kim, H.S. Algal Biopolymers as Sustainable Resources for a Net-Zero Carbon Bioeconomy. *Bioresour. Technol.* **2022**, *344*, 126397. [CrossRef]

5. Martins, J.T.; Cerqueira, M.A.; Bourbon, A.I.; Pinheiro, A.C.; Souza, B.W.S.; Vicente, A.A. Synergistic Effects between κ-Carrageenan and Locust Bean Gum on Physicochemical Properties of Edible Films Made Thereof. *Food Hydrocoll.* **2012**, *29*, 280–289. [CrossRef]
6. Beaumont, M.; Tran, R.; Vera, G.; Niedrist, D.; Rousset, A.; Pierre, R.; Shastri, V.P.; Forget, A. Hydrogel-Forming Algae Polysaccharides: From Seaweed to Biomedical Applications. *Biomacromolecules* **2021**, *22*, 1027–1052. [CrossRef] [PubMed]
7. Grand View Research. *Carrageenan Market Size, Share & Trends Analysis Report by Processing Technology (Semi-Refined, Gel Press, Alcohol Precipitation), by Function, by Product Type, by Application, by Region, and Segment Forecasts, 2023–2030*; Grand View Research: San Francisco, CA, USA, 2023.
8. Martínez-Sanz, M.; Gomez-Barrio, L.P.; Zhao, M.; Tiwari, B.; Knutsen, S.H.; Ballance, S.; Zobel, H.K.; Nilsson, A.E.; Krewer, C.; Östergren, K.; et al. Alternative Protocols for the Production of More Sustainable Agar-Based Extracts from Gelidium Sesquipedale. *Algal Res.* **2021**, *55*, 102254. [CrossRef]
9. Porse, H.; Rudolph, B. The Seaweed Hydrocolloid Industry: 2016 Updates, Requirements, and Outlook. *J. Appl. Phycol.* **2017**, *29*, 2187–2200. [CrossRef]
10. Verified Market Research. *Global Agar Market Size By Product Type (Strips, Powder, Flakes), by Application (Pharmaceutics, Cosmetics, Food And Beverages), by Geographic Scope And Forecast*; Verified Market Research: Washington, DC, USA, 2022.
11. Pereira, S.G.; Teixeira-Guedes, C.; Souza-Matos, G.; Maricato, É.; Nunes, C.; Coimbra, M.A.; Teixeira, J.A.; Pereira, R.N.; Rocha, C.M.R. Influence of Ohmic Heating in the Composition of Extracts from Gracilaria Vermiculophylla. *Algal Res.* **2021**, *58*, 102360. [CrossRef]
12. Abdul Khalil, H.P.S.; Lai, T.K.; Tye, Y.Y.; Rizal, S.; Chong, E.W.N.; Yap, S.W.; Hamzah, A.A.; Nurul Fazita, M.R.; Paridah, M.T. A Review of Extractions of Seaweed Hydrocolloids: Properties and Applications. *Express Polym. Lett.* **2018**, *12*, 296–317. [CrossRef]
13. Torres, M.D.; Flórez-Fernández, N.; Domínguez, H. Chondrus Crispus Treated with Ultrasound as a Polysaccharides Source with Improved Antitumoral Potential. *Carbohydr. Polym.* **2021**, *273*, 118588. [CrossRef]
14. Liu, Y.; An, D.; Xiao, Q.; Chen, F.; Zhang, Y.; Weng, H.; Xiao, A. A Novel κ-Carrageenan Extracting Process with Calcium Hydroxide and Carbon Dioxide. *Food Hydrocoll.* **2022**, *127*, 107507. [CrossRef]
15. Mandal, S.; Nagi, G.K.; Corcoran, A.A.; Agrawal, R.; Dubey, M.; Hunt, R.W. Algal Polysaccharides for 3D Printing: A Review. *Carbohydr. Polym.* **2023**, *300*, 120267. [CrossRef]
16. Xu, S.Y.; Huang, X.; Cheong, K.L. Recent Advances in Marine Algae Polysaccharides: Isolation, Structure, and Activities. *Mar. Drugs* **2017**, *15*, 388. [CrossRef]
17. Herrada-Manchón, H.; Fernández, M.A.; Aguilar, E. Essential Guide to Hydrogel Rheology in Extrusion 3D Printing: How to Measure It and Why It Matters? *Gels* **2023**, *9*, 517. [CrossRef] [PubMed]
18. Pinheiro, A.C.; Bourbon, A.I.; Rocha, C.; Ribeiro, C.; Maia, J.M.; Gonalves, M.P.; Teixeira, J.A.; Vicente, A.A. Rheological Characterization of κ-Carrageenan/Galactomannan and Xanthan/Galactomannan Gels: Comparison of Galactomannans from Non-Traditional Sources with Conventional Galactomannans. *Carbohydr. Polym.* **2011**, *83*, 392–399. [CrossRef]
19. He, H.; Ye, J.; Zhang, X.; Huang, Y.; Li, X.; Xiao, M. κ-Carrageenan/Locust Bean Gum as Hard Capsule Gelling Agents. *Carbohydr. Polym.* **2017**, *175*, 417–424. [CrossRef] [PubMed]
20. Brenner, T.; Wang, Z.; Achayuthakan, P.; Nakajima, T.; Nishinari, K. Rheology and Synergy of κ-Carrageenan/Locust Bean Gum/Konjac Glucomannan Gels. *Carbohydr. Polym.* **2013**, *98*, 754–760. [CrossRef] [PubMed]
21. Arda, E.; Kara, S.; Pekcan, Ö. Synergistic Effect of the Locust Bean Gum on the Thermal Phase Transitions of κ-Carrageenan Gels. *Food Hydrocoll.* **2009**, *23*, 451–459. [CrossRef]
22. Dionísio, M.; Grenha, A. Locust Bean Gum: Exploring Its Potential for Biopharmaceutical Applications. *J. Pharm. Bioallied Sci.* **2012**, *4*, 175–185. [CrossRef] [PubMed]
23. Dea, I.C.M.; Morrison, A. Chemistry and Interactions of Seed Galactomannans. *Adv. Carbohydr. Chem. Biochem.* **1975**, *31*, 241–312. [CrossRef]
24. Nunes, M.C.; Raymundo, A.; Sousa, I. Rheological Behaviour and Microstructure of Pea Protein/κ-Carrageenan/Starch Gels with Different Setting Conditions. *Food Hydrocoll.* **2006**, *20*, 106–113. [CrossRef]
25. Sousa, A.M.M.; Gonçalves, M.P. The Influence of Locust Bean Gum on Native and Alkali-Modified Agargels. *Food Hydrocoll.* **2015**, *44*, 461–470. [CrossRef]
26. Pereira, L.; Meireles, F.; Abreu, H.T.; Ribeiro-Claro, P.J.A. A Comparative Analysis of Carrageenans Produced by Underutilized versus Industrially Utilized Macroalgae (Gigartinales, Rhodophyta). *Mar. Algae Extr. Process. Prod. Appl.* **2015**, *1–2*, 277–294. [CrossRef]
27. Pereira, L.; Amado, A.M.; Critchley, A.T.; van de Velde, F.; Ribeiro-Claro, P.J.A. Identification of Selected Seaweed Polysaccharides (Phycocolloids) by Vibrational Spectroscopy (FTIR-ATR and FT-Raman). *Food Hydrocoll.* **2009**, *23*, 1903–1909. [CrossRef]
28. Gómez-Ordóñez, E.; Rupérez, P. FTIR-ATR Spectroscopy as a Tool for Polysaccharide Identification in Edible Brown and Red Seaweeds. *Food Hydrocoll.* **2011**, *25*, 1514–1520. [CrossRef]
29. Pereira, L.; Mesquita, J.F. Carrageenophytes of Occidental Portuguese Coast: 1-Spectroscopic Analysis in Eight Carrageenophytes from Buarcos Bay. *Biomol. Eng.* **2003**, *20*, 217–222. [CrossRef] [PubMed]
30. Pereira, L.; Van De Velde, F. Portuguese Carrageenophytes: Carrageenan Composition and Geographic Distribution of Eight Species (Gigartinales, Rhodophyta). *Carbohydr. Polym.* **2011**, *84*, 614–623. [CrossRef]

31. Martínez-Sanz, M.; Ström, A.; Lopez-Sanchez, P.; Knutsen, S.H.; Ballance, S.; Zobel, H.K.; Sokolova, A.; Gilbert, E.P.; López-Rubio, A. Advanced Structural Characterisation of Agar-Based Hydrogels: Rheological and Small Angle Scattering Studies. *Carbohydr. Polym.* **2020**, *236*, 115655. [CrossRef] [PubMed]
32. Sousa, A.M.M.; Borges, J.; Silva, A.F.; Gonçalves, M.P. Influence of the Extraction Process on the Rheological and Structural Properties of Agars. *Carbohydr. Polym.* **2013**, *96*, 163–171. [CrossRef] [PubMed]
33. Martínez-Sanz, M.; Gómez-Mascaraque, L.G.; Ballester, A.R.; Martínez-Abad, A.; Brodkorb, A.; López-Rubio, A. Production of Unpurified Agar-Based Extracts from Red Seaweed Gelidium Sesquipedale by Means of Simplified Extraction Protocols. *Algal Res.* **2019**, *38*, 101420. [CrossRef]
34. Silva-Brito, F.; Pereira, S.G.; Rocha, C.M.R.; da Costa, E.; Domingues, M.R.; Azevedo, A.; Kijjoa, A.; Abreu, H.; Magnoni, L. Improving Agar Properties of Farmed Gracilaria Gracilis by Using Filtered Sunlight. *J. Appl. Phycol.* **2021**, *33*, 3397–3411. [CrossRef]
35. Gomes-Dias, J.S.; Pereira, S.G.; Teixeira, J.A.; Rocha, C.M.R. Hydrothermal Treatments—A Quick and Efficient Alternative for Agar Extraction from Gelidium Sesquipedale. *Food Hydrocoll.* **2022**, *132*, 107898. [CrossRef]
36. Cebrián-Lloret, V.; Martínez-Abad, A.; López-Rubio, A.; Martínez-Sanz, M. Exploring Alternative Red Seaweed Species for the Production of Agar-Based Hydrogels for Food Applications. *Food Hydrocoll.* **2024**, *146*, 109177. [CrossRef]
37. Batista, A.P.; Nunes, M.C.; Fradinho, P.; Gouveia, L.; Sousa, I.; Raymundo, A.; Franco, J.M. Novel Foods with Microalgal Ingredients—Effect of Gel Setting Conditions on the Linear Viscoelasticity of Spirulina and Haematococcus Gels. *J. Food Eng.* **2012**, *110*, 182–189. [CrossRef]
38. Phillips, G.O.; Williams, P.A. *Handbook of Hydrocolloids*, 2nd ed.; Elsevier: Amsterdam, The Netherlands, 2009; ISBN 9781845695873.
39. Khoobbakht, F.; Khorshidi, S.; Bahmanyar, F.; Hosseini, S.M.; Aminikhah, N.; Farhoodi, M.; Mirmoghtadaie, L. Modification of Mechanical, Rheological and Structural Properties of Agar Hydrogel Using Xanthan and Locust Bean Gum. *Food Hydrocoll.* **2024**, *147*, 109411. [CrossRef]
40. Pereira, L.; Martins, M.; Mouga, T. Seasonal Nutritional Profile of Gelidium Corneum (Rhodophyta, Gelidiaceae) from the Center of Portugal. *Foods* **2021**, *10*, 2394.
41. Li, Y.; Zhao, M.; Gomez, L.P.; Senthamaraikannan, R.; Padamati, R.B.; O'Donnell, C.P.; Tiwari, B.K. Investigation of Enzyme-Assisted Methods Combined with Ultrasonication under a Controlled Alkali Pretreatment for Agar Extraction from Gelidium Sesquipedale. *Food Hydrocoll.* **2021**, *120*, 106905. [CrossRef]
42. Cebrián-Lloret, V.; Martínez-Abad, A.; López-Rubio, A.; Martínez-Sanz, M. Sustainable Bio-Based Materials from Minimally Processed Red Seaweeds: Effect of Composition and Cell Wall Structure. *J. Polym. Environ.* **2023**, *31*, 886–899. [CrossRef]
43. Ferro, R. *Production and Characterization of Protein-Rich Extracts from the Red Macroalga Gelidium Sesquipedale and Its Residues after Industrial Agar Extraction*; Instituto Superior Técnico, University of Lisbon: Lisbon, Portugal, 2021.
44. Lee, W.K.; Lim, Y.Y.; Leow, A.T.C.; Namasivayam, P.; Abdullah, J.O.; Ho, C.L. Factors Affecting Yield and Gelling Properties of Agar. *J. Appl. Phycol.* **2017**, *29*, 1527–1540. [CrossRef]
45. Xiao, Q.; Wang, X.; Zhang, J.; Zhang, Y.; Chen, J.; Chen, F.; Xiao, A. Pretreatment Techniques and Green Extraction Technologies for Agar from Gracilaria Lemaneiformis. *Mar. Drugs* **2021**, *19*, 617. [CrossRef]
46. Torres, M.D.; Flórez-Fernández, N.; Domínguez, H. Integral Utilization of Red Seaweed for Bioactive Production. *Mar. Drugs* **2019**, *17*, 314. [CrossRef]
47. Gomes-Dias, J.S.; Teixeira-Guedes, C.I.; Teixeira, J.A.; Rocha, C.M.R. Red Seaweed Biorefinery: The Influence of Sequential Extractions on the Functional Properties of Extracted Agars and Porphyrans. *Int. J. Biol. Macromol.* **2024**, *257*, 128479. [CrossRef]
48. Nova, P.; Pimenta-Martins, A.; Maricato, É.; Nunes, C.; Abreu, H.; Coimbra, M.A.; Freitas, A.C.; Gomes, A.M. Chemical Composition and Antioxidant Potential of Five Algae Cultivated in Fully Controlled Closed Systems. *Molecules* **2023**, *28*, 4588. [CrossRef] [PubMed]
49. Sousa, A.M.M.; Alves, V.D.; Morais, S.; Delerue-Matos, C.; Gonçalves, M.P. Agar Extraction from Integrated Multitrophic Aquacultured Gracilaria Vermiculophylla: Evaluation of a Microwave-Assisted Process Using Response Surface Methodology. *Bioresour. Technol.* **2010**, *101*, 3258–3267. [CrossRef]
50. Villanueva, R.D.; Sousa, A.M.M.; Gonçalves, M.P.; Nilsson, M.; Hilliou, L. Production and Properties of Agar from the Invasive Marine Alga, Gracilaria Vermiculophylla (Gracilariales, Rhodophyta). *J. Appl. Phycol.* **2010**, *22*, 211–220. [CrossRef]
51. Belattmania, Z.; Bhaby, S.; Nadri, A.; Khaya, K.; Bentiss, F.; Jama, C.; Reani, A.; Vasconcelos, V.; Sabour, B. Gracilaria Gracilis (Gracilariales, Rhodophyta) from Dakhla (Southern Moroccan Atlantic Coast) as Source of Agar: Content, Chemical Characteristics, and Gelling Properties. *Mar. Drugs* **2021**, *19*, 672. [CrossRef] [PubMed]
52. Rocha, C.M.R.; Sousa, A.M.M.; Kim, J.K.; Magalhães, J.M.C.S.; Yarish, C.; do, P. Gonçalves, M. Characterization of Agar from Gracilaria Tikvahiae Cultivated for Nutrient Bioextraction in Open Water Farms. *Food Hydrocoll.* **2019**, *89*, 260–271. [CrossRef]
53. Guerrero, P.; Etxabide, A.; Leceta, I.; Peñalba, M.; De La Caba, K. Extraction of Agar from Gelidium Sesquipedale (Rodhopyta) and Surface Characterization of Agar Based Films. *Carbohydr. Polym.* **2014**, *99*, 491–498. [CrossRef] [PubMed]
54. Oliveira, S.; Sousa, I.; Raymundo, A. Printability Evaluation of Chlorella Vulgaris Snacks. *Algal Res.* **2022**, *68*, 102879. [CrossRef]
55. Hernández, M.J.; Dolz, J.; Dolz, M.; Delegido, J.; Pellicer, J. Viscous Synergism in Carrageenans (κ and λ) and Locust Bean Gum Mixtures: Influence of Adding Sodium Carboxymethylcellulose. *Food Sci. Technol. Int.* **2001**, *7*, 383–391. [CrossRef]
56. Santinath Singh, S.; Aswal, V.K.; Bohidar, H.B. Internal Structures of Agar-Gelatin Co-Hydrogels by Light Scattering, Small-Angle Neutron Scattering and Rheology. *Eur. Phys. J. E Soft Matter* **2011**, *34*, 62. [CrossRef]

57. Gomes-Dias, J.S.; Romaní, A.; Teixeira, J.A.; Rocha, C.M.R. Valorization of Seaweed Carbohydrates: Autohydrolysis as a Selective and Sustainable Pretreatment. *ACS Sustain. Chem. Eng.* **2020**, *8*, 17143–17153. [CrossRef]
58. Joshi, M.; Adhikari, B.; Aldred, P.; Panozzo, J.F.; Kasapis, S. Physicochemical and Functional Properties of Lentil Protein Isolates Prepared by Different Drying Methods. *Food Chem.* **2011**, *129*, 1513–1522. [CrossRef]

Disclaimer/Publisher's Note: The statements, opinions and data contained in all publications are solely those of the individual author(s) and contributor(s) and not of MDPI and/or the editor(s). MDPI and/or the editor(s) disclaim responsibility for any injury to people or property resulting from any ideas, methods, instructions or products referred to in the content.

Article

A New Method for the Production of High-Concentration Collagen Bioinks with Semiautonomic Preparation

Jana Matejkova [1,*], Denisa Kanokova [1], Monika Supova [2] and Roman Matejka [1,*]

1. Department of Biomedical Technology, Faculty of Biomedical Engineering, Czech Technical University in Prague, 272 01 Kladno, Czech Republic; kanokden@fbmi.cvut.cz
2. Department of Composites and Carbon Materials, Institute of Rock Structure and Mechanics of The Czech Academy of Sciences, v.v.i., 182 09 Prague, Czech Republic; supova@irsm.cas.cz
* Correspondence: jana.matejkova@cvut.cz (J.M.); roman.matejka@cvut.cz (R.M.)

Abstract: It is believed that 3D bioprinting will greatly help the field of tissue engineering and regenerative medicine, as live patient cells are incorporated into the material, which directly creates a 3D structure. Thus, this method has potential in many types of human body tissues. Collagen provides an advantage, as it is the most common extracellular matrix present in all kinds of tissues and is, therefore, very natural for cells and the organism. Hydrogels with highly concentrated collagen make it possible to create 3D structures without additional additives to crosslink the polymer, which could negatively affect cell proliferation and viability. This study established a new method for preparing highly concentrated collagen bioinks, which does not negatively affect cell proliferation and viability. The method is based on two successive neutralizations of the prepared hydrogel using the bicarbonate buffering mechanisms of the 2× enhanced culture medium and pH adjustment by adding NaOH. Collagen hydrogel was used in concentrations of 20 and 30 mg/mL dissolved in acetic acid with a concentration of 0.05 and 0.1 wt.%. The bioink preparation process is automated, including colorimetric pH detection and adjustment. The new method was validated using bioprinting and subsequent cultivation of collagen hydrogels with incorporated stromal cells. After 96 h of cultivation, cell proliferation and viability were not statistically significantly reduced.

Keywords: bioprinting; bioink; collagen hydrogels; biofabrication; pH; neutralization; automation; stromal cells

1. Introduction

Bioinks—polymers with incorporated cells 'customized' for the preparation of tissues and organs—must meet many criteria to be structurally strong enough for the preparation of 3D structures, but also non-toxic to organisms, especially the cells that populate the gel. Ideally, cell adhesion and proliferation are also promoted. The resulting bioink should be nonimmunogenic and biodegradable [1]. Using natural or synthetic polymers is an option, where, in addition to properties, we must also consider the processability of the material, suitability for bioprinting, cost, and commercial availability. While synthetic polymers excel in mechanical durability, natural gels are highly biocompatible [2]. Using a natural material with a high concentration of polymers combines the advantages of both types of materials but has its limits for adhesion and cell growth [3].

A suitable natural polymer for bioink preparation is collagen, which is abundant in the extracellular matrix and thus is very natural to use and is also highly available in the connective tissues of organisms [4]. Collagen has a high affinity for adherent cells, which is the reason why these hydrogels are widely used in biomedical applications, from testing materials for 3D bioprinting to general tissue models for in vitro cell studies, drug testing, and specialized tissue models, especially when using cell-filled hydrogels [5].

Due to its good biocompatibility and low immunogenicity, collagen has been successfully used in clinical practice. However, low immunogenicity can only be achieved using high-purity collagen solutions when the protein is derived from collagen-containing tissues. The main weaknesses of this material are its low mechanical properties, difficulty in sterilization (e.g., heat sensitivity, degradation), and the commonly occurring shrinkage of collagen scaffolds in response to cellular activity [5]. Regarding 3D printing applications, collagen-based bioinks with potential for tissue regeneration have been developed to produce breast implants and artificial cartilage for joint reconstruction, among others [6]. Cell-based collagen bioinks also contribute to the growth of the 3D-printed alternative meat market, which companies are exploiting [7].

However, in the present studies, relatively low-concentration collagen polymers are used, as higher concentrations (10 mg/mL) reduce cell proliferation in the gel and differentiate ability, and the diffusion of waste products in the polymer is also impeded [8,9]. However, several factors influence the successful formation of the 3D structure. These include the properties of the bioink itself (composition, pH, temperature, viscosity, cell culture concentration, and crosslinking method) as well as the parameters of the printing process (temperature, pressure, and speed of parts of the printing system parts), as they have a significant impact on the resulting structural strength and viability of the incorporated cells [5].

Collagen hydrogels (COLs) are usually created from cold, acid-solubilized monomers (tropocollagen) by neutralization and heating to induce physical crosslinking. COL gelation is due to fibrillogenesis, a self-assembly process that proceeds from the self-association of triple helices [10]. This process is strongly affected by the nature of the COL monomers and many other physical factors [11]. Fibrillogenesis is induced by the neutralization of pH and by increasing temperature while increasing the ionic strength, which causes a decrease in the rate of fibrillogenesis [12]. The kinetics of COL fibrillogenesis consist of two stages: (1) a nucleation process and (2) the subsequent growth of nuclei and the aggregation of COL fibrils that result in the 3D network structure of the hydrogel [13]. Both the pH and the ionic strength affect the net surface charges of the COL molecules. Low pH causes a repulsive electrostatic interaction between COL molecules. Consequently, they are homogeneously dissolved [14], while neutral pH conditions cause a weakening of electrostatic repulsion. The kinetics of fibril self-assembly are also affected by the alteration of electrostatic, hydrophobic, and covalent interactions between monomers, resulting in the range of fibril size [11] and the final microstructure and the properties of the collagen gel [15,16]. These interactions are strongly influenced by concentrations of both COL and solvent [17]. The highest rates of fibrillogenesis occur between pH 6.9 and 9.2, with no significant changes in fibril diameters [18]. Another important factor is temperature, which affects the water-mediated hydrogen bonding between collagen molecules because water between triple helices may play a role in the association process [19].

Gelation occurs in two stages: first, triple helices in COL molecules are dissolved in dilute acid (typically acetic acid for several days at low temperatures) to disrupt weaker intermolecular hydrogen bonds and Schiff bases between collagen molecules [20], followed by the formation of 3D hydrogel structures induced by increasing the temperature and pH value (using dilute alkali solution or various simulated body fluid media). Therefore, COL hydrogel formation is a process of reconstruction of collagen molecules with water in a 3D space.

Studies that have used collagen concentrations of 10 mg/mL or more for bioink preparation have often used additives (riboflavin, Pluronic F127) to crosslink the polymer, resulting in increased mechanical strength of gels [21]. Still, the substances used are unnatural to cells. They can interfere with cell proliferation and differentiation, as well as the biocompatibility of the entire structure [22]. Crosslinking by mere modification of physical conditions (temperature, pH) is rarely represented due to high collagen concentrations [9].

In our previous work [23], we optimized the bioink properties with highly concentrated collagen (20 mg/mL dissolved in 0.1% acetic acid; the final collagen concentration was 10 mg/mL), especially the gel composition and adjustment of pH. When NaOH was added, we adjusted the final pH and gelation time. The composition and pH adjustment allowed the cells to grow and divide, and the higher mechanical modulus of these constructs also allowed the structures to be printed up to several millimeters in height with sufficient mechanical resistance. This experiment demonstrated that a high concentration of collagen gels is not necessarily a limiting factor for cell proliferation.

However, from an objective point of view, higher cell viability could be achieved in the resulting structures. Furthermore, the bioink preparation method also needed to be optimized regarding component volumes, as they varied up to a thousand times (from a few μL to several mL). This is particularly disadvantageous for mechanized, automated bioink preparation. Therefore, our paper aims to eliminate these weaknesses and obtain a 3D culture with high cell viability even after several days of cultivation.

This study builds on the results of a previous study to optimize the bioink composition so that cells achieve higher viability and proliferation even after several days of static culture. Furthermore, we also want to improve the method of mixing the bioink so that higher volumes of components can be handled, and a machine can then implement the mixing.

2. Results and Discussion

2.1. Cell Growth in Modified Culture Media

New formulations of culture media with a compensated additive concentration were compared to the formulation from our previous article and control growth culture medium. Growth and viability were evaluated after 3 days of static cultivation.

Cells in the nutritionally compensated media have proliferation similar to that of the control sample, although their shape is slightly elongated. In an uncompensated medium, proliferation is lower in all hatches. In this medium, the cells are not elongated are more isolated, and do not form a continuous structure as in the other samples. This indicates insufficient cell proliferation, which does not result in adequate formation of the potential tissue base. As illustrated in Figure 1, a medium lacking nutrients degraded cell growth.

Furthermore, nutrition-compensated media were tested to determine whether high concentrations of medium components during bioink preparation would result in cell damage and thus reduce viability. After 3 days of culture, there was no significant difference from control cells grown in the original culture medium. No dead cells are visible in any of the images; as in these 2D cultures, the cells are rinsed with PBS before analysis, and thus, the dead cells are removed from the culture.

2.2. Collagen FTIR Spectra

The FTIR spectra of isolated COL (original) and COL dissolved in 0.05 wt.% and in 0.1 wt.% acetic acid in a COL concentration of 20 mg/mL can be seen in Figure 2. All spectra contained five amidic bands typical for proteins. The band cantered at 3315 cm^{-1}, relating to a mutual band of hydrogen bonds among OH groups of water and humidity and amide A of COL, which belong to NH$_2$ stretching. The band at ~3077 cm^{-1} is mutual band of the N-H stretching vibrations in the secondary amides (amide B) and C-H stretching vibrations in the sp2 hybridization existing in aromatic amino acids. The amide I band describes the C=O stretching coupled with N-H bending vibrations. The amide II band originates from N-H bending coupled with C-N stretching vibrations. Another proof of the existence of a triple helical structure in the COL represents a band triplet (at ~1205, 1240, and 1280 cm^{-1}) of amide III together with the band at 1337 cm^{-1} [24,25]. Changes in the intensities and positions of individual amide bands could be connected with changes in the secondary structure of COL. As was proved in our previous study [23], the use of 70% ethanol for sterilization did not alter the secondary structure of COL. As shown in Figure 2, the use of acetic acid as a solvent, regardless of concentration, does not change the secondary structure of COL compared to original and commercial COL.

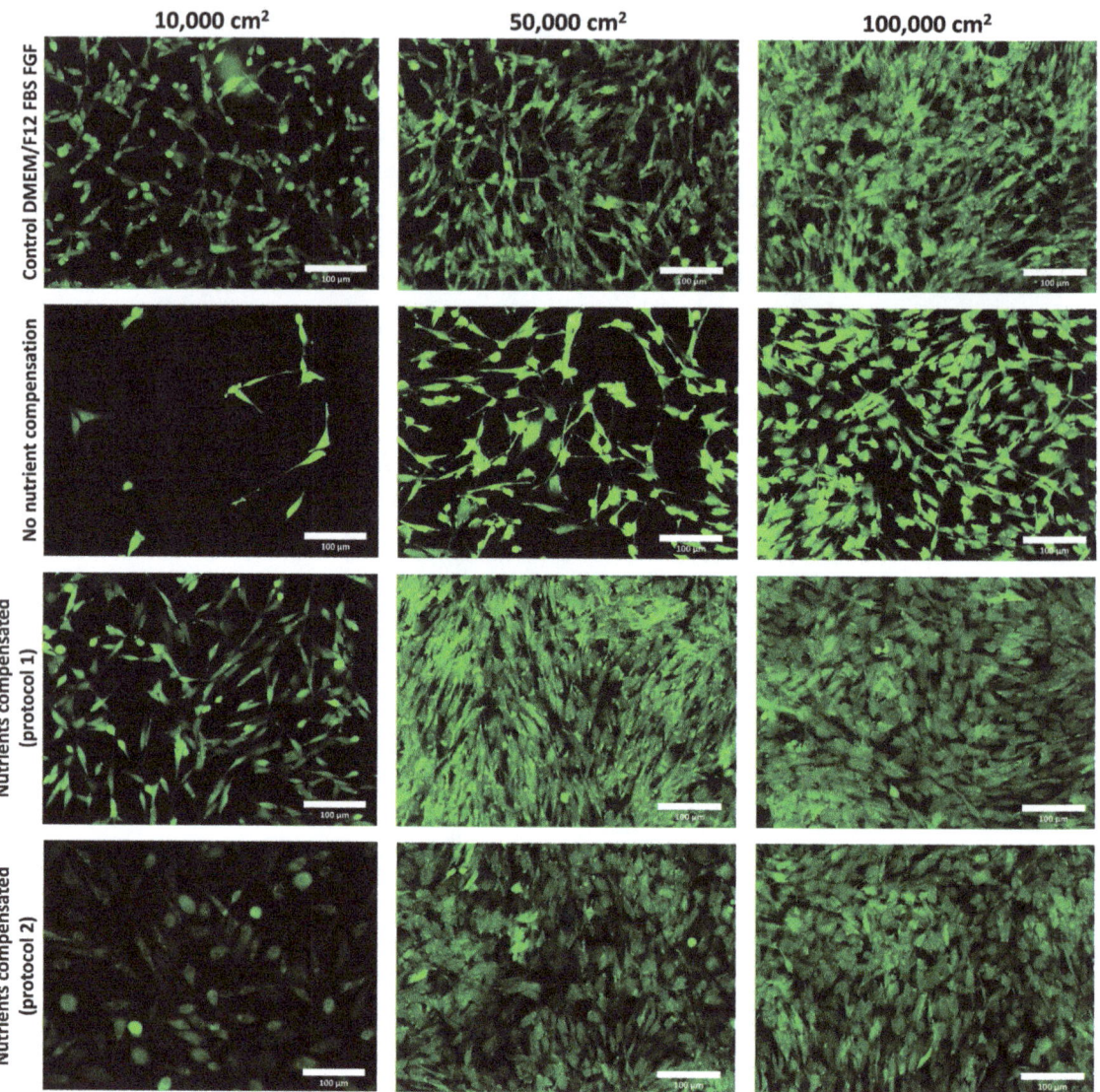

Figure 1. Live/dead images of cells cultured for 3 days in control culture medium, without compensation of nutrients and compensated (protocol 1, e.g., pH adjusted culture medium, and protocol 2, culture medium without pH adjustment). Live cells were stained with fluorescein diacetate (green), and dead cells (nuclei) were stained with propidium iodide (red), scale bar 100 µm—custom-built Thorlabs CERNA fluorescence microscope with Olympus Plan Fluorite 10× objective.

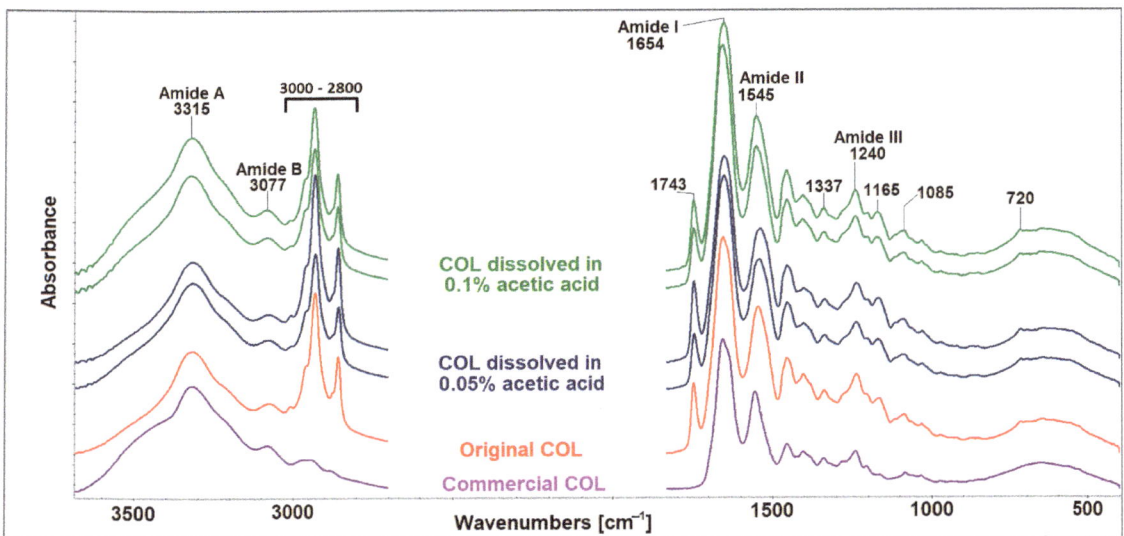

Figure 2. FTIR spectrum of porcine collagen lyophilizate after isolation (original COL) and examples of spectra of COL after dissolution in 0.05 wt.% and in 0.1 wt.% acetic acid in a COL concentration of 20 mg/mL and comparison with commercial COL.

Apparent changes in band intensities in spectral region 2800–3000 cm^{-1} and at 720 cm^{-1}, relating to C-H aliphatic bonds, and changes in bands at 1743 cm^{-1}, 1165 cm^{-1} and 1085 cm^{-1} ascribed to the C=O bonds in ester, are related to the presence of lipids. The different intensities of the infrared bands belonging to lipids in all spectra are caused by local inhomogeneity. Lipids are common residual impurities remaining in collagen after the isolation procedure [24]. As can be seen in Figure 2, the commercial COL does not contain such impurities.

The comparison of the infrared spectra of the original collagenous lyophilizate and the collagens subjected to acetic acid of two various concentrations (0.05 wt.%, 0.1 wt.%) proved that the spectra following dissolution in acetic acids did not exhibit significant changes compared to original COL.

2.3. Effect of Medium Preconditioning

The culture media used in this study use a bicarbonate and CO_2 buffering system that affects the resulting pH [26]. In our case, we prepared 2× enhanced media that is in an optimal state above the physiological range of approximately 7.6–7.8, depending on the batch and adjustments before sterilization. Figure 3 illustrates different pH levels of culture media preconditioned in a CO_2 atmosphere for 120 min and media left on air. The medium left in the air, due to the lack of CO_2, changed to an alkaline pH. The range of change was affected by the length of exposure, its volume, and also by the fact that our "2× enhanced" medium contains twice the bicarbonate compared to the standard proliferation medium (thus the large interval in the chart). The primary challenge is that the pH level of the culture medium has an impact on the neutralization of the collagen hydrogel. The pH of the collagen hydrogel then affects the gelling properties, cell viability, and printability. To ensure consistency and reproducible outcomes, we have used only preconditioned media in the following preparations of printable bioinks.

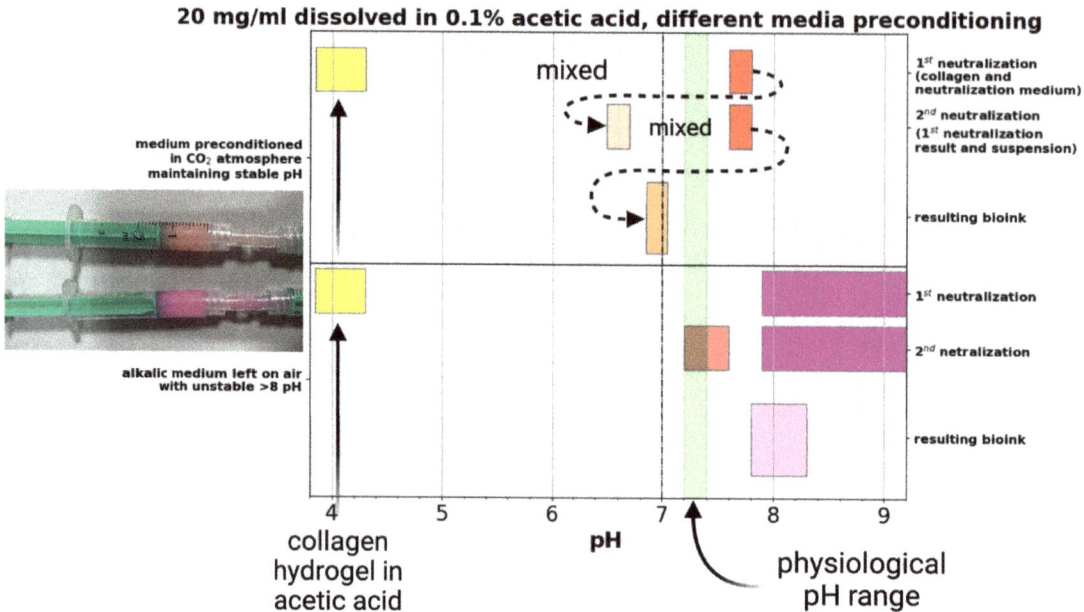

Figure 3. The difference in the pH of the resulting bioink is affected by the initial condition of the culture medium preconditioned in a CO_2 atmosphere and left in the air. Created with BioRender.com.

2.4. pH Response of Collagen Hydrogel to Medium Added

As described above, the use of lower concentrations of acetic acid (AA) in the dissolution of collagen yields almost an optimal pH after neutralization. Furthermore, some different adjustments of the culture media prior to sterilization (especially 10× concentrated) can affect the resulting parameters. To demonstrate this, we performed a simple volumetric neutralization in our mixing system. The mixing syringe was prefilled with 1400 µL of collagen (all variants—concentration of collagen 20 and 30 mg/mL and 0.05 and 0.1% dissolving AA). Then, the '2× enhanced' medium was dosed in 200 µL steps. After each addition, the collagen and the media were mixed, and the pH was captured using a colorimeter. This was performed until 1400 µL of media was dosed to achieve a 1:1 ratio of collagen to medium. As illustrated in Figure 4, it is obvious that 0.1% AA provides a more acidic and probably buffered collagen hydrogel, which may require more culture medium or NaOH modification. Differences in the preparation of the culture medium and its adjustments before sterilization can affect the neutralization behavior. In the case of 0.05% AA, there is a promising fact that with a different approach in media preparation, there is no need to perform any NaOH modification in culture media. Another approach can utilize continuous mixing and the estimation of pH in a designed mixing system, allowing for the change in the ratio between the neutralization and cell suspension media.

In our described experiments, the neutralization medium was modified using fixed steps of NaOH and a fixed 2:1:1 ratio (collagen:neutralization medium:suspension medium). These optimizations will be made in the following research.

Sterilization of the medium by filtration causes an increase in pH, and titration with this medium leads to an optimal pH much earlier. Collagen dissolved in less concentrated acid reaches physiological pH values with the addition of approximately 1 mL of culture medium so that further pH adjustment with NaOH is no longer necessary. This confirms the results of the previous analysis. By selecting the appropriate concentration of solvent and neutralizing with 2× enhanced medium, the pH of the mixture reaches a value of 7, at which point the cell suspension can be added without negatively affecting cell proliferation or viability. In some cases, it is unnecessary to add NaOH to the mixture, as we have implemented and published the results of the previous article [23] so far.

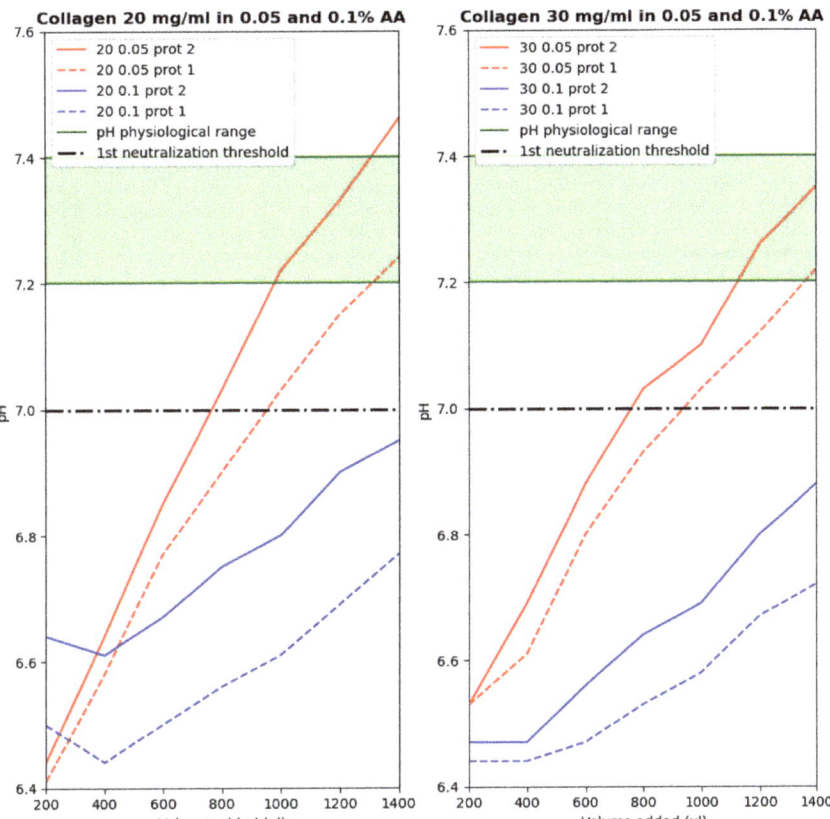

Figure 4. Volumetric titration of collagen hydrogels and response to changes in pH.

2.5. The pH Balancing

The pH of the prepared bioink varies according to the concentration of AA in which collagen dissolves, as shown in Figure 5. In the case of 0.1% acid, after the first neutralization, the pH is around 6.5, which is a fairly acidic environment for adding cells. After the second neutralization, the pH is below the physiological limit. Using a less concentrated solvent (0.05 wt.% AA) and no NaOH, the pH values are higher than with 0.1% acid but still below the physiological limit. The addition of NaOH at a concentration of 50 μL/mL leads to a change in pH below the physiological limit after the first neutralization. After the second neutralization, the pH is fully physiological for cell development.

To find the optimal neutralization medium and procedure, we find that two concentrations of collagen (20 and 30 mg/mL) and both dissolving AA concentrations (0.05 and 0.1 wt.%) were first neutralized using 2× enhanced medium with 0, 50, and 100 μL/mL NaOH. This initial neutralization step is crucial because it shifts the pH level closer to the physiological range, thereby making it more compatible and acceptable for cells. This prevents the cells from experiencing a shock from a rapid change in pH, which results in higher cell viability. The second neutralization step was performed with a 2× enhanced culture medium. The resulting pHs are illustrated in Figure 6. The 0.1% AA in both concentrations of collagen is obviously more acidic than 0.05%. Even when 100 µL/mL NaOH was used, collagens dissolved in 0.1% acidic were below the physiological range of pH. The 0.05% acidic acid as the dissolving agent provided easier control over the pH range. When neutralized without adding NaOH, the resulting pH after the first neutralization was slightly above 7 (6.8–7) and after the second neutralization reached the physiological range. However, it was still slightly below 0.1–0.15. For the following prints with living cells, 50 µL/mL NaOH was chosen as the optimal amount to add.

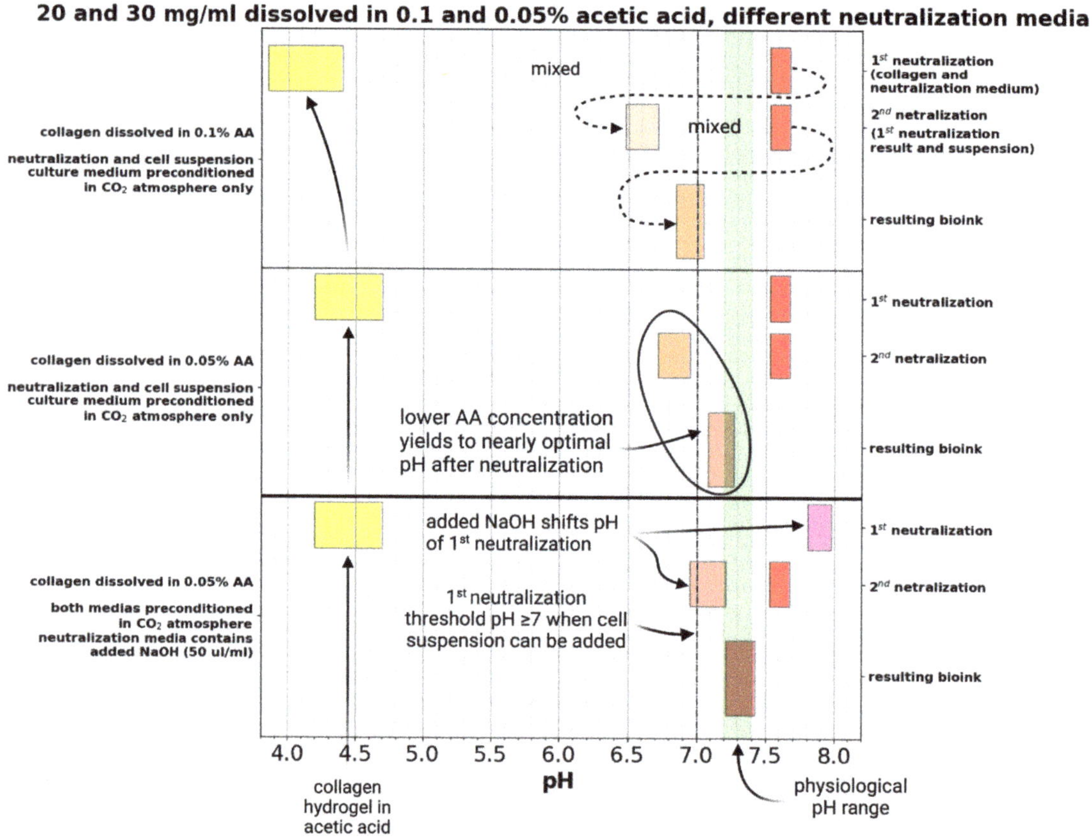

Figure 5. Two-step neutralization of collagens dissolved in 0.1 and 0.05 wt.% acetic acid using preconditioned and NaOH-modified 2× enhanced culture medium. A lower concentration of acetic acid provides a nearly optimal pH of the resulting bioink. However, depending on the batch of collagen, other pH adjustments using NaOH are optimal. Created with BioRender.com.

We have used two-step neutralization, where first neutralization shifts the pH of acid collagen into a natural pH of around 7. In addition, we chose this value as the working threshold for the second neutralization with culture media. As a simple alternative, we tested only single-step neutralization. In single-step neutralization, we mixed both neutralization media containing NaOH and cell suspension together. Then, it was mixed with acid collagen. The resulting pHs were similar to those of two-step neutralization; however, it resulted in poor cell viability. Skipping the two-step neutralization procedure leads to apoptosis of most cells in the hydrogel, as demonstrated in Figure 7.

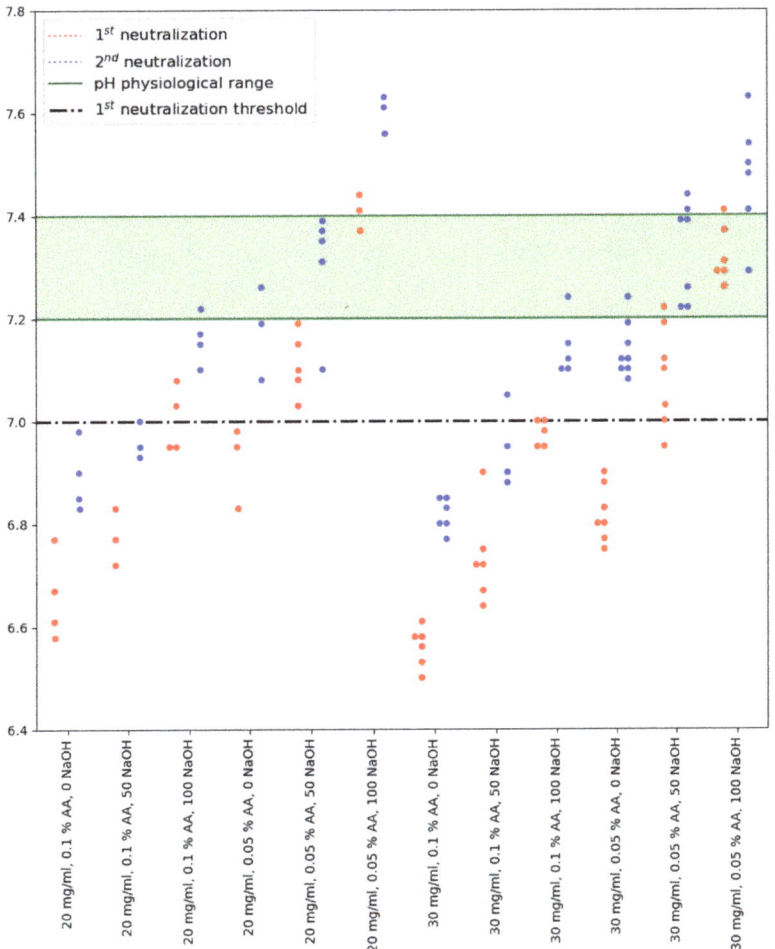

Figure 6. pH of the bioinks prepared after the first and second neutralization using a different neutralization medium (different concentrations of NaOH added).

Although it must be taken into account, the use of a less concentrated acid can result in less pH fluctuation during bioink preparation, and it can also not completely dissolve collagen fibers, which would have an adverse effect on subsequent printing.

In a previous article, we determined the optimal volumetric concentration of NaOH in collagen bioink to be 20 mg/mL in 0.1% AA–10 µL/mL, in terms of both mechanical strength and cell survival. Now, we must refine this conclusion. Several circumstances influence the resulting pH. Not only is the concentration of collagen in the bioink and the concentration of acid used for dissolution, but the pH is slightly different in each batch of production as a result of natural influences. Thus, it is not possible to generally determine the appropriate concentration of hydroxide to stabilize the pH, but in each individual experiment, the pH must be monitored and adjusted, if necessary, before cell suspension.

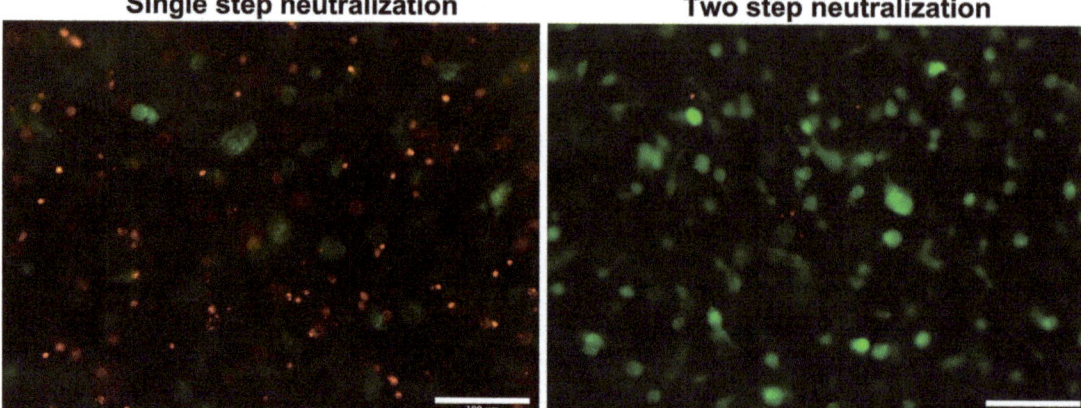

Figure 7. Difference in cell viability using single- or two-step neutralization. Live/dead images of cells grown for 3 h in a printed collagen bioink. Live cells stained with fluorescein diacetate (green) and dead cells (nuclei) stained with propidium iodide (red), scale 100 μm. Custom-built Thorlabs CERNA fluorescence microscope with Olympus Plan Fluorite 10× objective.

2.6. Cellular Behavior in Bioinks

It is well known that high-concentrated collagen hydrogels, especially their stiffness, porosity, pH level, chemical composition, and presence of adhesive ligands, affect cell morphology and behavior [27]. We analyzed the cell morphology within the four different categories of gels at three different time points. The stromal cells in the hydrogel showed high viability and density even in the inner parts of the materials after 96 h of culture, as shown in the confocal microscope images in Figure 8. For samples with different concentrations of AA, no difference in cell viability was evident because the gels were prepared by two-step neutralization, where the cell suspension was mixed into a gel at a pH of approximately 7. A slight but not statistically significant difference was observed for samples with different concentrations of 20 and 30 mg/mL (resulting in concentrations of 10 and 15 mg/mL). The 30 mg samples, regardless of the concentration of AA, showed lower cell viability at some time intervals. With the more concentrated collagens, the hydrogel does not mix completely with the cells, and the cells form clusters where more cells are clumped, so the gel is not completely uniformly populated with cells. Thus, these clusters are more susceptible to automatic and manual cell counting. Inaccuracies in counting also introduce some errors, as the cells are interspersed throughout the gel volume, and cell viability varies greatly from each section.

This confirms the validity of our method of gradual neutralization of collagen hydrogels during bioink preparation. None of the samples show reduced cell proliferation. This corresponds to our results that, depending on the actual batch and acidity of the collagen used, the process method needs to be adjusted to optimize the pH to physiological values suitable for cell growth.

A comparison of two bioink production methods (Figure 9) where 10× concentrated culture medium was (un)adjusted to pH 7.0 before mixing also showed no differences in cell morphology and proliferation. After 96 h of culture, the cells were viable, elongated, and formed a structure similar to that of physiological tissue.

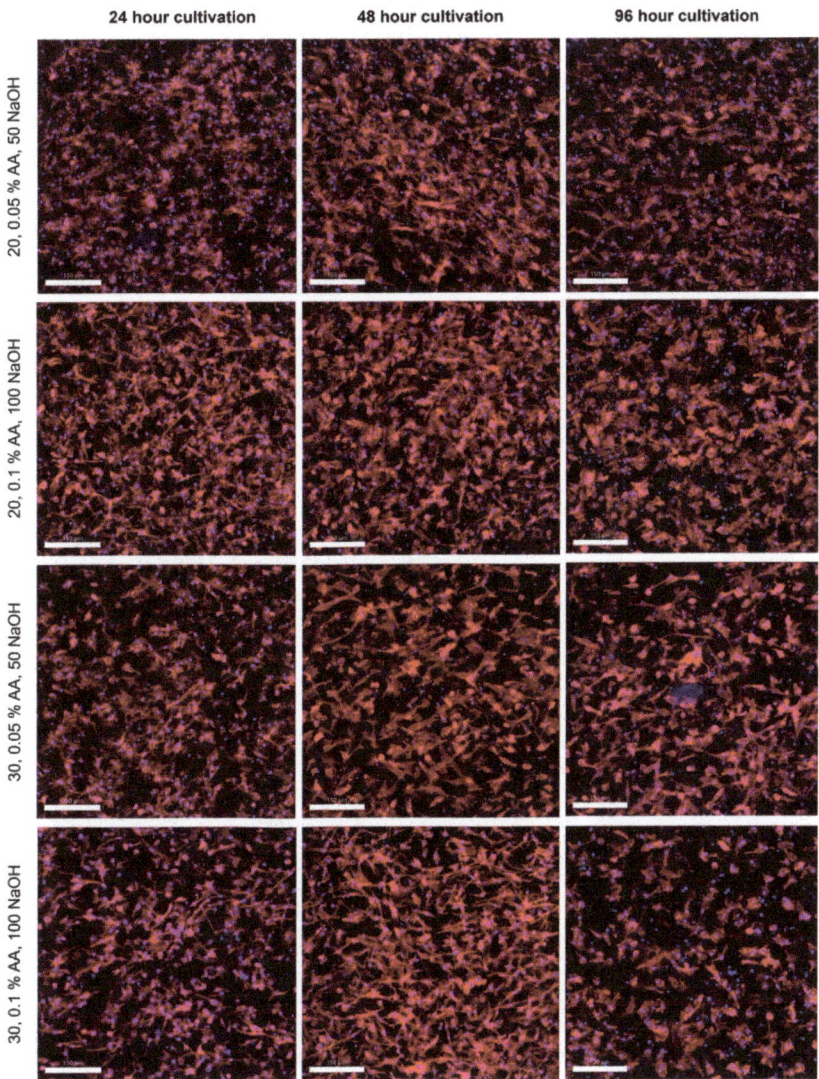

Figure 8. Confocal microscopy images of cultivated samples in 20 and 30 mg/mL collagen dissolved in 0.05 and 0.1 wt.% acetic acid. Cultivation for 24, 48, and 96 h. F-actin was stained using Alexa Fluor™ 594 Phalloidin (red), and cell nuclei were counterstained using DAPI (blue), scale 150 µm. Nikon CSU-W1 confocal microscope with CFI Plan Apo VC 20× dry objective. Orthographic projection rendered.

However, unlike cell morphology, cell viability differed between samples (Figure 10). The highest viability was observed for the 20 mg/mL, 0.1% AA, 100 NaOH sample and the 30 mg/mL, 0.05% and 50 NaOH samples; the other two samples showed higher cell mortality, which was obviously evident at other times. However, the comparison of multiple samples showed that the differences between samples were not statistically significant because the pH neutralization of the collagen hydrogels was emphasized in the preparation of the bioinks. Regardless of the initial concentration of collagen and acetic acid, the resulting bioink is chemically and structurally similar in all samples, and thus differences in cell growth and viability are not apparent. The comparison of the two production methods again showed small differences, especially after 24 h of cultivation, which were again not statistically significant (Figure 11). Several factors influence cell viability, particularly in 3D samples: the actual structure, homogeneity, and thickness of the layer in which the cells are incorporated. The higher density of collagen fibers can be concentrated in one location, preventing diffusion, and cells are not optimally nourished. Furthermore, the viability is affected by random effects during bioprinting, the temperature acting on the cells, the mechanical pressure, and the chemical effects of the components. After 96 h of culture, the images show the highest proportion of dead cells. As a result of the growth of cells throughout the gel volume, dead cells cannot be washed out during medium exchange, as in flat culture on the substrate. Thus, these cells accumulate there, and viability seems to be lower than in previous intervals. Also, wide-field fluorescence microscopy has its limits in taking images, especially in 96 h intervals from gel substrates. It is also expected that cells partially remodel the substrate. In images from confocal microscopy, there were no significant changes in the morphology or structure of cells in the 96 h interval compared to 48 h. Also, a limitation may be the substrate used (coverslip glass) that limits diffusion from one side of the 3D construct. The use of a mild culture perfusion in a bioreactor can also improve these problems.

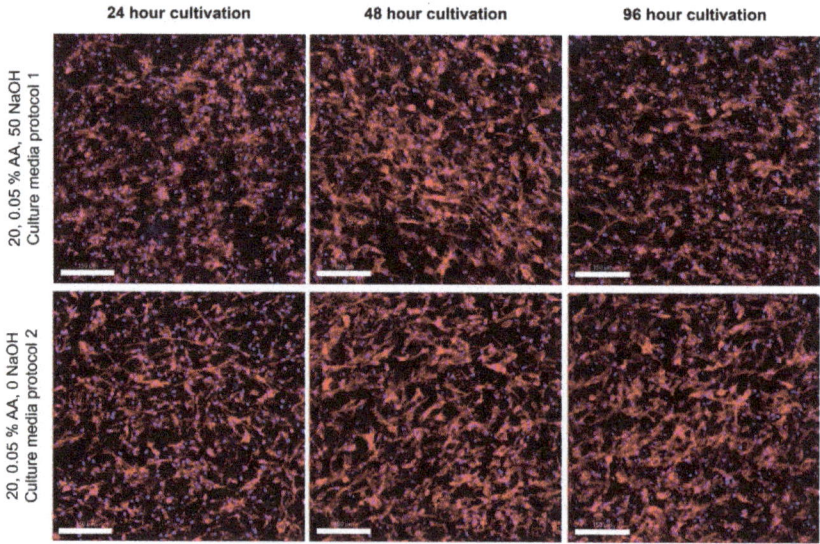

Figure 9. Confocal microscopy images of cultivated samples in 20 mg/mL collagen dissolved in 0.05 wt.% acetic acid prepared using two different culture media (protocol 1, e.g., pH adjusted culture medium, and protocol 2, culture medium with no pH adjustment). Cultivation for 24, 48, and 96 h. F-actin was stained using Alexa Fluor™ 594 Phalloidin (red), and cell nuclei were counterstained using DAPI (blue), scale 150 μm. Nikon CSU-W1 confocal microscope with CFI Plan Apo VC 20× dry objective. Orthographic projection rendered.

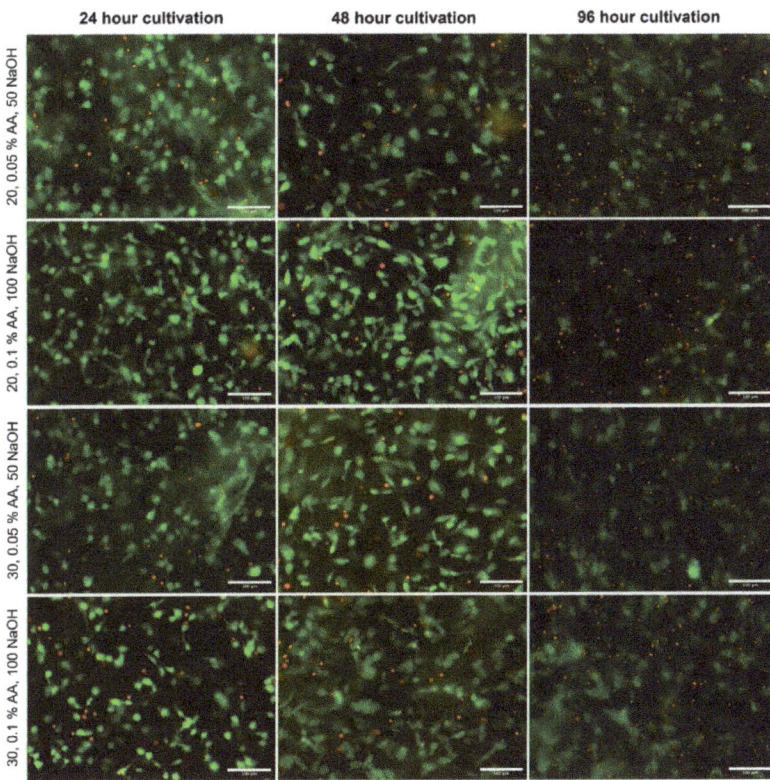

Figure 10. Live/dead images of cultivated samples in 20 and 30 mg/mL collagen dissolved in 0.05 and 0.1 wt.% acetic acid. Cultivation for 24, 48, and 96 h. Fluorescein diacetate-stained live cells (green) and dead cells (nuclei) stained with propidium iodide (red), scale 100 μm. Custom-built Thorlabs CERNA fluorescence microscope with Olympus Plan Fluorite 10× objective.

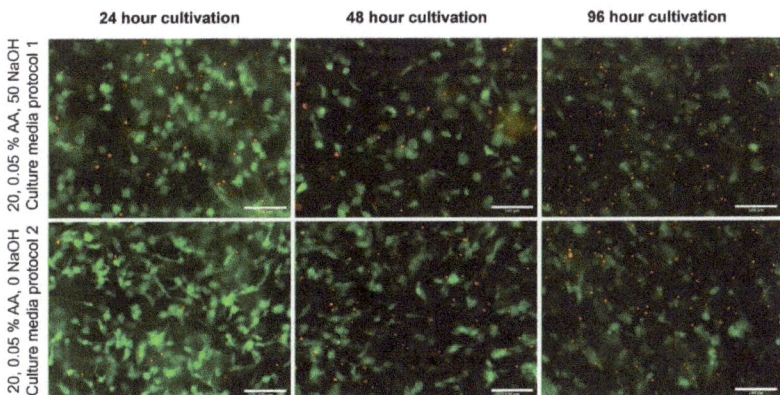

Figure 11. Live/dead images of cultivated samples in 20 mg/mL collagen dissolved in 0.05 wt.% acetic acid were prepared using two different culture media (protocol 1, e.g., pH adjusted culture medium, and protocol 2, culture medium with no pH adjustment). Fluorescein diacetate-stained live cells (green) and dead cells (nuclei) stained with propidium iodide (red), scale 100 μm. Custom-built Thorlabs CERNA fluorescence microscope with Olympus Plan Fluorite 10× objective.

2.7. Benefits and Risks of Current Bioink Formulation

The formulation of bioink in our previous work [23] was optimized to acquire the most suitable pH for the neutralization of the collagen hydrogel to promote its gelation, resulting in a stable bioink for 3D bioprinting and a printed substrate with cells. However, this formulation, which modifies the resulting pH, lacks nutrients and uses extremely low volumes of cell suspension, resulting in a complicated preparation.

In this study, we have modified our bioink based on a highly concentrated collagen hydrogel (20 and 30 mg/mL). Optimization of the formulation, especially in terms of FBS-derived growth factors, resulted in improved cell culture viability and cell growth and division during the 4-day period studied. Furthermore, the change in the volume ratio of the different bioink components led to the fact that the mixing process could be automated, which not only speeds up the process but also continuously monitors the pH during the bioink production process. Automated mixing also reduces the temperature shock to cells during bioink formation, which can negatively affect cell viability. The new method based on highly concentrated collagen may be a promising 3D bioprinting bioink for use in tissue engineering.

On the other hand, the cultivation of 3D scaffolds is practically limited by the thickness of the scaffold, with simple diffusion transferring nutrients and oxygen and removing waste products only for structures up to a maximum thickness of 2 mm. To nourish cells to deeper layers, a transport system, e.g., channels, must be added to permeate the entire volume of the scaffold. However, the use of type I collagen for bioprinting requires further research. In fact, some studies have shown that its use induces cell dedifferentiation, as demonstrated, for example, in chondrocytes in the study by Farjanel et al. [28]. The rapid expansion of chondrocytes in the monolayer leads to their dedifferentiation [29]. Furthermore, the synthesis of type III collagen to type I collagen occurred [29].

3. Conclusions

This study investigated the ability to control the collagen neutralization process that provides a printable bioink. These steps were utilized by our custom-built semiautomatic mixing system. The collagens used in this study were in two concentrations of 20 and 30 mg/mL dissolved in two concentrations of acetic acid 0.1 and 0.05 wt.%. In addition, phenol red was added to the dissolved collagen hydrogels, allowing colorimetric estimation of pH. To neutralize collagen hydrogels, a new formulation of culture media compositions was made—2× enhanced. A mixing ratio of 2:1:1 (collagen hydrogel:neutralization medium: cell suspension) was established as optimal to provide sufficient volumes for neutralization and the addition of cell suspension. The prepared collagens were titrated to determine the optimal volume of medium to neutralize the pH of the collagens by a bicarbonate buffering mechanism. Therefore, the method of 2 neutralizations was established to balance the pH of collagen hydrogels by supplying 0–100 uL/mL NaOH. After the first neutralization, the medium would be suitable for the addition of the cell suspension that constitutes the second neutralization. The amount of NaOH added cannot be fixed; it always depends on circumstances such as the buffering capacity of the culture medium, the pH of the batch of collagen hydrogels, etc. However, based on the established methodology of 2 neutralizations, the collagen bioinks were verified by 96 h culture. Cell viability and proliferation were not reduced after 96 h of culture.

4. Materials and Methods

4.1. Collagen Isolation

Type I collagen (COL), which was used for the preparation of hydrogels, was isolated from porcine skin according to our previous study [23]. Briefly, degreasing procedures were performed using 70 vol.% ethanol solution followed by the extraction of COL by application of 0.1 vol.% acetic acid solution (AA). After centrifugation, the COL in the collected supernatant was precipitated using a 0.1 M NaOH solution up to neutral pH.

The obtained COL pellets after centrifugation were then dissolved in 0.1 vol.% acetic acid solution, frozen to −30 °C, and lyophilized. All isolates were stored in a freezer at −20 °C.

4.2. Preparation of Collagen Hydrogels

Lyophilized COL I was sterilized by immersion in 70 vol.% ethanol for 2 h. After removal from ethanol, the COL was freely dried in a laminar flow box. COL was then dissolved in 0.05 wt.% and in 0.1 wt.% acetic acid in a COL concentration of 20 mg/mL and was stored at 4 °C for 10 days. Phenol red in the amount of 0.008 g was added to 1 L of acetic acid solutions to observe the changes in pH during further operations. The COL suspension was further homogenized using a disintegrator with 10,000 rpm for 2 min (IKA Dispersers, T 10 basic ULTRA-TURRAX®, Staufen, Germany) to obtain a homogenous COL solution.

4.3. Collagen Hydrogel Preparation

Lyophilized COL I was sterilized by immersion in 70 vol.% ethanol for 2 h. After removal from ethanol, the COL was dried freely in a laminar flow box. The COL was then dissolved in 0.05 wt.% and 0.1 wt.% AA in a concentration of COL of 20 or 30 mg/mL and stored at 4 °C for 10 days. Phenol red in the amount of 0.008 g was added to 1 L of AA solutions to observe changes in pH during future operations. The COL suspension was further homogenized using a disintegrator with 10,000 rpm for 2 min (IKA Dispersers, T 10 basic ULTRA-TURRAX®, Staufen, Germany) to obtain a homogenous COL solution.

4.4. Analysis of Porcine Collagen Properties

The secondary structure COL lyophilizates (original and dissolved in both concentrations of acetic acid) were evaluated by attenuated total reflection infrared spectrometry (ATR-FTIR) using an iS50 infrared spectrometer (Nicolet Instrument, Madison, WI, USA), the ATR device was equipped with a diamond crystal. Spectra were scanned in absorption mode in the range 4000–400 cm^{-1} at a resolution of 4 cm^{-1} 64 times. Infrared spectra were processed and evaluated using the OMNIC version 9 software. The spectra of studied samples were compared with commercial COL (lyophilized rat tail collagen; ROCHE, Basel, Switzerland), and they were scanned in a lyophilizate state several times to verify the homogeneity of the material.

4.5. Composition of Culture Media Used for Neutralization of Collagen Hydrogel

In the experiments, the control samples prepared according to the method of our previous work [23] and the samples prepared according to the new method were compared. For the current experiments, the culture medium for the bioink preparation was enriched with its components so that the resulting concentration of all components in the bioink (i.e., after mixing with collagen and cell suspension) was 1× that of the standard growth medium for stem cell culture, except for fetal bovine serum, which is 10 times more concentrated (in the original formulation only 3× more concentrated), as shown in Table 1.

The original culture growth medium (marked in the table as DMEM/F12 FBS FGF ABAM) is composed of a 1:1 mixture of Dulbecco's modified Eagle Medium and Ham's F-12 Medium (DMEM:F12, Gibco; supplemented with 2.438 g/L sodium bicarbonate), with 10% fetal bovine serum, 1% ABAM antibiotics (100 IU/mL penicillin, 100 µg/mL of streptomycin, and 0.25 µg/mL of Gibco Amphotericin B; Sigma-Aldrich, St. Louis, MO, USA) and 10 ng/mL FGF2.

Ten times (10×) concentrated DMEM/F12 was prepared using two protocols, differing in pH adjustment. The manufacturer's recommendations for 1× concentrated medium were followed in the preparation of the media. The medium prepared according to protocol 1 contained 1/10 water, and the pH of the medium was adjusted to pH 7.0 using HCl. Then, the medium was sterilized. In protocol 2, no HCl was added, and the pH of the medium was left at 7.5 and then filtered.

Table 1. Comparison of the composition culture medium prepared according to the previous protocol and the new protocol.

	Medium Composition According to [23]			New Composition		
	Partial Concentration (per 1 mL of Culture Media)			Partial Concentration (per 1 mL of Culture Media)		
	Volume Concentration	Ionic (DMEM:F12)	Nutrients (FBS)	Volume Concentration	Ionic (DMEM:F12)	Nutrients (FBS)
10× DMEM/F12	3.5%	0.7	0.0	3.5%	0.70	0.00
DMEM/F12 FBS FGF ABAM	16.5%	0.3	3.3	16.5%	0.33	3.30
H_2O	30.0%	0.0	0.0	26.5%	0.00	0.00
FBS	0.0%	0.0	0.0	3.5%	0.00	7.00
Concentration without collagen	50.0%	2.1	6.6	50%	2.06	20.60
Collagen	50.0%	0.0	0.0	50%	0.00	0.00
Resulting concentrtion	100.0%	1.03	3.3	100%	1.03	10.3

4.6. Cell Survival in Newly Prepared and Concentrated Media

The tests were performed in various cell concentrations in compensated and uncompensated culture medium and compared to standard proliferation media. Culture media were prepared according to Table 1. Culture media were tested with cells in which collagen was replaced with dH_2O. Cells were seeded in culture wells at a concentration of 10, 50, and 100,000 cells per cm^2. Cell viability was evaluated 3 days after seeding compared to the conventional growth medium.

Porcine stromal cells derived from porcine adipose tissue were used for all the experiments in this article. The entire procedure, including cell characterization, is described in a previous article by Matejka et al. [30]. Cells were harvested in 3–5 passages after cultivation in a cell growth medium. After centrifugation, cells were resuspended in a 2× enhanced culture medium.

4.7. Medium Preconditioning

The culture media used in this study use a bicarbonate buffer system with a 5% CO_2 atmosphere to maintain optimal pH. When fresh culture media is prepared, its optimal pH is in the physiological range of 7.2–7.4. However, manipulation in a normal atmosphere, especially in small volumes, tends to change the pH of the medium to alkaline [31]. Furthermore, in our case, we have prepared '2× enhanced' culture media consisting of twice bicarbonate compared to normal culture media. This increased bicarbonate shifts the prepared medium above the physiological range of 7.5–7.8 (according to the protocol used) and is more prone to change pH to alkaline in an atmosphere without CO_2.

To achieve an optimized pH of the culture medium used to mix the printable bioink that shifts the pH to the base, we have constructed a preconditioning setup. This setup consists of a standard GL45 bottle with a custom 3D printed pass-through plug with three fluidic ports. Two ports are used to recirculate the CO_2 atmosphere through two 220 nm PTFE sterile filters. The third port allows for the aseptic connection of a syringe to obtain a pH-optimized medium. When the medium is placed in the syringe without any air gaps, the change in pH is minimal. To speed up the process, a magnetic stirrer was also used. According to pH measurements and depending on volume, the culture medium stabilized to pH 7.6–7.7 in 120 min. This setup is illustrated in Figure 12.

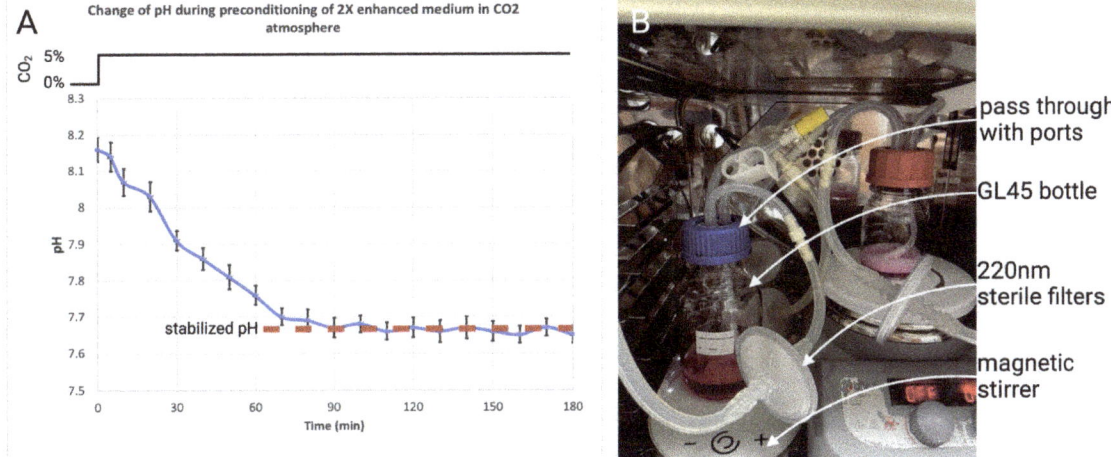

Figure 12. (**A**) Change in pH during preconditioning in CO_2 atmosphere; (**B**) setup for preconditioning of sterile culture medium.

4.8. Preparation of Printable Bioinks

In this study, we have formulated an optimal mixing ratio of 2:1:1 (collagen hydrogel:neutralization medium:cell suspension medium). This ratio was chosen to achieve a high concentration of the resulting collagen, the ability to control the pH of the resulting bioink using the neutralization step, and the ability to prepare cell suspensions with densities lower than in our previous work [23].

The preparation of the printable bioink is performed in two steps (two-step neutralization). In the first step, collagen gel dissolved in acetic acid (AA) is mixed with a neutralization medium. This neutralization medium is based on 2× enhanced culture media and can be mixed with the same medium with added NaOH. This step is crucial to neutralize acidic collagen to a neutral pH of around 7. In the second step, the cell suspension in the 2× enhanced medium is mixed into the pre-neutralized solution, which transforms into the final bioink. Throughout the process, the syringes containing all components are cooled to 4 °C to prevent premature gelation. These procedures are schematically shown in Figure 13.

To test the pH trend in the bioink, three different concentrations of 1 M NaOH were added to the 2× enhanced medium to form the neutralization medium: 0, 50, and 100 µL/mL. Four groups of samples were prepared, varying in collagen concentration and solvent concentration: 20 mg/mL and 0.1% AA, 30 mg/mL and 0.1% AA, 20 mg/mL and 0.05% AA, and finally 30 mg/mL and 0.05% AA. During the preparation of the bioinks, the pH of all components was measured colorimetrically. These initial tests to determine the optimal pH were performed without cell addition.

We have also tested the possibility of mixing neutralization and cell suspension media and then mixing them with collagen. There were similar results with the resulting pH of the bioink; however, cell viability was decreased as a result of the high pH in the culture medium, and the large difference was decreased. This procedure, called 'single-step neutralization', is illustrated in Figure 14 and was not used in the preparation process.

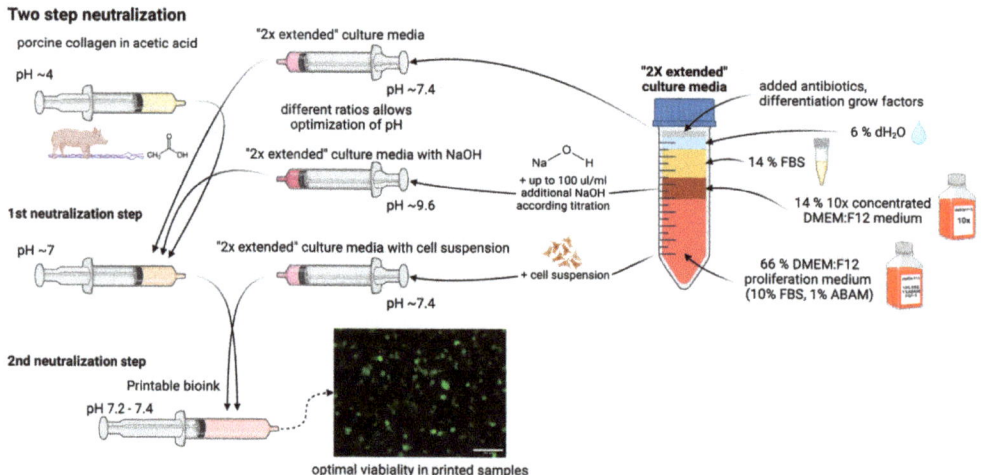

Figure 13. Schematic process of preparation of "2× enhanced" culture medium with its modification and two-step neutralization mixing procedure for preparation printable bioink.

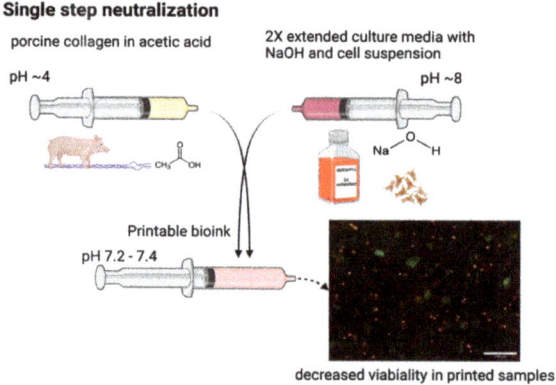

Figure 14. The schematic process of single-step neutralization provides poor results in cell viability.

4.9. Colorimetric pH Evaluation

The estimation of pH is crucial in the collagen neutralization process. First, it influences cell viability, and second, it affects printing properties [23]. Estimation with a conventional pH probe or touch pH probe has its limitations, which involve a long settling time, unsterile measurements, and a large sample volume. However, culture media and our prepared collagen hydrogels also contain phenol red as a pH indicator, allowing colorimetric estimation. For this purpose, we have created a unique cooled syringe holder equipped with a colorimetric sensor an EZO-RGB™ Embedded Color Sensor (Atlas Scientific, NY, USA). This colorimetric sensor provides output in the standardized CIE 1931 color space with outputs of x, y, and Y, where x and y are color space coordinates and Y is luminance. The white-point calibration of the colorimeter was set to the standardized D65 illuminant. This white point estimation was performed on the white and gray card, and then a syringe with white silicone was used. Calibration of pH was performed for both culture mediums (control, extended) and both collagen hydrogels. The standard flat-head touch electrode was used as a reference (THETA 90, Prague, Czech Republic).

The most important factor in determining pH was the CIE-y value, which changes in the pH range from 6.2 to 8.1. The measured data were fitted using a parabolic curve, and these values were also placed on a section of the CIE diagram. The measured CIE-x and Y values were affected by the type of media and hydrogel; however, they were not affected by the pH. These processing steps are illustrated in Figure 15.

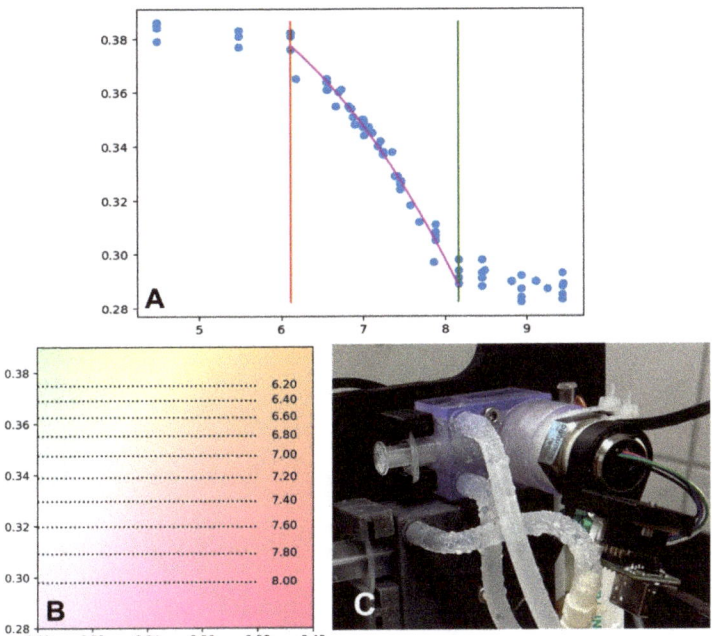

Figure 15. (**A**) calibrating values for estimation of pH from CIE-y coordinate shown by blue dots, red and green line define the measurable pH response of phenol red (**B**) pH scale over a section of CIE 1931 diagram, (**C**) colorimeter in mixing syringe-cooled block.

A major advantage of this approach is the fast response and measurement over the syringe wall, which ensures sterility. Based on fitted data and reverse evaluation, the resolution of the setup is better than 0.1 pH. The following measurements were made with this colorimetric setup.

4.10. Semiautomated System for Bioink Preparation

As mentioned above, the preparation of printable collagen bioink utilizes two steps: collagen hydrogel is first neutralized to a pH of around 7, and then cell suspension is added. In these procedures, the collagen hydrogel needs to be precisely dosed, and then a 2× enhanced culture medium with optimized pH is added. These components are dosed in hundreds of microliter volumes. Even small changes in volume can extensively affect the pH, printing, and gelling parameters of the prepared collagen bioink. Due to the low volumes needed and repeatability, we have constructed a custom semiautomated mixing system.

This system consists of three motorized linear actuators that hold three 10 mL syringes containing collagen, 2× enhanced culture medium, and 2× enhanced culture medium with the addition of 100 µL/mL NaOH. These three syringes are connected to a five-element, three-way stopper manifold. The upper two ports of the manifold connect the manual mixing syringe and the printhead syringe. All syringes are kept cooled below 10 °C by circulating water. The cooled mixing syringe holder also contains a colorimeter. The manifold is cleaned using sterile PBS and a connection to a vacuum.

The preparation of the bioink in this system consists of several steps. First, a neutralization medium with a given pH (added NaOH concentration) was prepared. This step consists of dosing the desired volumes of extended and NaOH-enhanced culture media from cooled syringes. In this study, we used fixed steps of NaOH concentrations (0, 25, 50, and 100). However, the system allows for proportional mixing and titration. The next steps utilize the dosing of collagen in the prepared medium, followed by manual mixing. This mixing is performed by a repeated manual push of the mixing and printing syringe plungers. A schematic of the semiautomated system with its components is illustrated in Figures 16 and 17.

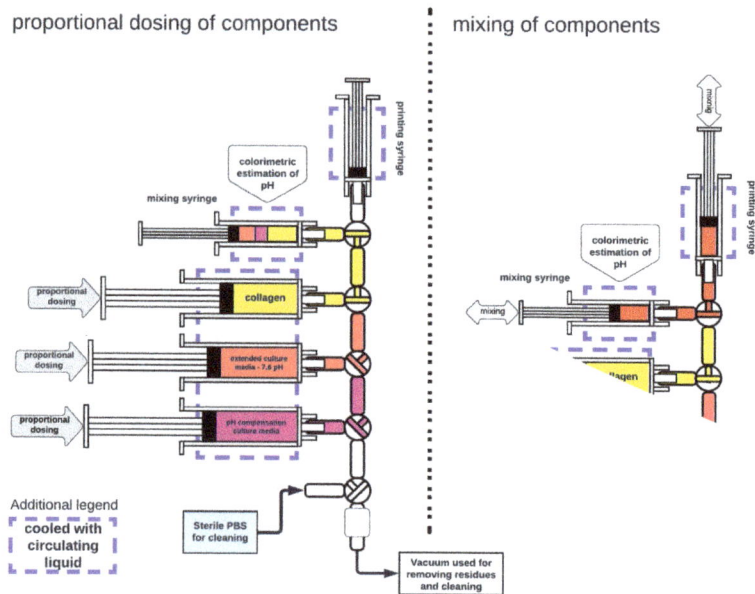

Figure 16. Schematic of the semiautomated mixing system.

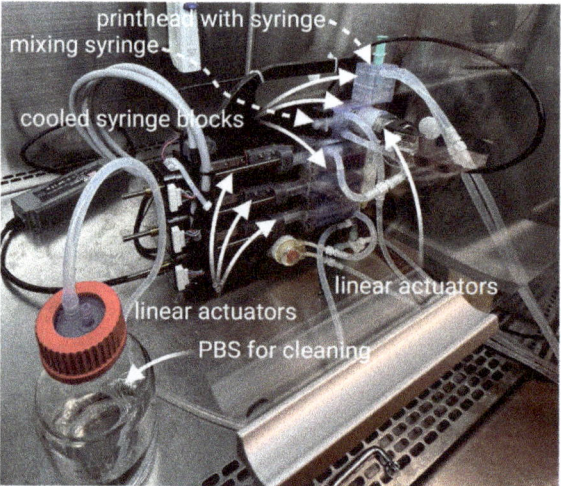

Figure 17. Mixing system installed in the biohazard box.

4.11. Bioprinting

Five groups of samples were printed, varying in collagen concentration and solvent concentration, and modified neutralization culture media: 20 mg/mL and 0.1% AA with 100 µL/mL neutralization medium, 30 mg/mL and 0.1% AA with 100 µL/mL neutralization medium, 20 mg/mL and 0.05% AA 50 µL/mL neutralization medium, 30 mg/mL and 0.05% AA 50 µL/mL neutralization medium and finally 20 mg/mL and 0.05% AA with medium protocol 2 without added NaOH in neutralization medium. The final cell density was set to 10 mil. cells/1 mL of bioink.

The prints were made in our modified bioprinter on tissue-treated coverslip glass (24 × 24 mm). The 15 × 15 × 1 mm square was set as a printing shape. Slicing was set to two layers with a G17 nozzle. The printing feed rate was set to 300 mm/min. The printing head was kept cool at <10 °C using cooled circulating water, while the print bed with the vacuum fixture was heated to 37 °C using heated circulating water. The print setup is illustrated in Figure 18.

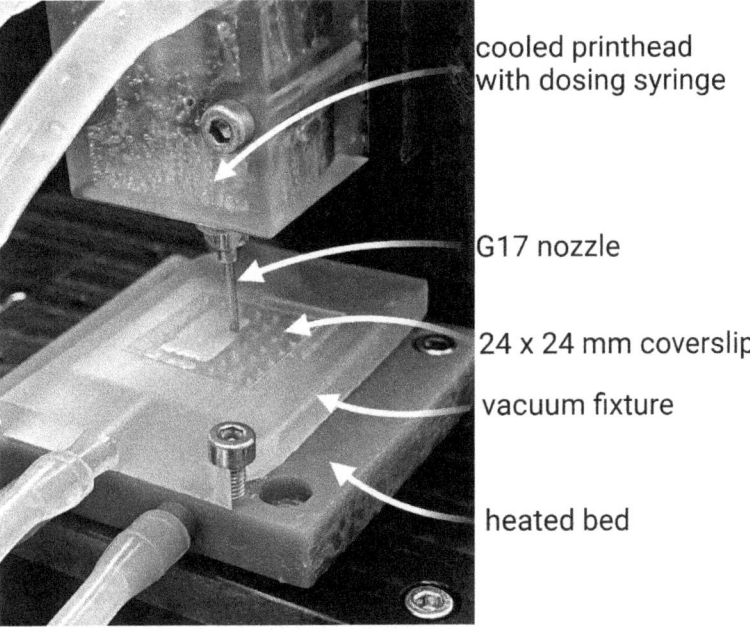

Figure 18. Bioprinting on 24 × 24 mm coverslip glass.

Printed samples were placed in six-well plates and flooded with culture media. The culture was held for 24, 48, and 96 h. Each culture interval was microscopically evaluated for cell viability and morphology.

4.12. Cell Viability Analysis

Cell viability was identified by live cell staining with fluorescein diacetate (FDA, 5 mg/mL; F1303, ThermoFisher, MA, USA) and dead cell staining with propidium iodide (PI, 2 mg/mL; P1304MP, ThermoFisher, MA, USA) after 24, 48 and 96 h of culture of printed samples. Samples were rinsed three times with PBS and stained with FDA and PI diluted in culture medium without fetal bovine serum for 5 min at 37 °C in a humidified incubator. According to the manufacturer's instructions, live and dead cells were labeled green and red. The images were then taken using a custom-built Thorlabs CERNA (Thorlabs, Bergkirchen, Germany) wide-field fluorescence microscope with Olympus Plan Fluorite

10× objectives (Olympus, Tokyo, Japan) and DCU223M CCD camera (The Imaging Source, Bremen, Germany).

4.13. Cell Morphology

The filamentous actin (F-actin) in the cell cytoskeleton was stained using Alexa Fluor™ 594 Phalloidin (Thermo-Fisher, Waltham, MA, USA, Cat. No. A12381, 200 U/mL) and cell nuclei using DAPI (Thermo-Fisher, D1306, 300 nM concentration). Staining was performed in humidified chambers for 120 min at room temperature. After staining, the samples were rinsed and stored in PBS. Confocal microscopic images were taken using a Nikon CSU-W1 inverted spinning disc confocal microscope based on the Nikon Eclipse Ti2 inverted microscope (Nikon, Tokyo, Japan) with the Yokogawa CSU-W1 spinning disc module (Yokogawa, Tokyo, Japan) and equipped with dual sCMOS PRIME BSI cameras (Teledyne Photometrics, Tucson, AZ, USA). Due to the manner of the samples, Nikon dry objectives CFI Plan Apo VC 20x were used with a 50 mm pinhole disc in Z stack mode (–0–250 mm depth). Images were then processed in Imaris 10.1.0 software (Oxford Instruments, Abingdon-on-Thames, UK).

Author Contributions: Conceptualization J.M., D.K., M.S. and R.M.; Investigation, J.M., D.K., R.M. and M.S.; Data Curation, R.M.; Software, R.M.; resources, J.M, R.M., D.K. and M.S.; writing—original draft preparation, J.M., D.K, M.S. and R.M.; writing—review and editing, J.M. and R.M.; visualization, D.K. and R.M. Funding acquisition, R.M. and M.S. All authors have read and agreed to the published version of the manuscript.

Funding: This research was funded by the Ministry of Health of the Czech Republic grant No. NV19-02-00068, and by the Grant Agency of the Czech Technical University in Prague, grant No. SGS22/201/OHK4/3T/17.

Institutional Review Board Statement: Porcine adipose-derived stromal cells used in this study were harvested during animal experimental surgeries approved by the Institutional Review Board of the Institute of Clinical and Experimental Medicine, Prague, Czech Republic (protocol code: 417; date of approval: 2 February 2021) and the Ministry of Health of the Czech Republic (protocol code: MZDR 8166/2021-4/OVZ, PID: MZDRX01EP4X5, 22/2021; date of approval: 3 March 2021).

Informed Consent Statement: Not applicable.

Data Availability Statement: All data and materials are available on request from the corresponding author. The data are not publicly available due to ongoing research using a part of the data.

Acknowledgments: We acknowledge the Imaging Methods Core Facility at BIOCEV, an institution supported by the MEYS CR (LM2023050 Czech-BioImaging), and for their support & assistance in this work. We also acknowledge the laboratory infrastructure at FBME CTU in Prague supported by MEYS CR and EU OPRDE (CZ.02.2.69/0.0/0.0/16_018/0002242 and CZ.02.1.01/0.0/0.0/16_017/0002244) and the Longterm Conceptual Development Research Organization under project no. RVO: 67985891.

Conflicts of Interest: The authors declare no conflict of interest.

References

1. Hospodiuk, M.; Dey, M.; Sosnoski, D.; Ozbolat, I.T. The bioink: A comprehensive review on bioprintable materials. *Biotechnol. Adv.* **2017**, *35*, 217–239. [CrossRef] [PubMed]
2. Abelardo, E. 7—Synthetic material bioinks. In *3D Bioprinting for Reconstructive Surgery*; Thomas, D.J., Jessop, Z.M., Whitaker, I.S., Eds.; Woodhead Publishing: Sawston, UK, 2018; pp. 137–144.
3. Delgado, L.M.; Bayon, Y.; Pandit, A.; Zeugolis, D.I. To crosslink or not to crosslink? Crosslinking associated foreign body response of collagen-based devices. *Tissue Eng. Part. B Rev.* **2015**, *21*, 298–313. [CrossRef] [PubMed]
4. Kim, A.; Lakshman, N.; Karamichos, D.; Petroll, W.M. Growth factor regulation of corneal keratocyte differentiation and migration in compressed collagen matrices. *Investig. Ophthalmol. Vis. Sci.* **2010**, *51*, 864–875. [CrossRef] [PubMed]
5. Stepanovska, J.; Supova, M.; Hanzalek, K.; Broz, A.; Matejka, R. Collagen Bioinks for Bioprinting: A Systematic Review of Hydrogel Properties, Bioprinting Parameters, Protocols, and Bioprinted Structure Characteristics. *Biomedicines* **2021**, *9*, 1137. [CrossRef] [PubMed]
6. Ashammakhi, N.; Ahadian, S.; Xu, C.; Montazerian, H.; Ko, H.; Nasiri, R.; Barros, N.; Khademhosseini, A. Bioinks and bioprinting technologies to make heterogeneous and biomimetic tissue constructs. *Mater. Today Bio* **2019**, *1*, 100008. [CrossRef]

7. Ahmad, K.; Lim, J.-H.; Lee, E.-J.; Chun, H.-J.; Ali, S.; Ahmad, S.S.; Shaikh, S.; Choi, I. Extracellular Matrix and the Production of Cultured Meat. *Foods* **2021**, *10*, 3116. [CrossRef]
8. Melchels, F.P.W.; Blokzijl, M.M.; Levato, R.; Peiffer, Q.C.; de Ruijter, M.; Hennink, W.E.; Vermonden, T.; Malda, J. Hydrogel-based reinforcement of 3D bioprinted constructs. *Biofabrication* **2016**, *8*, 035004. [CrossRef] [PubMed]
9. Gaudet, C.; Marganski, W.A.; Kim, S.; Brown, C.T.; Gunderia, V.; Dembo, M.; Wong, J.Y. Influence of type I collagen surface density on fibroblast spreading, motility, and contractility. *Biophys. J.* **2003**, *85*, 3329–3335. [CrossRef]
10. Silver, F.H.; Freeman, J.W.; Seehra, G.P. Collagen self-assembly and the development of tendon mechanical properties. *J. Biomech.* **2003**, *36*, 1529–1553. [CrossRef]
11. Fratzl, P. *Collagen: Structure and Mechanics*; Springer: New York, NY, USA, 2008.
12. Christiansen, D.L.; Huang, E.K.; Silver, F.H. Assembly of type I collagen: Fusion of fibril subunits and the influence of fibril diameter on mechanical properties. *Matrix Biol.* **2000**, *19*, 409–420. [CrossRef]
13. Wood, G.C. The formation of fibrils from collagen solutions. 2. A mechanism for collagen-fibril formation. *Biochem. J.* **1960**, *75*, 598–605. [CrossRef]
14. Matinong, A.M.E.; Chisti, Y.; Pickering, K.L.; Haverkamp, R.G. Collagen Extraction from Animal Skin. *Biology* **2022**, *11*, 905. [CrossRef] [PubMed]
15. Gobeaux, F.; Mosser, G.; Anglo, A.; Panine, P.; Davidson, P.; Giraud-Guille, M.M.; Belamie, E. Fibrillogenesis in Dense Collagen Solutions: A Physicochemical Study. *J. Mol. Biol.* **2008**, *376*, 1509–1522. [CrossRef]
16. Forgacs, G.; Newman, S.A.; Hinner, B.; Maier, C.W.; Sackmann, E. Assembly of Collagen Matrices as a Phase Transition Revealed by Structural and Rheologic Studies. *Biophys. J.* **2003**, *84*, 1272–1280. [CrossRef] [PubMed]
17. Zhang, X.; Xu, S.; Shen, L.; Li, G. Factors affecting thermal stability of collagen from the aspects of extraction, processing and modification. *J. Leather Sci. Eng.* **2020**, *2*, 19. [CrossRef]
18. Wood, G.C.; Keech, M.K. The formation of fibrils from collagen solutions 1. The effect of experimental conditions: Kinetic and electron-microscope studies. *Biochem. J.* **1960**, *75*, 588–598. [CrossRef] [PubMed]
19. Kar, K.; Amin, P.; Bryan, M.A.; Persikov, A.V.; Mohs, A.; Wang, Y.-H.; Brodsky, B. Self-association of Collagen Triple Helic Peptides into Higher Order Structures. *J. Biol. Chem.* **2006**, *281*, 33283–33290. [CrossRef]
20. Davison, P.F.; Cannon, D.J.; Andersson, L.P. The Effects of Acetic Acid on Collagen Crosslinks. *Connect. Tissue Res.* **1972**, *1*, 205–216. [CrossRef]
21. Adamiak, K.; Sionkowska, A. Current methods of collagen crosslinking: Review. *Int. J. Biol. Macromol.* **2020**, *161*, 550–560. [CrossRef]
22. Khor, E. Methods for the treatment of collagenous tissues for bioprostheses. *Biomaterials* **1997**, *18*, 95–105. [CrossRef]
23. Stepanovska, J.; Otahal, M.; Hanzalek, K.; Supova, M.; Matejka, R. pH Modification of High-Concentrated Collagen Bioinks as a Factor Affecting Cell Viability, Mechanical Properties, and Printability. *Gels* **2021**, *7*, 252. [CrossRef]
24. Jackson, M.; Choo, L.P.; Watson, P.H.; Halliday, W.C.; Mantsch, H.H. Beware of connective tissue proteins: Assignment and implications of collagen absorptions in infrared spectra of human tissues. *Biochim. Biophys. Acta* **1995**, *1270*, 1–6. [CrossRef] [PubMed]
25. Rabotyagova, O.S.; Cebe, P.; Kaplan, D.L. Collagen Structural Hierarchy and Susceptibility to Degradation by Ultraviolet Radiation. *Mater. Sci. Eng. C Mater. Biol. Appl.* **2008**, *28*, 1420–1429. [CrossRef] [PubMed]
26. Levraut, J.; Labib, Y.; Chave, S.; Payan, P.; Raucoules-Aime, M.; Grimaud, D. Effect of sodium bicarbonate on intracellular pH under different buffering conditions. *Kidney Int.* **1996**, *49*, 1262–1267. [CrossRef] [PubMed]
27. Breslin, S.; O'Driscoll, L. Three-dimensional cell culture: The missing link in drug discovery. *Drug Discov. Today* **2013**, *18*, 240–249. [CrossRef] [PubMed]
28. Farjanel, J.; Schürmann, G.; Bruckner, P. Contacts with fibrils containing collagen I, but not collagens II, IX, and XI, can destabilize the cartilage phenotype of chondrocytes. *Osteoarthr. Cartil.* **2001**, *9* (Suppl. S1), S55–S63. [CrossRef]
29. Mayne, R.; Vail, M.S.; Mayne, P.M.; Miller, E.J. Changes in type of collagen synthesized as clones of chick chondrocytes grow and eventually lose division capacity. *Proc. Natl. Acad. Sci. USA* **1976**, *73*, 1674–1678. [CrossRef]
30. Matějka, R.; Koňařík, M.; Štěpanovská, J.; Lipenský, J.; Chlupáč, J.; Turek, D.; Pražák, Š.; Brož, A.; Šimůnková, Z.; Mrázová, I.; et al. Bioreactor Processed Stromal Cell Seeding and Cultivation on Decellularized Pericardium Patches for Cardiovascular Use. *Appl. Sci.* **2020**, *10*, 5473. [CrossRef]
31. Michl, J.; Park, K.C.; Swietach, P. Evidence-based guidelines for controlling pH in mammalian live-cell culture systems. *Commun. Biol.* **2019**, *2*, 144. [CrossRef]

Disclaimer/Publisher's Note: The statements, opinions and data contained in all publications are solely those of the individual author(s) and contributor(s) and not of MDPI and/or the editor(s). MDPI and/or the editor(s) disclaim responsibility for any injury to people or property resulting from any ideas, methods, instructions or products referred to in the content.

Article

Alginate–Gelatin Hydrogel Scaffolds; An Optimization of Post-Printing Treatment for Enhanced Degradation and Swelling Behavior

Christina Kaliampakou [1], Nefeli Lagopati [2,3,*], Evangelia A. Pavlatou [4] and Costas A. Charitidis [1,*]

1. RNanoLab, Research Unit of Advanced, Composite, Nano Materials & Nanotechnology, School of Chemical Engineering, Zografos Campus, National Technical University of Athens, 9 Heroon, Polytechniou St., 15780 Athens, Greece; kaliabakou@chemeng.ntua.gr
2. Laboratory of Biology, Department of Basic Medical Sciences, Medical School, National and Kapodistrian University of Athens, 11527 Athens, Greece
3. Biomedical Research Foundation, Academy of Athens, 11527 Athens, Greece
4. Laboratory of General Chemistry, School of Chemical Engineering, National Technical University of Athens, Zografou Campus, 15772 Athens, Greece; pavlatou@chemeng.ntua.gr
* Correspondence: nlagopati@med.uoa.gr (N.L.); charitidis@chemeng.ntua.gr (C.A.C.); Tel.: +30-210-7462362 (N.L.); +30-210-7724046 (C.A.C.)

Citation: Kaliampakou, C.; Lagopati, N.; Pavlatou, E.A.; Charitidis, C.A. Alginate–Gelatin Hydrogel Scaffolds; An Optimization of Post-Printing Treatment for Enhanced Degradation and Swelling Behavior. *Gels* 2023, 9, 857. https://doi.org/10.3390/gels9110857

Academic Editors: Enrique Aguilar and Helena Herrada-Manchón

Received: 1 October 2023
Revised: 26 October 2023
Accepted: 26 October 2023
Published: 28 October 2023

Copyright: © 2023 by the authors. Licensee MDPI, Basel, Switzerland. This article is an open access article distributed under the terms and conditions of the Creative Commons Attribution (CC BY) license (https://creativecommons.org/licenses/by/4.0/).

Abstract: The generation of 3D structures comprises three interlinked phases: material development, the printing process, and post-printing treatment. Numerous factors control all three phases, making the optimization of the entire process a challenging task. Until now, the state of the art has mainly focused on optimizing material processability and calibration of the printing process. However, after the successful Direct Ink Writing (DIW) of a hydrogel scaffold, the post-printing stage holds equal importance, as this allows for the treatment of the structure to ensure the preservation of its structural integrity for a duration that is sufficient to enable successful cell attachment and proliferation before undergoing degradation. Despite this stage's pivotal role, there is a lack of extensive literature covering its optimization. By studying the crosslinking factors and leveling the post-treatment settings of alginate–gelatin hydrogel, this study proposes a method to enhance scaffolds' degradation without compromising the targeted swelling behavior. It introduces an experimental design implementing the Response Surface Methodology (RSM) Design of Experiments (DoE), which elucidated the key parameters influencing scaffold degradation and swelling, and established an alginate ratio of 8% and being immersed for 15 min in 0.248 M $CaCl_2$ as the optimal level configuration that generates a solution of 0.964 desirability, reaching a degradation time of 19.654 days and the swelling ratio of 50.00%.

Keywords: optimization DoE; post-printing treatment; scaffolds; degradation; swelling

1. Introduction

Direct Ink Writing (DIW) is an Additive Manufacturing (AM) method that enables the creation of complex 3D structures with intricate designs and various material compositions for biomedical applications [1]. The rapid progress in the development of innovative biomaterials that are processable for extrusion-based AM methods like DIW allows for strong biological interactions, eventually facilitating the cultivation of 3D artificial tissues [2]. In DIW, a viscoelastic ink is extruded through a deposition nozzle, layer by layer, to construct scaffolds [3]. Unlike other manufacturing methods, DIW-based scaffold development allows for the precise and controlled placement of biomaterials, highlighting its potential as a means to produce innovative and adaptable grafts which are suitable for various types of tissues [4]. The cost-effectiveness, ease of use, and capability to blend multiple materials in a single manufacturing step have gathered significant interest from various research teams worldwide, leading to extensive advancements in this cutting-edge technology. The precision of DIW can address the need of Tissue Engineering (TE) for artificial tissues with embedded vascular networks and establish directed vascularization, increasing the compatibility and longevity of the printed structures [5]. Up until now, much effort has been made in achieving a high precision in geometry fidelity by optimizing materials' ink properties and the printing process. However, one of the most crucial factors in developing functional 3D systems is their ability to preserve their structural integrity for a sufficient duration, as guided by the specific requirements of the target tissue. In the case of guided vascularization, studies suggest that a three-week (21-day) cell culture is the appropriate timeframe for vessel formation [6].

This stage, which is referred to as the post-printing treatment, is mainly governed by the crosslinking process which, in turn, is determined by the ink's composition. In the present study, based on the printing approach and the need for in situ gelation, to ensure that the printed products maintain their structural integrity and stability after printing, alginate–gelatin ink was selected. This hydrogel ink possesses the desired properties for efficient use in DIW, as presented in previous results, and shows great compatibility with biological systems [7]. Alginate-based biomaterial inks have gained popularity, as alginate is a biodegradable and biocompatible polysaccharide derived from brown algae, and is capable of forming a gel through ionic crosslinking [8]. It has been employed for printing vascular tissue, bone, and cartilage-like structures. Gelatin is another well-researched biomaterial used as ink for DIW. Gelatin provides arginine–glycine–aspartic acid (RGD) cell adhesion motifs in alginate–gelatin composite hydrogel, imparting bioactivity and enhancing cell adhesion capabilities. In addition, these oligopeptide sequences containing RGD peptide also result in more favorable scaffold degradation compared to pure alginate [9,10]. Thus, although alginate lacks the desired bioactive properties and gelatin has limitations in terms of mechanical strength, the right combinations of these materials can lead to biomaterial inks that are suitable for scaffolds that promote directed vascularization [11].

Post-printing treatment is necessary for both alginate and gelatin to acquire good mechanical properties and stability. First, by cooling down the printed structures (20 °C), the physical crosslinking of the temperature-dependent gelatin is activated. Then, these pre-crosslinked samples are soaked in Ca^+ solution. The interaction between multivalent cations and the glucuronic blocks in alginate leads to ionic crosslinking of the alginate particles, resulting in the formation of a 3D network and enhancing the mechanical properties of the created scaffold [12]. The crosslinking of hydrogels has the effect of preserving their shape even though they can absorb significant amounts of fluids and swell. Thus, a necessary equilibrium must be established between the degradation rate and the ratio of swelling when introducing a scaffold in in vitro and in vivo applications [13].

However, when studying the post-printing treatment, and aiming to decrease the degradation rate, the optimization process should take into consideration that the scaffold's functionality might be decreased due to four different reasons related to the handling of the scaffold following the printing. Firstly, when sodium alginate-based hydrogels, crosslinked with calcium ions, are exposed to physiological conditions, a process occurs where divalent calcium ions are gradually replaced by monovalent sodium ions present in the degradation medium [13,14]. Secondly, when alginate–gelatin scaffolds are incubated in culture media, temperature-dependent gelatin is gradually dissolved and released from the scaffold due to high temperature (37 °C), leading to hypoxia affecting cell viability. However, the inclusion of crosslinked sodium alginate matrices, as well as the interaction between alginate and gelatin, enhances the thermal stability of gelatin and slows down the release of gelatin from the printed scaffolds when maintained at 37 °C [15]. Furthermore, densely arranged hydrogels could pose a substantial obstacle to the diffusion of various substances (such as protein molecules, gases, growth factors, and metabolic waste) between the enclosed cells and the surrounding culture medium. This confinement of cells could lead to diminished cell viability. For this reason, the swelling capacity of the scaffolds is highly important [16]. Last, but not least, after implantation, a distinct reduction in size can lead to detachment, and an uncontrolled increase in the substrate could result in severe inflammation or patient discomfort. The shaping process can be accurately managed through specific parameters during fabrication. Therefore, it is of high importance to address the deformation that occurs after the treatment [17].

Thus, through the examination of the factors related to crosslinking, an approach that improves the degradation of scaffolds while maintaining their functionality needs to be developed. In the existing literature, when aiming to decrease the scaffold's degradation rate in order to reach the desired time interval without losing its structural integrity, emphasis is mainly given to studying the parameters of the crosslinking stage (alginate ratio, crosslinker concentration, immersion time) in correlation with other scaffold properties, such as its stiffness and mechanical properties [18]. For instance, Naghieh at al. showed that both the duration of the immersion and the concentration of the crosslinker employed are critical factors that significantly influence the mechanical properties in 3D bioplotted alginate scaffolds [19]. Bahrami et al. demonstrated that the 4% alginate scaffold experienced substantial weight loss in various solutions, and its dissolution rate was significantly greater than that of the 8% and 16% alginate scaffolds in all of the tested solutions, while 16% alginate scaffold, due to its density, presented a very low degradation rate [20]. Furthermore, Sonaye et al. showed that increasing the concentration of crosslinking and the amount of alginate in the printed scaffolds improved the swelling capacity and reduced the degradation rate [21]. Specifically, they identified that an optimal crosslinking concentration of 500 mM $CaCl_2$ and an alginate content of 12% (w/v) resulted in scaffolds with high swelling (70%) and low degradation rates (28%). This research team established a connection between crosslinking parameters and scaffold characteristics, such as stiffness, swelling capacity, and degradation rate, to create scaffolds suitable for durable skeletal muscle tissue constructs in tissue engineering applications. Finally, Lei et al. suggested treating alginate/gelatin scaffolds with $CaCl_2$ and $CaCl_2$–EDC solutions when targeting minimal swelling (50%) throughout the post-treatment period. Their findings confirm that the deformation occurring after treatment can be controlled through a crosslinking process [17].

In addition, until now, the state of the art, including our team's previous work [22], in Direct Ink Writing (DIW) has primarily relied on the Design of Experiments (DoE) to optimize the initial two stages, which encompass the material processability and printing process parameters. There is also a limited number of studies optimizing some scaffold properties, like porosity, based on material composition [23]. Notably, the existing literature lacks any evidence of a similar methodology applied to the subsequent post-printing stage. However, it is imperative to extend the application of the DoE to optimize the post-printing treatment phase, given that it is influenced by numerous factors which are often overlooked (UV treatment, culture media), resulting in inefficient resource utilization through trial-and-error approaches [24]. Furthermore, the implementation of an experimental design becomes crucial to identify the most functional compromise within the 3D system, focusing on key responses such as the degradation behavior and swelling ratio. Such a design would not only illuminate the interactions among the post-printing treatment factors but also expedite the development of a predictive model for enhanced efficiency.

The primary objective of this research is to bridge this gap and offer insights into optimizing the post-printing treatment of alginate–gelatin hydrogel, building upon our team's prior published research [22], which focused on optimizing ink processability and the printing process. Specifically, this study seeks to improve the degradation characteristics of scaffolds while maintaining their swelling capacity. To address these objectives, an experimental design is introduced that employs Statistical Response Surface Methodology (RSM DoE). This approach aims to systematically uncover the key parameters that govern post-printing treatment and establish their optimal levels. In this context, we developed alginate–gelatin scaffolds through DIW to explore and optimize the effect of crosslinking on their biodegradability and deformation. The ink formulation and printing parameters were adjusted based on the results of the previous work through the corresponding three-step optimization method [22]. The combination of the levels that we judged to be optimal for each parameter has resulted in the printing of rectangular scaffolds with a high geometry fidelity. Following, via RSM DoE, we studied which post-printing factors play the most crucial role in scaffolds' functionality and, after defining the optimal levels, we investigated the properties of the scaffolds that were treated with the conditions identified as optimal by the experimental design. Thus, the current paper presents an innovative approach that addresses the critical yet underexplored stage of scaffold post-printing treatment. By implementing a novel approach that utilizes RSM design, this study concurrently optimizes two vital scaffold properties: degradation and swelling behavior. These responses are inherently challenging to co-currently optimize, as maximizing one often leads to a decrease in the other and vice versa. Reaching the targeted values for both responses would improve the total functionality of alginate–gelatin scaffolds.

2. Results and Discussion
2.1. Scaffolds Development

After the development of the three distinct alginate–gelatin ink configurations (4/4 (%), 6/4(%) and 8/4 (%)), the successful Direct Ink Writing of the scaffolds for use in the suggested screening tests was subsequently undertaken. The four samples were printed in triplicate and each of them was then methodically treated in accordance with the specified conditions, as determined by the selected leveling of the post-printing factors, in order to record the variance in the degradation time and swelling ratio results. For the assessment of the scaffolds' shape fidelity and structural integrity, precise measurements were taken, including their dimensions and the diameter of their individual strands, via ImageJ (National Institutes of Health, Bethesda, MD, USA) software. The term 'material strand' refers to the diameter of the deposited material. In this context, it is compared to the theoretical value, which is defined by the tip's diameter (0.41 mm in this case). This comparison serves to define the printability of the ink based on the following equation (Equation (1)):

$$\text{Strand Printability}: 1 - \frac{Ds - Dexp}{Ds} \quad (1)$$

where D_s is the theoretical strand diameter and D_{exp} is the experimental strand diameter [22].

Furthermore, their dry weight was determined and recorded for further use in the deformation analysis. The results for Samples 1–4 are presented in Table 1, providing a comprehensive overview of the screening scaffolds' characteristics. An ideal printability value of 1 signifies a perfect match between the experimental strand and the theoretical model, demonstrating the precise deposition of material. The results presented in Table 1 reveal a printability trend for samples with a fixed gelatin ratio of 4% and varying alginate content within the 4–8% range. Specifically, it becomes evident that, as the alginate ratio increases, as shown by Sample 4 with 8% alginate, printability significantly improves. This observation is consistent with prior findings in the field [22,25]. As presented, higher printability values correlate with a reduced scaffold mass and smaller dimensions, indicating a more efficient printing process. This is attributed to the deposition of less material in each layer, resulting in superior shape fidelity.

Table 1. Scaffolds' properties for the screening tests.

Samples	Length (mm)	Width (mm)	Height (mm)	Strand (mm)	Mass (gr)	Printability
Sample 1	18.121 ± 0.037	18.332±	4.061 ± 0.020	0.493 ± 0.003	0.919 ± 0.023	1.202 ± 0.035
Sample 2	17.793 ± 0.023	18.015±	3.465 ± 0.022	0.484 ± 0.002	0.860 ± 0.012	1.180 ± 0.019
Sample 3	17.924 ± 0.031	17.777±	3.791 ± 0.019	0.489 ± 0.002	0.854 ± 0.007	1.192 ± 0.017
Sample 4	16.767 ± 0.015	16.597±	3.374 ± 0.011	0.471 ± 0.001	0.693 ± 0.009	1.148 ± 0.011

2.2. Screening Tests Characterizations Results

2.2.1. Degradation Behavior

Figure 1 displays the FT-IR (Fourier-transform infrared spectroscopy) spectra of the pellets that were collected after the centrifugation of the media in which the screening scaffolds (Samples 1–4) were immersed for 7 days. As confirmed in Figure 1, the amino group in gelatin and the carboxyl acid group in alginate exhibited electrostatic attraction, leading to the formation of the alginate–gelatin network backbone [26]. The small absorption peaks at 3263 cm^{-1} are characteristic of sodium alginate for –OH stretching [27]. However, with regard to this absorption band, concerning the stretching vibration of the N–H group bonded to O–H group, the increase in intensity is reported to be characteristic of an increase in intramolecular bonding attributed to $CaCl_2$ crosslinking [17,28]. Another small peak at 2922 cm^{-1} which is partially overlapped with the peak for –OH stretching is assigned to N–H stretching from gelatin or C–H stretching from alginate [29]. The distinctive peak at 1643 cm^{-1} is associated with the stretching vibrations of the asymmetric and symmetric -COO- groups, as found in sodium alginate infrared spectra (1635 cm^{-1}) peaks at 1640 cm^{-1} for C=O, from gelatin [27]. Another characteristic peak of alginate at 560 cm^{-1} clearly indicates its dissolution in the culture media. The absorption peaks at 1028 cm^{-1} are strongly indicative of dissolute gelatin (Amide III) [26] and have coupled with stretching vibrations of C-O from pure sodium alginate, which is recorded at 1043 cm^{-1} [29]. Moreover, the strong peaks exhibited at 3363 and 1643 cm^{-1}, as well as the wide and small peak at 2089 cm^{-1}, are characteristic peaks of the water spectrum. According to the obtained spectra, it can be concluded that Sample 1, with strong characteristic peaks of alginate and gelatin, was totally dissolved in the culture medium by Day 7, demonstrating a high degradation rate. Samples 2 and 3 exhibited peaks that were smaller but also indicative of alginate and gelatin, leading to the conclusion that those samples demonstrated a medium degradation rate. Finally, the spectrum of Sample 4 is characterized by water-indicative peaks, which means that no dissolute alginate or gelatin was detected in the culture medium. Thus, Sample 4 exhibits a small degradation rate, verifying that the selected levels for the optimization DoE indicate a sufficient range of degradation rates.

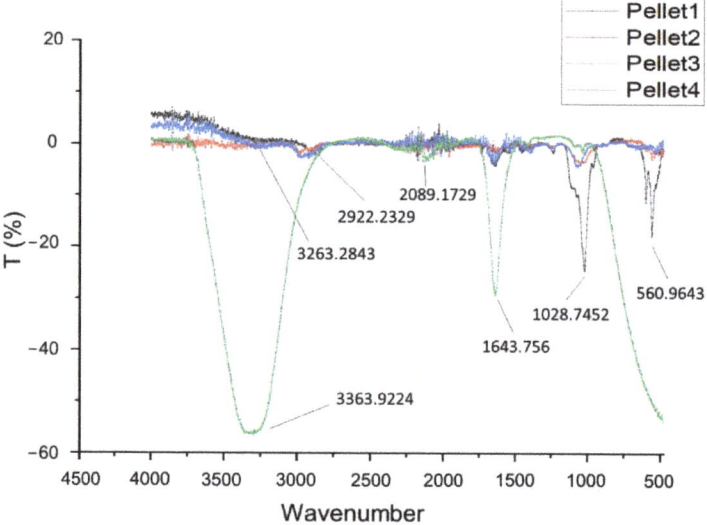

Figure 1. FT-IR spectra (4500–450 cm^{-1}) of centrifuged culture media for the four screening samples.

These findings are also confirmed in Figure 2, where there is evidence that the configuration of low levels led to rapid degradation until Day 2, whereas the configuration of high levels for the relevant factors enabled the scaffold to maintain its structure until Day 23. In addition, the observed slight deviation between the degradation times of Sample 2 and Sample 3 can be attributed to a single differentiating factor, which is the UV exposure, while all other post-printing parameters remained consistent in their handling. Sample 3, which did not undergo UV treatment, displayed a slightly faster degradation rate compared to Sample 2. These findings underline the significance of considering UV exposure as a post-printing treatment factor in the DoE framework.

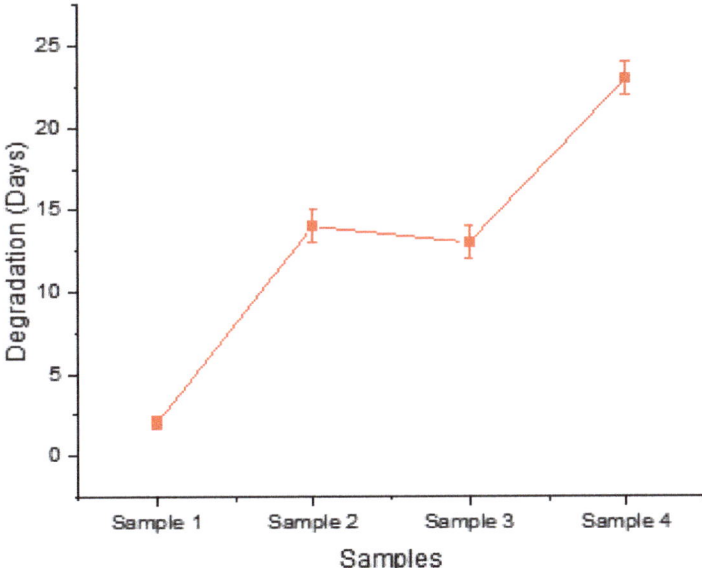

Figure 2. Time needed for each screening sample to maintain its structural integrity before dissolving in culture medium.

2.2.2. Swelling Ratio

The scaffolds' swelling behavior first reached a high absorbency and subsequently showed a decline in the swelling percentage. This constitutes a typical phenomenon for an ionically crosslinked hydrogel. When the ionic strength of the swelling medium is raised, the swelling capacity of ionic hydrogels decreases, resulting in poor anion–anion electrostatic repulsion and a decreased osmotic pressure [30]. In line with this, the plateau of the scaffolds' maximum swelling was observed up to Day 3, regardless of the crosslinker's concentration or the immersion time. As is evident from Figure 3a, displaying the combination of the factors, lower levels led to greater media absorption, gradually increasing the scaffolds mass. A lower alginate ratio, combined with less immersion time and a lower crosslinker concentration, lead to a limited degree of crosslinking which, in turn, resulted in a less dense hydrogel network that is able to absorb water, as recorded in the mass diagram of Sample 1 (Figure 3a), which exhibits the highest swelling ratio. On the other hand, Sample 4 represents a configuration characterized by the high levels of these factors, resulting in extensive crosslinking of the alginate blocks. This higher-level combination scaffold (Sample 4) displayed minimal deformation due to its density, which hindered media absorbance. By Day 3, Samples 1–4 displayed maximum swelling ratios of 96.543%, 79.833%, 57.120%, and 12.090%, respectively, as presented in Figure 3b,c. The value range covers the targeted swelling rate of 50.00% and is evaluated as suitable for the process optimization. The fluctuation observed in the scaffolds' mass recording is partly attributed to temperature, which results in gelatin dissolution, and partly to the change in culture medium at Day 2. After the change in the culture medium, the process of monovalent sodium ions replacing divalent calcium ions reoccurs, leading to decreased crosslinking degree and increased swelling ability until the reestablishment of the ion equilibrium [14].

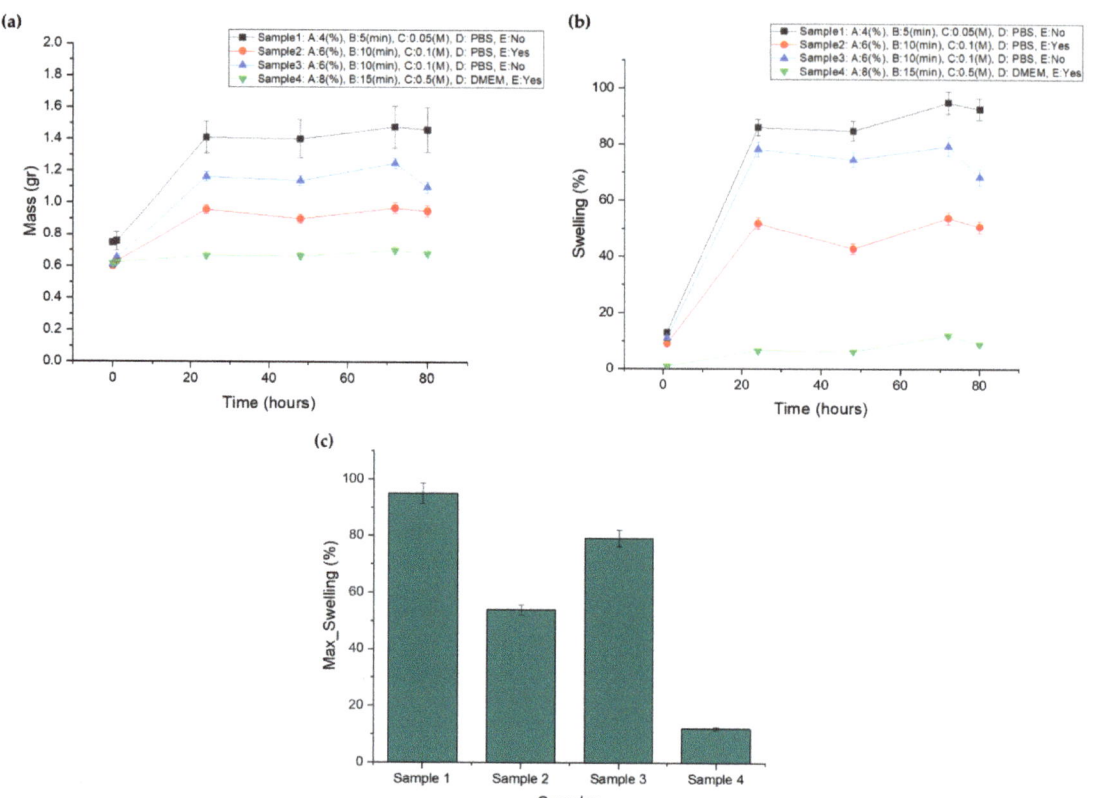

Figure 3. (a) Screening samples' mass plot after immersion in culture media, (b) screening samples' swelling curve after immersion in culture media, (c) maximum swelling ratio for each screening sample.

2.3. Design of Experiments

2.3.1. Model Summary and Regression Analysis for Response 1: Degradation Time and Response 2: Swelling Ratio

The structure of the RSM's robust design and the degradation and swelling results for each sample are presented in Table 2. The set of experiments was conducted two times and the mean value was obtained for both responses. The results acquired from the run of 34 experimental sets, suggested by the I-optimal type of RSM model, were analyzed by the Design Expert© 11.0 (Stat-Ease, Minneapolis, MN, USA) software. For the degradation time results, the day that the scaffolds were dissolved in the media, losing their original structure, was recorded while, for the swelling ratio results, the maximum swelling ratio of each run was recorded. As shown in Table 2, a maximum degradation time of 23 days was achieved with runs 24 and 29. The maximum swelling ratio of the hydrogel scaffolds was 120.3% (run 19). However, the targeted swelling ratio is 50.00%. Swelling ratios of 52.5% and 49.9% were observed in runs 9 and 27 respectively.

Table 2. RSM Design of Experiments structure.

Run	Factor 1 A: Alginate %	Factor 2 B: Time (CrossLinking) min	Factor 3 C: Concetration (Crosslinker) M	Factor 4 D: Media [1] -	Factor 5 E: UV -	Response 1 Degradation Days	Response 2 Swelling %
1	8	10	0.5	PBS	No	16.5	16.749
2	6	10	0.5	DMEM	No	16	9.4675
3	8	15	0.05	PBS	No	7	112.544
4	4	15	0.5	DMEM	Yes	13	0.000
5	8	5	0.05	DMEM	No	7	97.262
6	6	10	0.1	DMEM	No	11	80.254
7	6	10	0.1	DMEM	No	10.5	85.993
8	6	15	0.05	DMEM	Yes	10	100.873
9	4	5	0.5	PBS	Yes	10	52.548
10	6	15	0.1	PBS	No	3.5	110.401
11	4	10	0.1	DMEM	Yes	7	39.036
12	4	10	0.5	DMEM	Yes	12	13.559
13	4	15	0.5	PBS	No	13	3.934
14	8	5	0.5	DMEM	Yes	16	23.195
15	4	5	0.05	DMEM	Yes	6	89.373
16	6	10	0.05	PBS	Yes	4.5	95.019
17	8	15	0.1	DMEM	Yes	15	82.043
18	4	10	0.05	PBS	No	2	88.401
19	6	10	0.05	PBS	Yes	4	120.275
20	4	5	0.5	DMEM	No	11	39.992
21	6	10	0.5	PBS	Yes	14.5	2.579
22	6	10	0.1	DMEM	No	11	91.872
23	6	5	0.5	PBS	No	13	25.956
24	8	15	0.5	DMEM	No	23	3.008
25	6	5	0.1	PBS	No	5	67.782
26	6	15	0.1	PBS	No	5.5	95.145
27	4	15	0.1	PBS	Yes	4	49.895
28	4	15	0.05	DMEM	No	9.5	71.652
29	8	15	0.5	PBS	Yes	23	12.945
30	8	10	0.1	PBS	No	6.5	105.106
31	6	15	0.05	DMEM	Yes	10.5	95.814
32	6	10	0.1	DMEM	Yes	10.5	92.279
33	4	10	0.1	DMEM	Yes	8.5	43.810
34	8	5	0.1	PBS	Yes	8	111.501

[1] **PBS**: Phosphate buffered saline, **DMEM**: Dulbecco Modified Eagle Medium.

To evaluate whether the design is well-balanced and study the possibility of multi-collinearity, the Standard Error, the Variance Inflations Factor, and the Ri^2 values were obtained and assessed, as displayed in Table 3. As shown, the similarity and small values of standard errors indicate a balanced design and precise measurements, while VIF values of about 1 suggest no presence of multicollinearity. In addition, an Ri^2 close to zero indicates that the terms included in the model are not interrelated, a phenomenon which could lead to suboptimal model performance [22,23].

To assess which terms and interactions among the variables should be included in the model, two-way non-parametric Kruskal–Wallis tests (ANOVA equivalent) were run. According to the p-values (<0.05) (Table 4), the interaction terms AB, AC, AE, BC, BD, and CD will be included in the degradation analysis model, and the interaction terms AC and BC will be, respectively, included in the swelling ratio analysis model [23].

Table 3. Model terms and interactions.

Term	Standard Error *	VIF	R_i^2	Power
A	0.2374	1.12095	0.1079	98.0%
B	0.2344	1.07312	0.0681	98.1%
C	0.2290	1.43552	0.3034	99.6%
D	0.1831	1.136	0.1197	99.9%
E	0.1844	1.15616	0.1351	99.9%
AB	0.2836	1.12591	0.1118	90.9%
AC	0.2546	1.14246	0.1247	95.6%
AD	0.2421	1.14503	0.1267	97.1%
AE	0.2446	1.16822	0.1440	96.8%
BC	0.2623	1.18698	0.1575	94.5%
BD	0.2380	1.13322	0.1176	97.5%
BE	0.2398	1.14976	0.1303	97.3%
CD	0.1977	1.12462	0.1108	99.7%
CE	0.1990	1.14333	0.1254	99.7%
DE	0.1835	1.12884	0.1141	99.9%
A^2	0.4137	1.40955	0.2906	99.5%
B^2	0.4057	1.35545	0.2622	99.6%
C^2	1.07	1.42195	0.2967	42.0%

* p-value < 0.0001 is considered as significant.

With regard to the degradation analysis, the model is described as significant (p-value < 0.0001) (Table 5). A model F-value of 75.040 can also be considered as an indicator of the model's significance. The Lack of Fit F-value of 1.819 suggests that the Lack of Fit is not statistically significant when compared to the pure error and it indicates that the model is a good fit for the data (Table 5) [23]. The R-squared value of the degradation analysis model is 0.975, indicating a very strong level of correlation. The Predicted R^2 of 0.933 closely aligns with the Adjusted R^2 of 0.963, with a difference of less than 0.2 between them (Table 5) [23,31,32].

As supported by the p-values (Table 5), among the design's selected factors, Alginate ratio, Time, Concentration, and Media all have a significant effect on the scaffold's degradation behavior and should be taken into account to achieve the targeted functionality. The UV Exposure term (p-value > 0.05) could be excluded from the analysis, leading to model reduction. However, not including this term did not increase the R^2 value. In addition, in order to define the required steps for the optimal post-printing handling of the scaffolds, the UV exposure effect was decided to be further investigated and, thus, included in the degradation analysis model.

Furthermore, in order to identify any patterns and outliers in the residuals associated with different factor levels or combinations, the Residuals vs. Factor plots were studied. The horizontal axis of the plots in Figure 4 represents the factors that were identified as significant and their levels. As shown in the colored dot plots (Figure 4a–c), the samples that maintained their structural integrity for more days when immersed in the culture medium (marked as red points) were those that were treated with the maximum levels of the three statistically significant numerical factors. In addition, the degradation time was maximized when the samples were immersed in DMEM instead of PBS, as shown in Figure 4d. The fact that the residuals fall between the two horizontal red lines suggests that the residuals have a relatively uniform variability and are evenly distributed around zero, and that the patterns in the data are sufficiently captured.

Regarding the analysis of the scaffolds' swelling ratios, the statistical model shows strong significance with a very low p-value (<0.0001) (Table 6). The Lack of Fit F-value of 2.68 implies that the Lack of Fit is not significant relative to the pure error, indicating a good fit of the model [23]. The R-squared value for the swelling ratio analysis model is 0.898, demonstrating a strong level of correlation. In addition, the Predicted R^2 of 0.829 is in reasonable agreement with the Adjusted R^2 of 0.870, as the difference is less than 0.2 (Table 6) [23,31,32].

Table 4. Analysis of variance for interactions.

Source		Alginate Ratio × Time	Alginate Ratio × Concentration	Alginate Ratio × Media	Alginate Ratio × UV Exposure	Time Alginate Ratio × UV Exposure Concentration	Time Alginate Ratio × UV Exposure Media	Time Alginate Ratio × UV Exposure UV Exposure	Concentration Alginate Ratio × UV Exposure Media	Concentration Alginate Ratio × UV Exposure UV Exposure	Media Alginate Ratio × UV Exposure UV Exposure
	Terms	AB	AC	AD	AE	BC	BD	BE	CD	CE	DE
p-Value	Degradation	0.003	0.001	0.397	0.003	0.003	0.009	0.795	0.000	0.783	0.023
	Swelling	0.143	0.003	0.879	0.276	0.050	0.902	0.126	0.702	0.999	0.624

Table 5. Analysis of variance table for Response 1 and statistical summary for the linear model for the five factors affecting degradation behavior.

Source	Sum of Squares	df	Mean Square	F-Value	p-Value
Model	825.460	11	75.040	80.010	0.000
A-Alginate	124.100	1	124.100	132.300	0.000
B-Time (crosslinking)	49.550	1	49.550	52.830	0.000
C-Concentration (CaCl$_2$)	474.300	1	474.300	505.670	0.000
D-Media	74.980	1	74.980	79.940	0.000
E-UV	0.659	1	0.659	0.703	0.410
AB	11.640	1	11.640	12.410	0.001
AC	14.170	1	14.170	15.110	0.000
AE	11.680	1	11.680	12.450	0.001
BC	12.140	1	12.140	12.950	0.001
BD	7.330	1	7.330	7.820	0.010
CD	27.220	1	27.220	29.020	0.000
Residual	20.640	22	0.938		
Lack of Fit	17.090	16	1.070	1.810	0.238
Pure Error	3.540	6	0.590		
Cor Total	846.100	33			
Std. Dev.	0.968		R^2	0.975	
Mean	10.220		Adjusted R^2	0.963	
C.V.%	9.480		Predicted R^2	0.933	
			Adeq. Precision	35.370	

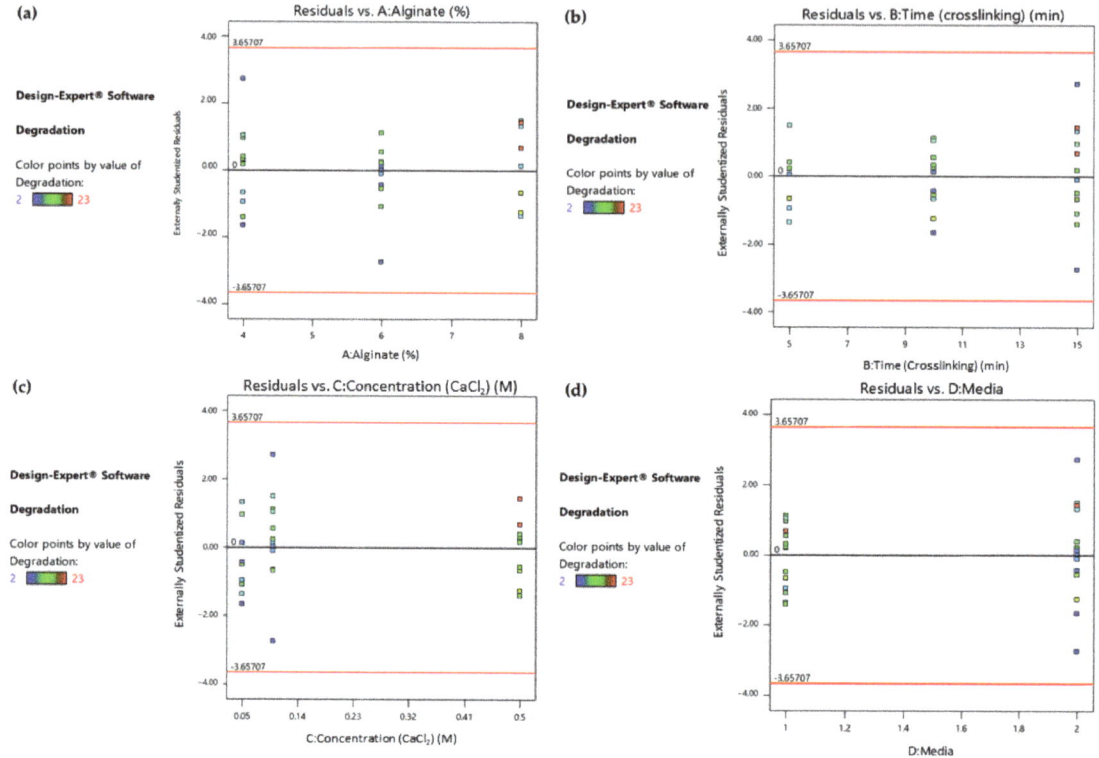

Figure 4. (**a**) Response 1: Residual vs. Alginate ratio plot, (**b**) Response 1: Residual vs. Time (crosslinking) plot, (**c**) Response 1: Residual vs. Concentration (CaCl$_2$) plot, (**d**) Response 1: Residual vs. Culture media plot.

Table 6. Analysis of variance table for Response 2 and statistical summary for the linear model for the five factors affecting swelling ratio.

Source	Sum of Squares	df	Mean Square	F-Value	*p*-Value
Model	46,294.00	7	6613.530	32.800	0.000
A-Alginate	1384.170	1	1384.170	6.870	0.014
B-Time (crosslinking)	806.930	1	806.930	4.000	0.056
C-Concentration (CaCl$_2$)	39,930.220	1	39,930.220	198.050	0.000
D-Media	528.260	1	528.260	2.620	0.117
E-UV	13.720	1	13.720	0.068	0.796
AC	2068.080	1	2068.080	10.260	0.003
BC	1150.540	1	1150.540	5.710	0.024
Residual	5242.070	26	201.620		
Lack of Fit	4715.090	20	235.750	2.680	0.112
Pure Error	526.980	6	87.830		
Cor Total	51,536.770	33			
Std. Dev.	14.200		R^2	0.898	
Mean	62.650		Adjusted R^2	0.870	
C.V.%	22.660		Predicted R^2	0.829	
			Adeq. Precision	18.048	

Studying the *p*-values of the terms included in the model (Table 6), it is evident that the factors A, alginate ratio, and C, concentration (crosslinker), have a statistically significant impact on the swelling behavior of the samples. However, the factors B, D, and E could not be omitted from the model as the main goal of this study is to level all post-printing treatment conditions by taking into consideration the optimization of both the degradation and swelling behavior in a simultaneous manner.

As depicted in the colored dot plot (Figure 5), when studying the statistically significant factors separately, it can be deduced that the samples exhibiting a moderate swelling ratio (marked as green points) are those that had lower levels of their alginate ratio (Figure 5a) or those which were treated with a higher-level concentration of $CaCl_2$. Moreover, both diagrams are indicators of a uniform variability and even distribution.

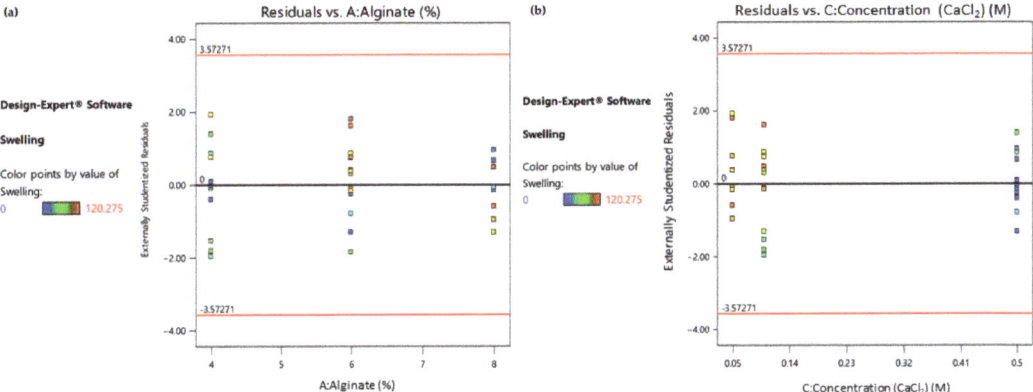

Figure 5. (**a**) Response 2: Residual vs. Alginate ratio plot; (**b**) Response 2: Residual vs. Concentration ($CaCl_2$) plot.

Finally, in order to understand how the values of both responses vary, a Reduced Two-Factor Interaction Model was employed to propose a possible relationship between the five most-critical post-printing variables affecting the scaffold's functionality and the statistically significant term-interactions. Based on this, two reduced 2FI models (Table 7) were generated using the Design Expert software to forecast the degradation time (days) and swelling ratio (%). Both equations are expressed in terms of the actual factors. For this reason, the coefficients in the equation are adjusted to account for the units of each factor and should not be utilized to assess the relative influence of each factor.

As the culture medium and UV exposure treatment are two categorical predictors, the model consists of four equations, one for each of the combinations of the two categorical variables. In addition, as presented in Figure 6, as the line is not significantly skewed either left or right, it can be assumed that there is no significant deviation from the normal distribution for the observations for both responses.

In order to evaluate the equations' accuracy, a confirmation experiment was piloted selecting the medium levels of the factors: A: alginate ratio = 6%, B: time (crosslinking) = 10 min, C: concentration ($CaCl_2$) = 0.1 M, D: media = DMEM, UV exposure = Yes. The measurements were conducted in triplicate and the average values were obtained for both responses (degradation time and swelling ratio). The results and the comparison with the values predicted by the actual reduced 2FI model equations are presented in Table 8. Both experimental values fall into the Confidence Interval, indicating the equations' high predictability.

Table 7. Degradation time equations and swelling ratio equations of the reduced 2FI model for each combination of media and UV exposure.

Degradation		Swelling	
Media	DMEM	Media	DMEM
UV	No	UV	No
+9.856		+26.131	
−0.608	Alginate	+10.985	Alginate
−0.334	Time (crosslinking)	+0.724	Time (crosslinking)
−5.645	Concentration ($CaCl_2$)	+47.471	Concentration ($CaCl_2$)
+0.095	Alginic × Time (crosslinking)	−24.374	Alginic × Concentration ($CaCl_2$)
+2.050	Alginic × Concentration ($CaCl_2$)	−7.390	Time (crosslinking) × Concentration ($CaCl_2$)
+0.781	Time (crosslinking) × Concentration ($CaCl_2$)		
Media	DMEM	Media	DMEM
UV	Yes	UV	Yes
+5.331		+24.838	
+0.194	Alginate	+10.985	Alginate
−0.334	Time (crosslinking)	+0.724	Time (crosslinking)
−5.645	Concentration ($CaCl_2$)	+47.471	Concentration ($CaCl_2$)
+0.095	Alginic × Time (crosslinking)	−24.374	Alginic × Concentration ($CaCl_2$)
+2.050	Alginic × Concentration ($CaCl_2$)	−7.390	Time (crosslinking) × Concentration ($CaCl_2$)
+0.781	Time (crosslinking) × Concentration ($CaCl_2$)		
Media	PBS	Media	PBS
UV	No	UV	No
+6.744		+34.178	
−0.608	Alginate	+10.985	Alginate
−0.585	Time (crosslinking)	+0.724	Time (crosslinking)
+3.440	Concentration ($CaCl_2$)	+47.471	Concentration ($CaCl_2$)
+0.095	Alginic × Time (crosslinking)	−24.374	Alginic × Concentration ($CaCl_2$)
+2.050	Alginic × Concentration ($CaCl_2$)	−7.390	Time (crosslinking) × Concentration ($CaCl_2$)
+0.781	Time (crosslinking) × Concentration ($CaCl_2$)		
Media	PBS	Media	PBS
UV	Yes	UV	Yes
+2.219		+32.885	
+0.194	Alginate	+10.985	Alginate
−0.585	Time (crosslinking)	+0.724	Time (crosslinking)
+3.440	Concentration ($CaCl_2$)	+47.471	Concentration ($CaCl_2$)
+0.095	Alginic × Time (crosslinking)	−24.374	Alginic × Concentration ($CaCl_2$)
+2.050	Alginic × Concentration ($CaCl_2$)	−7.390	Time (crosslinking) × Concentration ($CaCl_2$)
+0.781	Time (crosslinking) × Concentration ($CaCl_2$)		

Table 8. Reduced 2FI model equations' predicted responses for numerical factors' medium levels and observed results from confirmation tests.

Response	Predicted Mean	Predicted Median	Observed	Std. Dev	SE Mean	95% CI Low for Mean	95% CI High for Mean	95% TI Low for 99% Pop	95% TI High for 99% Pop
Degradation	12.990	12.990	14.000	0.968	0.543	11.866	14.115	8.725	17.252
Swelling	97.813	97.813	93.872	14.199	6.587	84.272	111.355	39.055	156.52

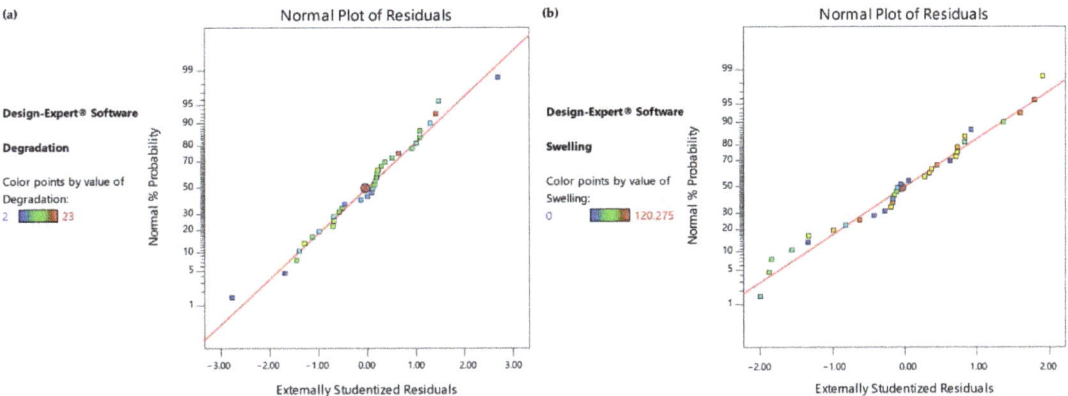

Figure 6. (**a**) Normal Probability Plot for degradation time observations, (**b**) Normal Probability Plot for swelling ratio observations.

2.3.2. Parameter Optimization by Response Surface Methodology for Scaffold's Post-Printing Treatment

When optimizing the post-printing treatment for scaffolds, specific criteria must be met to ensure their suitability for the targeted application. For instance, in the case of in vitro micro-vessel network formation, a critical evaluation of 3D cultures typically occurs at the 14-day mark. However, it is essential to note that optimal scaffold performance in facilitating vessel formation often demands structural integrity extending beyond this period, ideally reaching the 21-day threshold. Additionally, the swelling ratio of 3D systems is of pivotal importance, as excessive swelling can lead to inflammation and discomfort, while minimal deformation may result in detachment post-implantation. As a guideline, a moderate swelling ratio of 50% could strike the targeted balance. These criteria help guide the effective optimization of post-printing treatment for scaffolds, ensuring their suitability for the intended applications [6,17,21].

As displayed in the contour diagrams (Figure 7), for factor B: time (crosslinking) = 15 min, the higher the alginate ratio and the crosslinker concentration, the higher the time interval that the scaffolds maintain their structure when immersed in culture medium (contoured as red region) (Figure 7a). However, for the same conditions, the swelling ratio is rather low (contoured as blue region) (Figure 7b), not reaching the targeted value of 50% (contoured as green region) (Figure 7b). For this reason, an optimization process was piloted to define the optimum compromise between the two expected functionalities.

The numerical optimization [23,33,34] process was conducted with the primary objective of addressing two specific functionalities of the scaffold: achieving a low degradation rate and maintaining a moderate swelling ratio. In order to achieve these goals, the optimization criteria were precisely defined. The targeted values for the two key responses were set at 21 days for the degradation time and 50% for the swelling ratio. Through the optimization modeling process, a total of 100 solutions were generated, with their desirability decreasing progressively. Remarkably, the initial solution, which defined the optimal levels for both the numerical and categorical factors, yielded an impressive desirability score of 0.964. Implementing these recommended factor settings resulted in the model predicting a degradation time of 19.654 days and a swelling ratio of 50.00%. This achievement not only met one of the primary criteria (swelling ratio) but also exhibited nearly ideal degradation behavior. The predicted optimal values for each variable are comprehensively presented in the Ramps diagram (Figure 8) and summarized in Table 9.

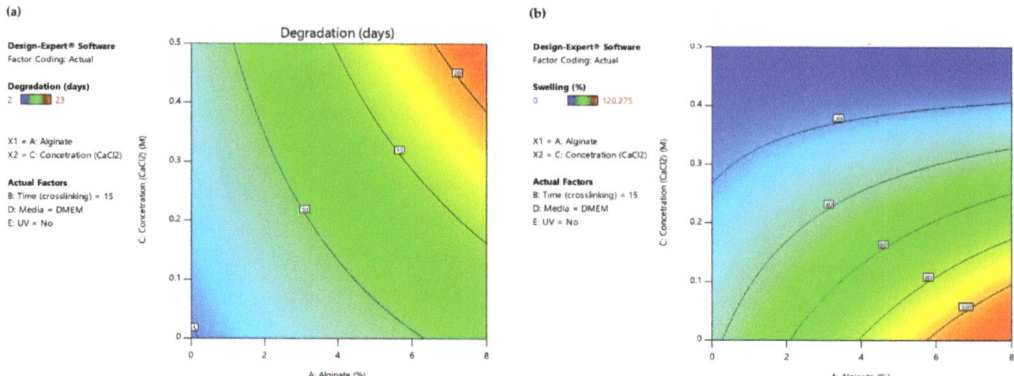

Figure 7. (**a**) AC Contour model graph for degradation analysis with term B fixed at 15 min, (**b**) AC Contour model graph for swelling ratio analysis with term B fixed at 15 min (Design-Expert 11 was employed).

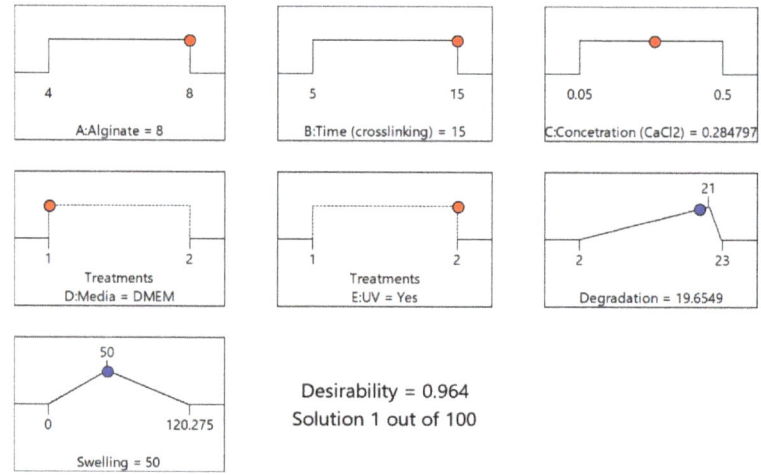

Figure 8. Solution's Ramps graph (1/100; Desirability = 0.964) for optimization of both Response 1 and Response 2.

Table 9. RSM's numerical optimization variable levels (Solution 1/100; Desirability = 0.964).

Factors	A: Alginate Ratio (%)	B: Time (Crosslinking)(min)	C: Concentration (CaCl$_2$) (M)	D: Media (-)	E: UV Exposure (-)
Optimal Levels	8.000	15.000	0.284	DMEM	Yes

The "All Factors" diagram (Figure 9) plots each individual response on the y-axis against each design factor on the x-axis, providing a structured representation of the relationship between variables and responses. Additionally, to encapsulate the co-optimization objective, a third response is introduced in the diagram by plotting desirability on the y-axis. It becomes evident that, for the majority of the responses, an apparent linearity prevails. Notably, the Desirability-Concentration (CaCl$_2$) plot deviates from this pattern, revealing a non-linear relationship where the peak response is displayed at C = 0.284 M.

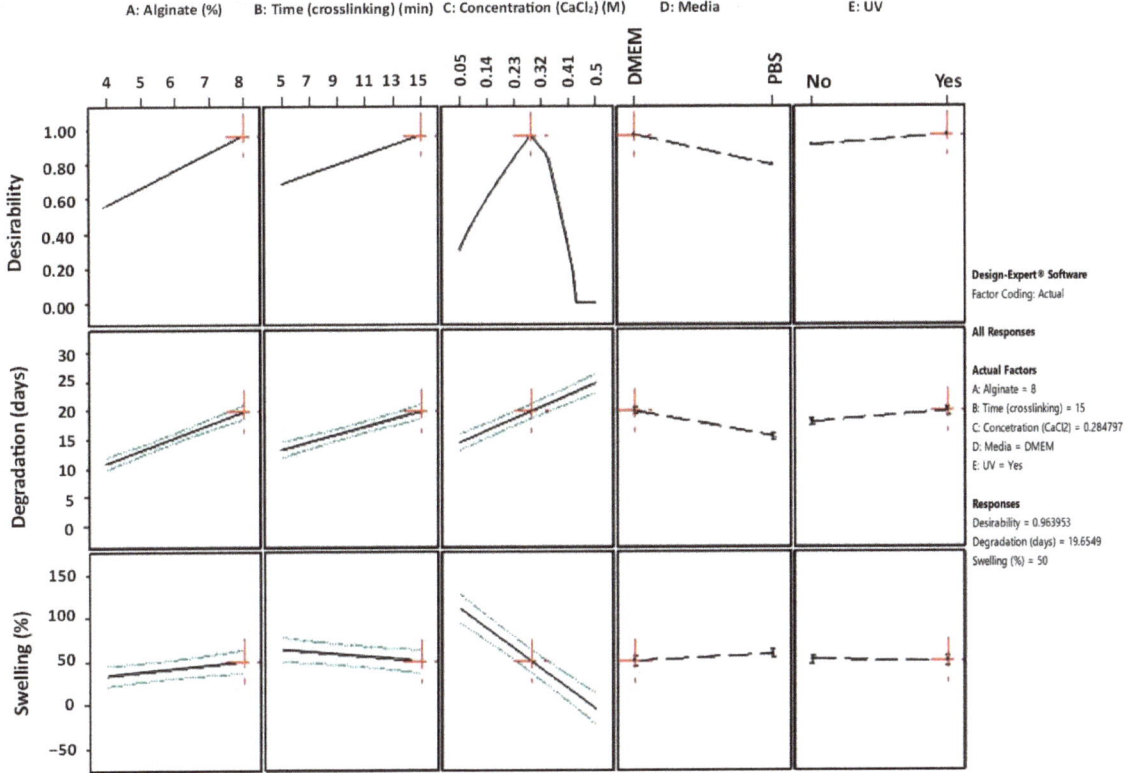

Figure 9. All Factors graph with annotation for 1/100 numerical optimization's solution (Design-Expert 11 was employed).

Furthermore, the slope of these lines demonstrates the magnitude of a variable's impact on the response; a steeper slope signifies a more pronounced effect. As is comprehensively presented, the optimization's objective can be reached by using a high alginate ratio for a high level of immersion time in the medium crosslinker concentration. In regard to the conditions, UV exposure is suggested. Immersing the scaffolds in PBS might negatively affect the degradation rate of the scaffold. These results are also verified by the 2D contour plots (Figure 10), where the highest desirability is observed (contoured as red region) for high alginate ratios and medium levels of crosslinker concentration when the immersion time in DMEM is 15 min.

2.3.3. Model Validation

In order to validate the obtained models and predicted optimal values, a parallel analysis in Minitab 2021 was conducted utilizing the levels' configuration generated by the RSM design, along with the corresponding results for the degradation and swelling ratio. The results of this analysis in Minitab 2021 totally mirrored those obtained in Design Expert (DE). Specifically, the same terms and interactions were identified as statistically significant for both the degradation and swelling ratios, as presented in Figure 11a,b respectively. Moreover, regarding the generated models' analysis of variance as well as the models' summary, the calculated values are identical to those obtained from DE as presented in Table 10, describing the observed data effectively. Moreover, the desirability score was again calculated as 0.963, affirming that the validation test confirmed the existence of a solution that optimizes both responses without significant compromise. Notably, as shown in Figure 12, all of the predicted optimal levels are completely identical with those obtained from DE. Thus, the results obtained from this validation approach completely align with the outcomes obtained using Design Expert, further reinforcing the accuracy of the software and models used in the study.

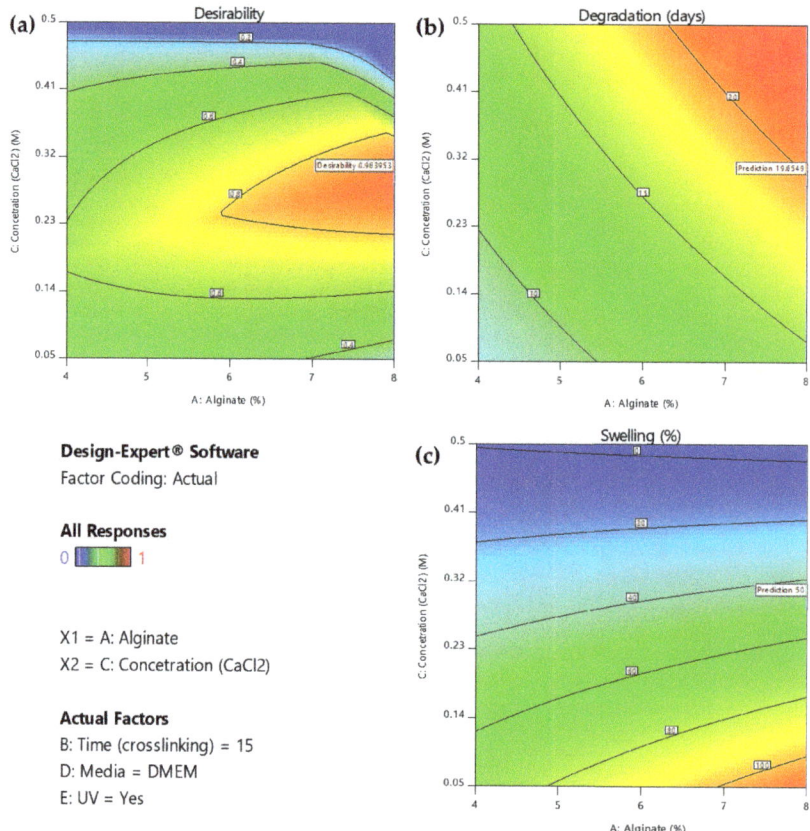

Figure 10. 2D Contour plot for effect of the interaction of A: alginate ratio and C: concentration (CaCl$_2$) factors on (**a**) desirability, (**b**) degradation, and (**c**) swelling respectively for B = 10 min. (Design-Expert 11 was employed).

Table 10. Analysis of variance and model summary comparison as obtained by Design Expert and Minitab software, respectively.

	Source		Sum of Squares	df	Mean Square	F-Value	p-Value
Degradation Analysis	Design Expert	Model	825.460	11	75.040	80.010	0.000
	Minitab		825.460	11	75.040	80.010	0.000
Swelling ratio Analysis	Design Expert	Model	46,294.700	7	6613.530	32.800	0.000
	Minitab		46,294.700	7	6613.530	32.800	0.000
Degradation Analysis			Std. Dev.	R^2	Adjusted R^2		Predicted R^2
	Design Expert		0.968	0.975	0.963		0.933
	Minitab						
Swelling ratio Analysis	Design Expert		0.968	0.975	0.963		0.933
	Minitab		14.200	0.898	0.870		0.829

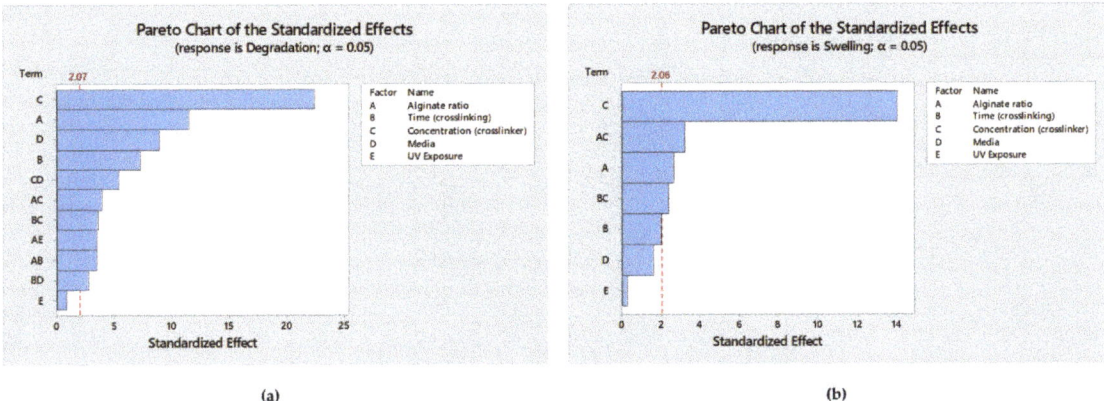

Figure 11. (**a**) Pareto Chart displaying the terms from the largest effect to the smallest effect on degradation response with a reference line for statistical significance; (**b**) Pareto Chart displaying the terms from the largest effect to the smallest effect on swelling ratio response with a reference line for statistical significance.

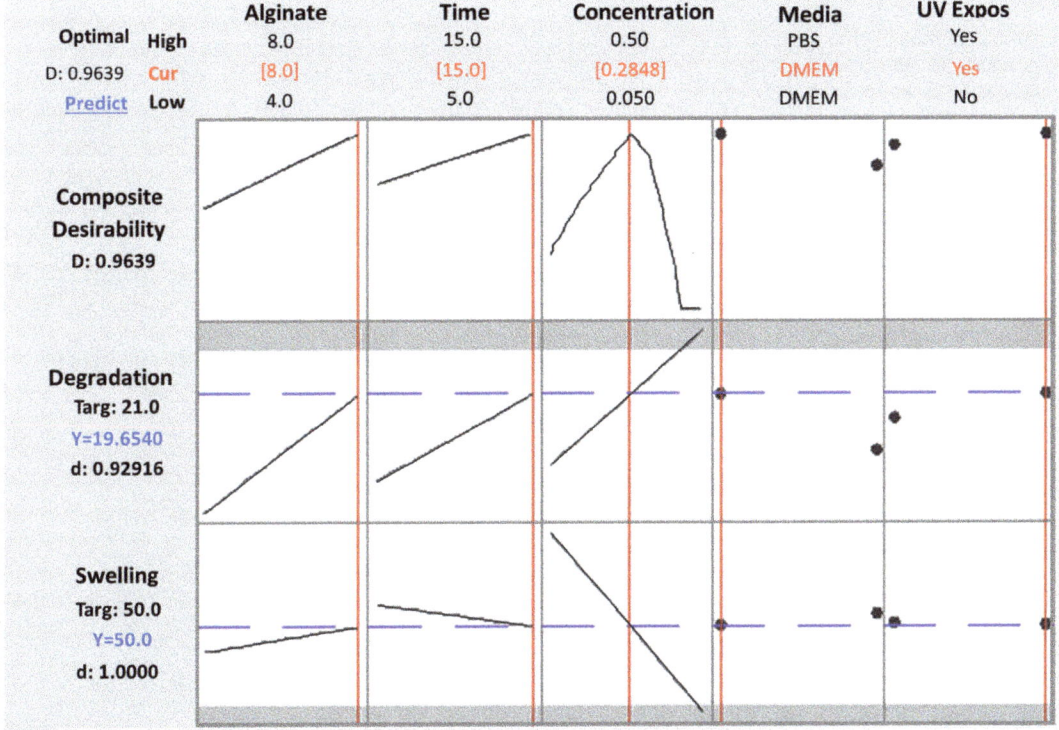

Figure 12. All Factors graph obtained by Minitab 2021 software, validating the predicted levels for each parameter as defined by Design Expert software for desirability = 0.9639.

2.4. Confirmation Tests

2.4.1. RSM

In order to compare the experimental results with the RSM-predicted values, confirmation experiments were then conducted by using the suggested levels of the numerical variables and by applying the optimal conditions for the post-printing treatment as indicated by the optimization process. The as-suggested treated scaffold maintained its structural integrity for 18.5 days, while reaching a swelling ratio of 54.120%. Both results are within the given limits of [18.37, 20.897] and [36.840, 63.571], respectively, according to C.I., with 95% confidence (z =1.96) (Table 11). The experimental results deviate by 5.621 (%) and 8.842 (%), respectively, from the predicted values, verifying the model's suitability for the post-printing stage's optimization.

Table 11. RSM's DoE point prediction for optimum variable levels and observed results from confirmation tests.

Optimal Response	Predicted Mean	Predicted Median	Observed	Std Dev	SE Mean	95% CI Low for Mean	95% CI High for Mean	95% TI Low for 99% Pop	95% TI High for 99% Pop
Degradation	19.654	19.654	18.500	0.968	0.607	18.376	20.897	15.262	24.011
Swelling	50.00	50.000	54.120	14.189	6.502	36.840	63.571	−8.406	108.818

Therefore, the RSM was used to define the optimum alginate ratio, time of immersion, and crosslinker concentration while determining that UV exposure facilitates the achievement of the targeted functionalities. The model also indicated that the degradation rate is higher when the scaffold is immersed in PBS, suggesting cautious and rapid procedures when treating it with PBS.

2.4.2. Characterization Results for the Confirmation Tests

Degradation Behavior

FT-IR Spectroscopy

Figure 13 displays the FT-IR spectra of the pellets that were collected after centrifugation of the media in which the optimized scaffold was immersed for 7 days and 14 days, respectively. The wide peak at 3358 is a characteristic peak of sodium alginate for –OH stretching [27]. The stronger peaks at 1034 (Amide III) and 1412 (O-C-C stretching) are indicative of dissolute gelatin and sodium alginate, demonstrating that, on Day 7, the scaffold exhibited lower degradation [29]. The narrow peaks at 3317 and 1647 cm^{-1}, as well as at 562 and 2132 cm^{-1}, are water characteristics, and this might indicate that the pellet was not totally dried. The scaffold was dissolved in the media in 18.5 days, almost reaching the ideal degradation interval of 21 days.

Shape retention

In order to more comprehensively present the degradation behavior of the optimized scaffold from Day 0 of its post-printing treatment until Day 18.5 when the scaffold lost its structural integrity, photos were taken at various time intervals as displayed in Figure 14. As is shown, the scaffold retained its infill's structure at a sufficient level up to Day 18. For the shape retention measurement, the shape of the scaffold was also recorded at various time intervals (3, 7, 14, and 18 days) during its 18.5-day culturing (Figure 15). Representative images of the scaffold were also acquired at Day 1 to ensure the culture conditions and the sustainability of the produced 3D structure (Figure 16). By using a face contrast microscope (Zeiss, Axiovert S100, Oberkochen, Germany, 5×), images of the scaffold were obtained and analyzed via Image J software. A commonly used criterion to evaluate the shape fidelity after treatment of the scaffold is based on the measurement of the diameter of the deposited filament (strand) and its comparison with the theoretical strand diameter (0.41 mm based on the nozzle diameter).

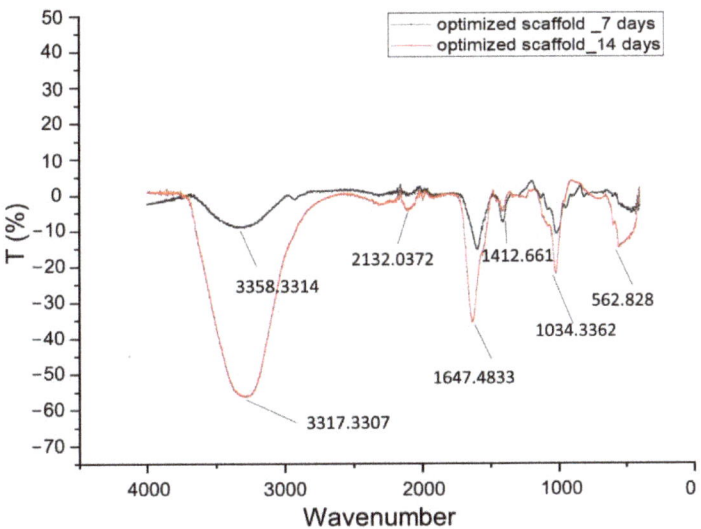

Figure 13. FT-IR spectra (4500–0 cm^{-1}) of centrifuged culture media for optimized scaffold at Day 7 and Day 14.

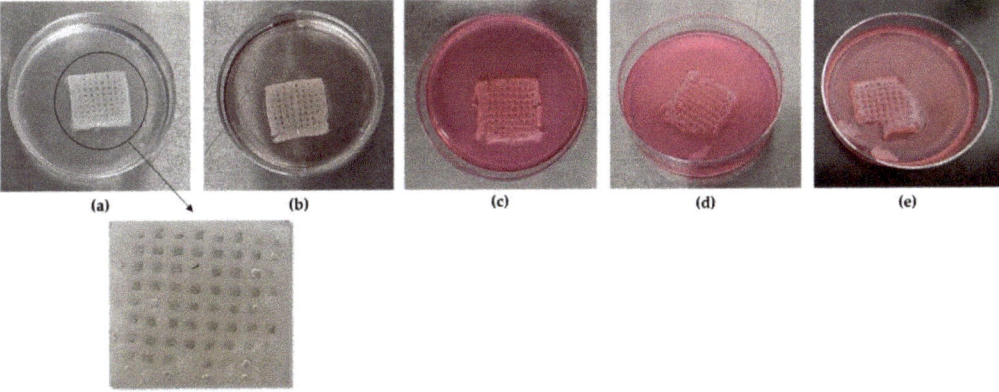

Figure 14. Representative photos of the scaffold were acquired at several time points. (**a**) Optimized scaffold aft post-printing treatment at Day 0, (**b**) optimized scaffold at Day 3 after immersion in DMEM, (**c**) optimized scaffold at Day 7 after immersion in DMEM, (**d**) optimized scaffold at Day 15 after immersion in DMEM, (**e**) optimized scaffold at Day 18 after immersion in DMEM.

For the diameter of each strand, multiple measurements were taken and the mean value was calculated. For instance, the vertical strand's diameters are displayed in Figure 15, indicating the scaffold's deformation throughout the degradation study. As was expected, until Day 3, where the plateau in the swelling curve occurs, the scaffold was expanding due to water uptake, and an increase by 2.05% in the scaffold's strand diameter was recorded. From Day 3 to Day 18.5 (Figure 17), the scaffold gradually exhibited degradation, leading to the reduction in the strand's diameter by 39.4% before its complete dissolution in the culture media.

Morphology

SEM characterization was applied as a supplementary method for assessing the infill morphology of alginate–gelatin scaffolds over time, following their immersion in

DMEM. In line with the shape retention results, the optimized scaffold demonstrated high stability, maintaining its infill structure integrity from Day 0 through to Day 14 and Day 18, as presented by the preserved primary shape of the pores in Figure 18a–c. However, on Day 14, a critical time point for the cell culture, a slight crack appeared in the pore's perimeter, signifying the initial stages of gradual alteration of the scaffold's geometry. Comparing the morphology of the Day 0 scaffold (Figure 18d) with that of Day 18 (Figure 18f), it is apparent that the pores significantly retained their geometry, while the scaffold's macroscopic structure exhibited some deformation, likely due to gelatin dissolution. Additionally, a representative EDS analysis conducted on Day 14 ensured the detection of calcium in the infill, responsible for crosslinking the alginate blocks, despite its release into the culture media and replacement by sodium ions (Figure 18e).

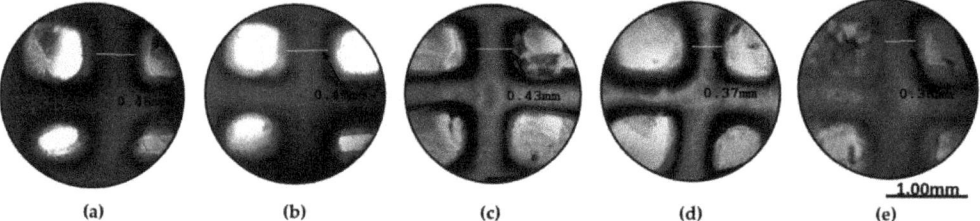

Figure 15. Representative images of the scaffold were acquired at several time points. (**a**) Strand retention measurements at Day 0, (**b**) strand retention measurements at Day 3, (**c**) strand retention measurements at Day 7, (**d**) strand retention measurements at Day 14, (**e**) strand retention measurements at Day 18 (Zeiss, Axiovert S100, Oberkochen, Germany) (×5 magnification).

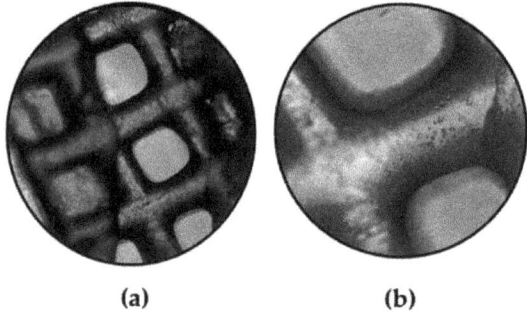

Figure 16. Representative images of the scaffold were also acquired also at Day 1 to ensure the scaffold's sustainability. (**a**) Scaffold at Day 1 (×4 magnification), (**b**) scaffold at Day 1 (×10 magnification) (Zeiss, Axiovert S100, Oberkochen, Germany).

Swelling Behavior

Swelling Ratio

As was expected, after the optimized level combination, the scaffold exhibited moderate deformation (Figure 19a), reaching a maximum swelling ratio of 54.120% (Figure 19b) and reaching the critical equilibrium of strong water uptake that facilitates cell viability and nutrient diffusion but not exhibiting excessive deformation that could cause inflammation or discomfort to the host after implantation. The fluctuation observed in the swelling ratio of the optimized scaffold is a result of multiple factors. Firstly, temperature variations can induce gelatin dissolution, leading to an initial degradation which was recorded at Day 2. However, the change in culture media at Day 2 triggers, again, the ion exchange, leading to a temporary decrease in the crosslinking degree and an increase in the scaffold's swelling ability. These changes persist until the ion equilibrium is reestablished, resulting in the subsequent stabilization of the swelling ratio [14,30].

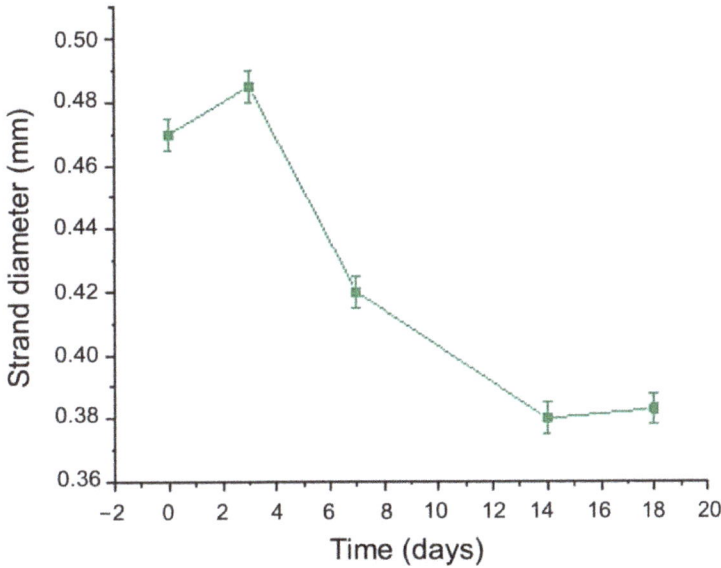

Figure 17. Mean value of optimized scaffold's strand diameter versus time.

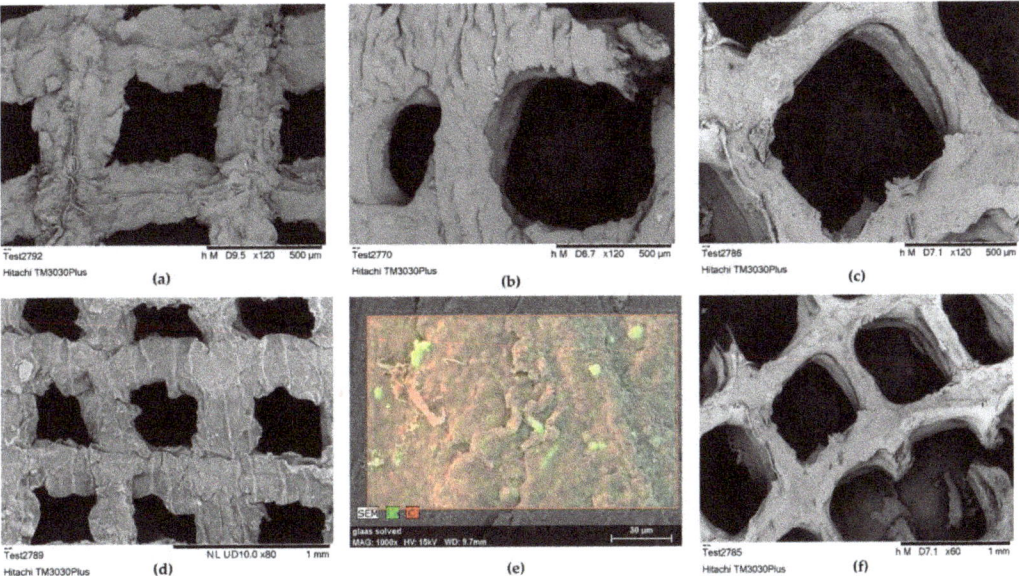

Figure 18. (**a**) Optimized scaffold's infill structure integrity morphology assessment at Day 0 at an acceleration voltage of 5 kV (×120 magnification); (**b**) optimized scaffold's infill structure integrity morphology assessment at Day 14 after immersion in DMEM, at an acceleration voltage of 5 kV (×120 magnification); (**c**) optimized scaffold's infill structure integrity morphology assessment at Day 18 after immersion in DMEM, at an acceleration voltage of 5 kV (×120 magnification); (**d**) optimized scaffold's infill morphology at Day 0 at an acceleration voltage of 15 kV (×80 magnification); (**e**) EDS analysis of scaffold's infill selected region, at Day 14 after immersion in DMEM at an acceleration voltage of 15 kV (×1000 magnification); (**f**) optimized scaffold's infill morphology at Day 18 at an acceleration voltage of 5 kV (×60 magnification).

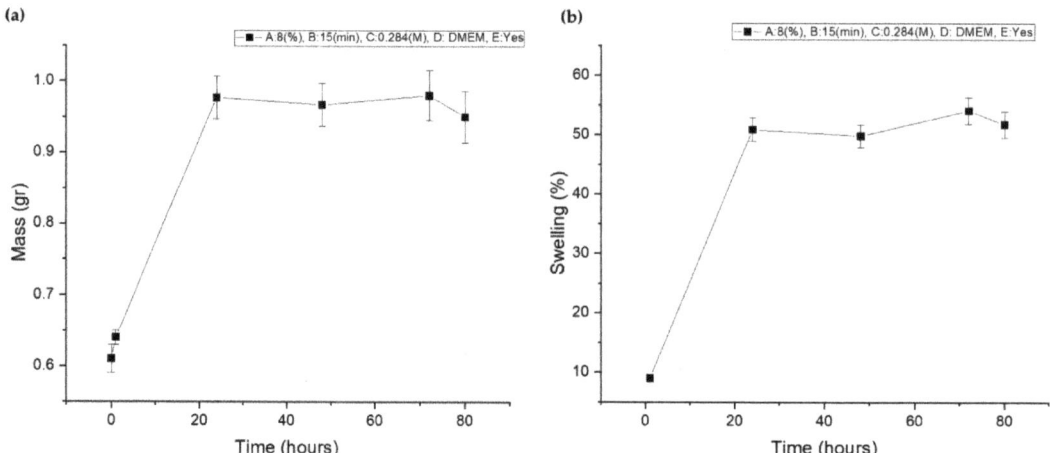

Figure 19. (a) Optimized scaffold's mass plot after immersion in culture media; (b) optimized scaffold's swelling curve after immersion in culture media.

3. Conclusions

This study, which constitutes a follow-up to our previously published work [22], represents a systematic and data-driven exploration of post-printing treatment optimization which aimed to enhance the entire implementation of hydrogels' DIW. The current work is driven by the use of RSM DoE to uncover the optimal combination of conditions for the desired degradation and swelling behavior of the printed structure, further advancing the field of bioprinting and tissue engineering. By employing RSM DoE, this study identified the individual and interactive effects of five post-printing treatment factors on scaffolds' degradation time and swelling behavior, and also suggested an optimized methodology to guide the selection of post-printing treatment conditions tailored to the scaffold's targeted functionalities.

Thus, this study managed to address a critical gap in the existing literature by concurrently optimizing the degradation and swelling behavior of DIW scaffolds. The developed model has not only identified the key factors influencing the degradation time and swelling ratio but has also yielded a solution that achieves a remarkable desirability score of 0.964, closely approaching the desired benchmarks of 21 days for degradation and a 50% swelling ratio. The optimal conditions, including an 8% alginate ratio, 15 min of crosslinking time, a crosslinker concentration of 0.284 M, DMEM as the immersion medium, and the incorporation of UV exposure, have been successfully predicted with a high degree of accuracy, with experimental values confirming the model's reliability. Furthermore, this optimization process has shed light on crucial post-printing handling procedures for scaffolds, notably revealing that UV exposure enhances structural integrity while decreasing the degradation rate and emphasizing the need for caution when treating the scaffolds with PBS, which accelerates degradation. This innovative approach highlights the applicability of the Response Surface Methodology in simultaneous response optimization, offering a promising possibility for future research aimed at co-optimizing scaffold-targeted responses, and ultimately enhancing scaffold functionality without compromising other critical properties. This approach aims for not only improved scaffold functionality but also enhanced efficiency and resource utilization by reducing reliance on trial-and-error methods.

The present research is focused on the post-printing treatment of a specific composite hydrogel based on its common use in tissue engineering applications. Although the widespread use of alginate–gelatin hydrogel makes this study's findings as relevant to a significant portion of ongoing research in this field, the study's scope is limited to the materials and conditions utilized in this specific experiment and may not cover a broad range of possible materials and conditions relevant to all applications. However, the proposed approach, which entails the simultaneous optimization of two critical responses, offers a versatile and customized methodology. This adaptability enables future studies to tailor the DoE's variables to match the specific materials and conditions in their optimization scenarios, thereby broadening the current study's applicability across a diverse range of hydrogel-based 3D bioprinting systems. This paper introduces an innovative methodology that addresses the intricate balance between these pivotal properties and offers a novel dimension to scaffold optimization through the application of RSM.

4. Materials and Methods

4.1. Design of Experiments

The responses that this study aims to optimize are two-fold: degradation time and swelling ratio. Degradation time is a crucial factor in scaffold functionality, influencing the rate at which the scaffold provides support before naturally degrading as the vessel network is formatting [6]. Simultaneously, maintaining strong but not excessive water uptake enhances cell migration and infiltration, as well as nutrient diffusion. Achieving an optimal balance between scaffold degradation time and swelling ratio is a complex and critical task [35]. For this challenge to be addressed, this study adopts a systematic and data-driven approach utilizing Statistical Response Surface Methodology (RSM DoE). RSM DoE is an experimental design technique used to optimize processes, as it offers a comprehensive means to navigate the interdependent factors that influence a system's responses.

Thus, three numerical factors were selected for investigation: alginate ratio, crosslinker concentration, and immersion time. These factors are crucial elements of the scaffolds' final properties, affecting their structural integrity, deformation, and biocompatibility. Additionally, two categorical factors are be explored, namely UV treatment and culture media, recognizing their substantial impact on post-printing treatment outcomes.

The experimental design encompasses 4 centroids within the factor space, yielding a total of 34 runs. Each run represents a carefully orchestrated combination of these factors, allowing for a comprehensive study of the selected parameter range.

Prior to the development of the Statistical Response Surface Methodology step, a crucial screening stage was essential to qualitatively define the most influential factors affecting the response variables: *degradation time* and *swelling*. This initial phase also aimed to establish a range of values for each factor within which the optimal levels were more likely to be found. The selection of the alginate–gelatin blend composition range was guided by our previous work, which was grounded in the outcomes of the material processability optimization process [22]. In this context, the gelatin ratio was fixed at 4%, while the alginate ratio varied between 4%, 6%, and 8%. Regarding the ionic crosslinking of sodium alginate, the screening tests indicated that the two most critical factors were the concentration of the crosslinker and the immersion time of the scaffolds in the crosslinking solution. Based on the existing literature, for the initial observations of scaffold degradation behavior, the designated value range for $CaCl_2$ concentration was set at [0.01–0.8] M. However, scaffolds that were treated with $CaCl_2$ solutions in the range of [0.01–0.04] were difficult to handle and exhibited a very high degradation rate. On the other hand, hydrogels treated with $CaCl_2$ solutions in the range of [0.6–0.8] were too dense and exhibited a very low degradation rate. For these reasons, the final range for $CaCl_2$ concentration was set at [0.05–0.5]. Although the release of calcium ions from the printed structures during their culture could potentially have a cytotoxic effect, the selected high level of 0.5 M $CaCl_2$ is recorded as not cytotoxic with good biocompatibility results [21]. In line with the findings of the existing literature and after screening trials, the immersion time was set to range from 5 to 15 min. Furthermore, it was of paramount importance to investigate the impact of culture media and UV treatment on the scaffold, as these two factors represent major conditions that need to be considered during the post-printing treatment of the scaffolds. With regard to culture media, DMEM is a well-established and widely utilized basal medium known for its capacity to facilitate the proliferation of diverse mammalian cell lines, offering great suitability as the selected medium for this experimental setup. The future perspectives of this study include the co-culture of cells and scaffolds in DMEM. PBS is also a commonly used buffer in everyday 3D cell-culture routine, for scaffold washes and during procedures such as trypsinization (cell splits). Thus, these two media were selected to be studied. In addition, UV treatment was used for sterilization reasons. Towards addressing these considerations, these two factors were incorporated into the experimental design as categorical variables, with different levels determined based on the respective conditions, such as "PBS or DMEM" and "UV or No UV" treatments. The levels for each numerical factor that were classified as "low", "medium", and "high", and the two levels for the categorical variables, are displayed in Tables 12 and 13, respectively.

Table 12. Defined levels for all three numerical factors of the post-printing treatment of alginate–gelatin scaffolds.

Post-Printing Treatment Factor	A: Alginate Ratio (%)	B: Time (CrossLinking) (min)	C: Concentration (CrossLinker) (M)
Low Level	4	5	0.05
Medium Level	6	10	0.1
High Level	8	15	0.5

Table 13. Defined levels for both vategorical variables of the post-printing treatment of alginate-gelatin scaffolds.

Post-Printing Treatment Factors	D: Culture Media	E: UV Exposure
Level 1	DMEM	NO
Level 2	PBS	YES

4.2. Screening Tests of DoE's Selected Levels

In the proposed methodology, an initial screening stage was incorporated consisting of experiments to evaluate the effectiveness of the levels chosen for the RSM DoE. Aiming to determine whether these selected levels warranted further optimization, a total of four samples were prepared. These samples encompassed various combinations of factors at different levels: Sample 1 involved the low-level combination of factors, Samples 2 and 3 represented medium-level combinations, and, finally, Sample 4 encompassed the high-level combination of factors. A detailed breakdown of these combinations can be found in Table 14. The samples were then evaluated in terms of degradation rate and swelling ratio.

Table 14. Factors' level combinations for screening tests samples.

Post-Printing Treatment Factors	A: Alginate Ratio (%)	B: Time (Crosslinking) (min)	C: Concentration ($CaCl_2$) (M)	D: Media (-)	E: UV Exposure (-)
Sample 1	4	5	0.05	PBS	No
Sample 2	6	10	0.1	PBS	Yes
Sample 3	6	10	0.1	PBS	No
Sample 4	8	15	0.5	DMEM	Yes

4.3. Materials

Sodium alginate (alginic acid sodium salt, low viscosity, (Alfa Aesar) Thermo Fisher Scientific, TechnoBiochem, Athens, Greece), gelatin (general purpose grade, Fisher Scientific, Waltham, MA, USA), PBS (Phosphate buffered saline tablets, Fisher Bioreagents, TechLine, Athens, Greece), and calcium chloride ($CaCl_2$) (Calcium Chloride Dihydrate, Riedel de Haen, Germany) were procured and used without modifications. Also, DMEM (Dulbecco's Modified Eagle Medium) (Gibco BRL, Life Technologies, ThermoScientific, Paisley, UK) culture medium was supplemented with 10% fetal bovine serum (FBS), 1% L-glutamine, 1% sodium pyruvate, and antibiotics (Gibco BRL, Life Technologies, Thermo Scientific, Paisley, UK) [36].

4.4. Hydrogel Synthesis

As indicated by the Design of Experiments, alginate and gelatin were combined in the various ratios of 4/4, 6/4, and 8/4 (%) for the generation of the three hydrogel blends displayed in Table 15. For instance, the following procedures were followed to prepare 50 mL of alginate/gelatin hydrogel 4/4 (%). A total of 2 g of gelatin was dissolved in 50 mL of PBS under continuous mechanical stirring at 60 °C. Then, an amount of 2 g of alginic acid sodium salt was added to the solution followed by stirring at 5600 rpm for 1.5 h at 50 °C to ensure homogeneity. The developed composite alginate–gelatin hydrogel inks were left on a hot plate to achieve a temperature of 30 °C before the Direct Ink Writing of the scaffolds.

Table 15. Different ratios of developed alginate–gelatin hydrogel blends.

Hydrogel	Alginate (%)	Gelatin (%)
1	4	4
2	6	4
3	8	4

4.5. Direct Ink Writing of the Scaffolds

The 3D printing of the scaffolds was accomplished using the Bioprinter Regemat 3D BIO V1 (REGEMAT 3D S.L., Granada, Spain). The composite alginate–gelatin hydrogel inks were loaded into 5 mL syringes. Subsequently, the material was extruded from the syringe nozzle to form a continuous filament, aided by the motor-assisted piston mechanism. The hydrogel filament was deposited layer by layer, following the designed 3D blueprint [22]. The designated scaffold dimensions were specified as L:17 mm × W:17 mm × H:3.5 mm, featuring an infill pattern of orthogonal pores measuring L:1.7 mm × W:1.7 mm. In order to ensure high printability and geometric accuracy, the optimal levels of four primary (Temperature, Extrusion Speed, Nozzle Diameter, Layer Height) and four secondary (Perimeter speed, Infill Speed, Retract Speed, Travel Speed) printing parameters were implemented. These optimal parameter settings had been previously determined through a robust Taguchi design, as detailed in our earlier work on optimizing the 3D printing process. A comprehensive overview of the nine printing parameters and their respective optimal levels can be found in Table 16.

Table 16. Optimal levels for 8 selected printing settings for high printing accuracy.

Printing Parameters	Temperature (°C)	Extrusion Speed (mm/s)	Nozzle Diameter (mm)	Layer Height (mm)	Perimeter Speed (mm/s)	Infill Speed (mm/s)	Retract Speed (mm/s)	Travel Speed (mm/s)
Printability Window	24	2	0.41	0.25	3	2	30	50

4.6. Crosslinking Process

A major requirement for the ink used in Direct Ink Writing of scaffolds is for it to be easily stabilized after its extrusion from the nozzle tip for the printed structure to sufficiently maintain its geometry. For this reason, the main step that governs the post-printing treatment is the crosslinking process [37]. This step is essential not only for the enhancement of the rigidity of the hydrogel-based printed scaffolds but also for the maintenance of the scaffolds' deformation at minimal levels. In the initial screening phase, different combinations of calcium chloride ($CaCl_2$) concentrations and scaffolds' immersion time into the crosslinking solution were investigated to identify the range of the most optimal crosslinking conditions suitable for the scaffolds. Based on the results of the screening phase, following the printing process, the scaffolds were submerged in a $CaCl_2$ solution with varying concentrations (0.05 M, 0.1 M, or 0.5 M) for different time intervals (5 min, 10 min, or 15 min). Subsequently, the scaffolds were removed from the $CaCl_2$ solution and tapped dry. Additionally, to activate the temperature-dependent gelatin's physical crosslinking, the scaffolds were left at room temperature (22 °C) overnight before their immersion in culture medium at 37 °C.

4.7. UV-Exposure

A supplementary step in the post-printing treatment is the UV exposure of the crosslinked scaffolds. As documented by Carranza et al., despite the sterilization effect, UV samples also exhibited a higher degree of dimensional stability [38]. The samples sterilized by UV radiation in particular presented the highest dimensional stability. In the direction of studying whether the UV-exposure affects the degradation behavior of the scaffolds, some of the samples were exposed to UV light before their immersion in the culture medium, while others not. For the preparation of UV scaffolds, the samples were placed in petri dishes, exposed to 254 nm UV light in a cabinet (MSC-AdvantageTM) with type-II laminar flow for 30 min, according to common sterilization protocols [39].

In consideration of environmental sustainability, the experimental procedure was designed to minimize UV light utilization, but if UV-exposure was totally avoided, then possible contaminations in the incubator and the laboratory abductor would lead to pointless repetitions of the scaffold printing that would eventually burden the environment with a higher energy footprint and waste. To accomplish this, all samples were first printed and those selected for UV exposure based on DoE runs were collectively subjected to UV light, limiting the frequency of UV lamp usage. Moreover, safety protocols were implemented to prevent any researcher exposure to UV light during the experimentation process, utilizing automatic lock system of the laboratory door during the exposure time.

4.8. Degradation Test

The printed scaffolds were crosslinked and treated as indicated by the matrix runs of the DoE. In order to collect data for the monitoring of the design's selected response, degradation time, the scaffolds were then immersed in culture media after being rinsed with ethanol (90%) for sterilization. The selected media were DMEM and PBS as it is of high importance to reassure that the structure will not be dissolute at standard cell cultures conditions. DMEM is a very commonly used medium and PBS is used in everyday routine, for cell washes and during procedures such as trypsinization (cell splits) [40]. Thus, some structures were immersed in DMEM and PBS, respectively, at 37 °C, 95% humidity, in a 5% CO_2 incubator and steady pH conditions of 7.4 [41] for varying durations (3, 7,14, and 21 days). During the degradation study, it was ensured that the scaffolds stayed submerged in the culture media. At the end of the time points, samples were retrieved and shape-retention testing of the crosslinked hydrogel scaffolds was conducted via Image J (National Institutes of Health, Bethesda, MD, USA) software. In addition, the supernatant media was centrifuged (HITACHI High Speed Refrigerated Centrifuge, CR22, Eppendorf Himac Technologies, Takeda, Japan) and the pellets' FT-IR spectra were obtained.

4.9. Swelling Test

In vitro cell culture studies depend significantly on the scaffolds' capacity to swell, as this property permits the flow of cell nutrients inside the scaffold, increasing the cell's longevity. Moreover, a swelling test is critical as it provides an estimate of the scaffolds' maximum volume following implantation. In this study, the swelling tests were conducted by immersing dry samples in DMEM or PBS at 37 °C. The samples were taken out at various time intervals, they were tapped dry to remove excess water and then weighed until water absorption reached saturation. The following equation was used to determine the swelling ratio of scaffold sample (Equation (2)):

$$\text{Swelling ratio} = \frac{W_1 - W_0}{W_0} \quad (2)$$

where W_1 is the weight of the wet scaffold after immersion in the culture media (DMEM/PBS), and W_0 is the weight of the dry scaffold before water uptake.

4.10. FT-IR Spectroscopy

FTIR spectroscopy was conducted with the FTIR spectrometer (Cary 630 FTIR, Agilent Technologies, Santa Clara, CA, USA). In parallel with morphological observation through a face contrast microscopy and Scanning Electron Microscopy (SEM), FTIR spectroscopy was the characterization method that was employed in order to study the degradation rate of scaffolds, with regard to the chemical composition of the scaffold over time [42–44]. Aiming for maintaining aseptic conditions, this non-invasive approach to monitor scaffold dissolution was applied.

The FTIR spectra of the pellets of the centrifuged culture media for varying durations (7 and 14 days) enabled the identification of the dissolution of alginate and gelatin in the culture media. Resolution of 2 cm^{-1} was maintained in all cases. This method not only allowed for effective tracking of scaffold dissolution but also eliminated the risk of potential contamination associated with direct scaffold handling during mass or dimensional measurements.

4.11. Scanning Electron Microscopy

Scanning Electron Microscopy (SEM) was applied to assess the optimized scaffold's morphology at various time points of immersion in culture media. Alginate–gelatin hydrogel scaffolds were desiccated by passing the samples through a gradation of alcohol dehydration series followed by vacuum drying and characterized using SEM at an accelerating voltage of 5 kV. Scanning electron microscopy was performed with a Hitachi TM3030 (Thermo Fisher Scientific Waltham, MA, USA) tabletop microscope, equipped with an energy dispersive X-ray spectrophotometer (EDX) system (QUANTAX 70) for the coupled analysis of chemical structure.

4.12. Flow Diagram

In order to clarify the aforementioned methodology, the following flow diagram (Figure 20) summarizes the main steps. This study focuses on the steps that are enclosed in the light blue frame.

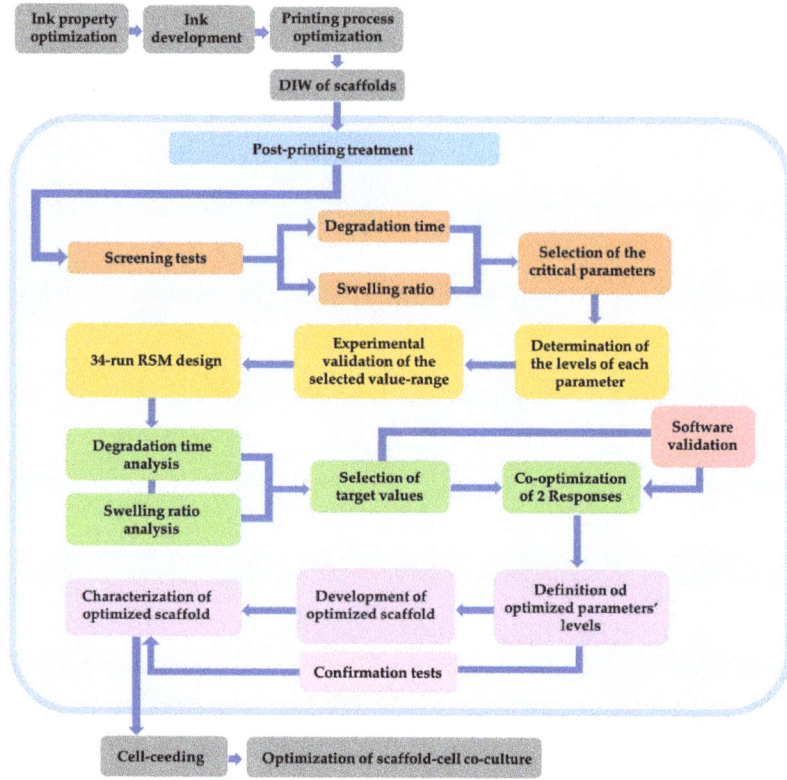

Figure 20. Flow diagram of the experimental procedure. The light blue area includes the main steps of this study.

Author Contributions: Conceptualization, C.K. and C.A.C.; methodology, C.K. and N.L.; software, C.K.; validation, C.K., E.A.P., N.L. and C.A.C.; formal analysis, C.K., E.A.P., N.L. and C.A.C.; investigation, C.K.; resources, N.L. and C.A.C.; data curation, C.K.; writing—original draft preparation, C.K. and N.L.; writing—review and editing, C.K., E.A.P., N.L. and C.A.C.; visualization, C.K. and N.L.; supervision, N.L. and C.A.C.; project administration, N.L. and C.A.C.; funding acquisition, C.A.C. All authors have read and agreed to the published version of the manuscript.

Funding: This research was funded by an NTUA-supported PhD Scholarship for C.K. (SARG-NTUA: No 65219100).

Institutional Review Board Statement: Not applicable.

Informed Consent Statement: Not applicable.

Data Availability Statement: The data presented in this study are openly available in article.

Acknowledgments: The authors would like to thank BIOG3D for the access to their scientific equipment (printer BIOV1).

Conflicts of Interest: The authors declare no conflict of interest.

References

1. Saadi, M.A.S.R.; Maguire, A.; Pottackal, N.T.; Thakur, M.S.H.; Ikram, M.M.; Hart, A.J.; Ajayan, P.M.; Rahman, M.M. Direct Ink Writing: A 3D Printing Technology for Diverse Materials. *Adv. Mater.* **2022**, *34*, e2108855. [CrossRef] [PubMed]
2. Gu, Z.; Fu, J.; Lin, H.; He, Y. 3D Bioprinting: A Novel Avenue for Manufacturing Tissues and Organs. *Engineering* **2019**, *5*, 777–794. [CrossRef]

3. Jiang, Z.; Diggle, B.; Li Tan, M.; Viktorova, J.; Bennett, C.W.; Connal, L.A. Extrusion 3D Printing of Polymeric Materials with Advanced Properties. *Adv. Sci.* **2020**, *7*, 2001379. [CrossRef] [PubMed]
4. Zielińska, A.; Karczewski, J.; Eder, P.; Kolanowski, T.; Szalata, M.; Wielgus, K.; Szalata, M.; Kim, D.; Shin, S.R.; Słomski, R.; et al. Scaffolds for drug delivery and tissue engineering: The role of genetics. *J. Control Release* **2023**, *359*, 207–223. [CrossRef]
5. Saini, G.; Segaran, N.; Mayer, J.L.; Saini, A.; Albadawi, H.; Oklu, R. Applications of 3D Bioprinting in Tissue Engineering and Regenerative Medicine. *J. Clin. Med.* **2021**, *10*, 4966. [CrossRef]
6. Yavvari, P.; Laporte, A.; Elomaa, L.; Schraufstetter, F.; Pacharzina, I.; Daberkow, A.D.; Hoppensack, A.; Weinhart, M. 3D-Cultured Vascular-Like Networks Enable Validation of Vascular Disruption Properties of Drugs In Vitro. *Front. Bioeng. Biotechnol.* **2022**, *10*, 888492. [CrossRef]
7. Di Giuseppe, M.; Law, N.; Webb, B.; Macrae, R.A.; Liew, L.J.; Sercombe, T.B.; Dilley, R.J.; Doyle, B.J. Mechanical behaviour of alginate–gelatin hydrogels for 3D bioprinting. *J. Mech. Behav. Biomed. Mater.* **2018**, *79*, 150–157. [CrossRef]
8. Lagopati, N.; Pavlatou, E.A. Advanced Applications of Biomaterials Based on Alginic Acid. *Am. J. Biomed. Sci.* **2020**, *9*, 47–53. [CrossRef]
9. Anitua, E.; Zalduendo, M.; Troya, M.; Erezuma, I.; Lukin, I.; Hernáez-Moya, R.; Orive, G. Composite alginate–gelatin hydrogels incorporating PRGF enhance human dental pulp cell adhesion, chemotaxis and proliferation. *Int. J. Pharm.* **2022**, *617*, 121631. [CrossRef]
10. Lagopati, N.; Pippa, N.; Gatou, M.-A.; Papadopoulou-Fermeli, N.; Gorgoulis, V.G.; Gazouli, M.; Pavlatou, E.A. Marine-Originated Materials and Their Potential Use in Biomedicine. *Appl. Sci.* **2023**, *13*, 9172. [CrossRef]
11. Tomić, S.L.; Babić Radić, M.M.; Vuković, J.S.; Filipović, V.V.; Nikodinovic-Runic, J.; Vukomanović, M. Alginate-Based Hydrogels and Scaffolds for Biomedical Applications. *Mar. Drugs* **2023**, *21*, 177. [CrossRef]
12. Ketabat, F.; Maris, T.; Duan, X.; Yazdanpanah, Z.; Kelly, M.E.; Badea, I.; Chen, X. Optimization of 3D Printing and in Vitro Characterization of Alginate/Gelatin Lattice and Angular Scaffolds for Potential Cardiac Tissue Engineering. *Front. Bioeng. Biotechnol.* **2023**, *11*, 1161804. [CrossRef]
13. Wierzbicka, A.; Bartniak, M.; Rosińska, K.; Bociąga, D. Optimization of the preparation process stages of the bioink compositions based on sodium alginate and gelatin to improve the viability of biological material contained in hydrogel 3D printouts. *Eng. Biomater.* **2022**, *165*, 7–16. [CrossRef]
14. Chawla, D.; Kaur, T.; Joshi, A.; Singh, N. 3D Bioprinted Alginate–gelatin Based Scaffolds for Soft Tissue Engineering. *Int. J. Biol. Macromol.* **2020**, *144*, 560–567. [CrossRef]
15. Mondal, A.; Gebeyehu, A.; Miranda, M.; Bahadur, D.; Patel, N.; Ramakrishnan, S.; Rishi, A.K.; Singh, M. Characterization and printability of Sodium alginate -Gelatin hydrogel for bioprinting NSCLC co-culture. *Sci. Rep.* **2019**, *9*, 19914. [CrossRef]
16. Sarker, M.; Izadifar, M.; Schreyer, D.; Chen, X. Influence of ionic crosslinkers (Ca^{2+}/Ba^{2+}/Zn^{2+}) on the mechanical and biological properties of 3D Bioplotted Hydrogel Scaffolds. *J. Biomater. Sci. Polym. Ed.* **2018**, *29*, 1126–1154. [CrossRef] [PubMed]
17. Shi, L.; Xiong, L.; Hu, Y.; Li, W.; Chen, Z.; Liu, K.; Zhang, X. Three-Dimensional Printing Alginate/Gelatin Scaffolds as Dermal Substitutes for Skin Tissue Engineering. *Polym. Eng. Sci.* **2018**, *58*, 1782–1790. [CrossRef]
18. Amr, M.; Dykes, I.; Counts, M.; Kernan, J.; Mallah, A.; Mendenhall, J.; Van Wie, B.; Abu-Lail, N.; Gozen, B.A. 3D Printed, Mechanically Tunable, Composite Sodium Alginate, Gelatin and Gum Arabic (SA-GEL-GA) Scaffolds. *Bioprinting* **2021**, *22*, e00133. [CrossRef]
19. Naghieh, S.; Karamooz-Ravari, M.R.; Sarker, M.D.; Karki, E.; Chen, X. Influence of crosslinking on the mechanical behavior of 3D printed alginate scaffolds: Experimental and numerical approaches. *J. Mech. Behav. Biomed. Mater.* **2018**, *80*, 111–118. [CrossRef]
20. Bahrami, N.; Farzin, A.; Bayat, F.; Goodarzi, A.; Salehi, M.; Goodarzi, A.; Salehi, M.; Karimi, R.; Mohamadnia, A.; Parhiz, A.; et al. Optimization of 3D Alginate Scaffold Properties with Interconnected Porosity Using Freeze-drying Method for Cartilage Tissue Engineering Application. *Arch. Neurosci.* **2019**, *6*, e85122. [CrossRef]
21. Sonaye, S.Y.; Ertugral, E.G.; Kothapalli, C.R.; Sikder, P. Extrusion 3D (Bio)Printing of Alginate–gelatin-Based Composite Scaffolds for Skeletal Muscle Tissue Engineering. *Materials* **2022**, *15*, 7945. [CrossRef] [PubMed]
22. Kaliampakou, C.; Lagopati, N.; Charitidis, C.A. Direct Ink Writing of Alginate–Gelatin Hydrogel: An Optimization of Ink Property Design and Printing Process Efficacy. *Appl. Sci.* **2023**, *13*, 8261. [CrossRef]
23. Gupta, P.; Nayak, K.K. Optimization of Keratin/Alginate Scaffold Using RSM and Its Characterization for Tissue Engineering. *Int. J. Biol. Macromol.* **2016**, *85*, 141–149. [CrossRef]
24. Pepelnjak, T.; Stojšić, J.; Sevšek, L.; Movrin, D.; Milutinović, M. Influence of Process Parameters on the Characteristics of Additively Manufactured Parts Made from Advanced Biopolymers. *Polymers* **2023**, *15*, 716. [CrossRef] [PubMed]
25. Gao, T.; Gillispie, G.J.; Copus, J.S.; Pr, A.K.; Seol, Y.J.; Atala, A.; Yoo, J.J.; Lee, S.J. Optimization of gelatin-alginate composite bioink printability using rheological parameters: A systematic approach. *Biofabrication* **2018**, *10*, 034106. [CrossRef] [PubMed]
26. Shan, Y.; Li, C.; Wu, Y.; Li, Q.; Liao, J. Hybrid Cellulose Nanocrystal/Alginate/Gelatin Scaffold with Improved Mechanical Properties and Guided Wound Healing. *RSC Adv.* **2019**, *9*, 22966–22979. [CrossRef]
27. Aljohani, W.; Ullah, M.W.; Li, W.; Shi, L.; Zhang, X.; Yang, G. Three-Dimensional Printing of Alginate–gelatin-Agar Scaffolds Using Free-Form Motor Assisted Microsyringe Extrusion System. *J. Polym. Res.* **2018**, *25*, 62. [CrossRef]
28. Li, Y.; Jia, H.; Cheng, Q.; Pan, F.; Jiang, Z. Sodium alginate–gelatin polyelectrolyte complex membranes with both high water vapor permeance and high permselectivity. *J. Membr. Sci.* **2011**, *375*, 304–312. [CrossRef]

29. Helmiyati; Aprilliza, M. Characterization and properties of sodium alginate from brown algae used as an ecofriendly superabsorbent. *IOP Conf. Ser. Mater. Sci. Eng.* **2017**, *188*, 012019. [CrossRef]
30. Gupta, N.V.; Shivakumar, H.G. Investigation of Swelling Behavior and Mechanical Properties of a pH-Sensitive Superporous Hydrogel Composite. *Iran. J. Pharm. Res.* **2012**, *11*, 481–493.
31. Talaei, A.; O'Connell, C.D.; Sayyar, S.; Maher, M.; Yue, Z.; Choong, P.F.; Wallace, G.G. Optimizing the composition of gelatin methacryloyl and hyaluronic acid methacryloyl hydrogels to maximize mechanical and transport properties using response surface methodology. *J. Biomed. Mater. Res. B Appl. Biomater.* **2023**, *111*, 526–537. [CrossRef]
32. Coşkun, S.; Akbulut, S.O.; Sarıkaya, B.; Çakmak, S.; Gümüşderelioğlu, M. Formulation of Chitosan and Chitosan-nanoHAp Bioinks and Investigation of Printability with Optimized Bioprinting Parameters. *Int. J. Biol. Macromol.* **2022**, *222*, 1453–1464. [CrossRef] [PubMed]
33. El Magri, A.; El Mabrouk, K.; Vaudreuil, S.; Ebn Touhami, M. Experimental Investigation and Optimization of Printing Parameters of 3D Printed Polyphenylene Sulfide through Response Surface Methodology. *J. Appl. Polym. Sci.* **2021**, *138*, 49625. [CrossRef]
34. Vates, U.K.; Kanu, N.J.; Gupta, E.; Singh, G.K.; Daniel, N.A.; Sharma, B.P. Optimization of FDM 3D Printing Process Parameters on ABS Based Bone Hammer Using RSM Technique. *IOP Conf. Ser. Mater. Sci. Eng.* **2021**, *1206*, 012001. [CrossRef]
35. Dhote, V.; Vernerey, F.J. Mathematical model of the role of degradation on matrix development in hydrogel scaffold. *Biomech. Model. Mechanobiol.* **2014**, *13*, 167–183. [CrossRef] [PubMed]
36. Lagopati, N.; Kotsinas, A.; Veroutis, D.; Evangelou, K.; Papaspyropoulos, A.; Arfanis, M.; Falaras, P.; Kitsiou, P.V.; Pateras, I.; Bergonzini, A.; et al. Biological Effect of Silver-modified Nanostructured Titanium Dioxide in Cancer. *Cancer Genom. Proteom.* **2021**, *18* (Suppl. 3), 425–439. [CrossRef] [PubMed]
37. Fu, Q.; Saiz, E.; Tomsia, A.P. Direct ink writing of highly porous and strong glass scaffolds for load-bearing bone defects repair and regeneration. *Acta Biomater.* **2011**, *7*, 3547–3554. [CrossRef] [PubMed]
38. Carranza, T.; Zalba-Balda, M.; Baraibar, M.J.B.; de la Caba, K.; Guerrero, P. Effect of sterilization processes on alginate/gelatin inks for three-dimensional printing. *Int. J. Bioprint.* **2022**, *9*, 645. [CrossRef]
39. Yen, S.; Sokolenko, S.; Manocha, B.; Blondeel, E.J.; Aucoin, M.G.; Patras, A.; Daynouri-Pancino, F.; Sasges, M. Treating cell culture media with UV irradiation against adventitious agents: Minimal impact on CHO performance. *Biotechnol. Prog.* **2014**, *30*, 1190–1195. [CrossRef]
40. Katifelis, H.; Nikou, M.-P.; Mukha, I.; Vityuk, N.; Lagopati, N.; Piperi, C.; Farooqi, A.A.; Pippa, N.; Efstathopoulos, E.P.; Gazouli, M. Ag/Au Bimetallic Nanoparticles Trigger Different Cell Death Pathways and Affect Damage Associated Molecular Pattern Release in Human Cell Lines. *Cancers* **2022**, *14*, 1546. [CrossRef]
41. Papadopoulou-Fermeli, N.; Lagopati, N.; Pippa, N.; Sakellis, E.; Boukos, N.; Gorgoulis, V.G.; Gazouli, M.; Pavlatou, E.A. Composite Nanoarchitectonics of Photoactivated Titania-Based Materials with Anticancer Properties. *Pharmaceutics* **2023**, *15*, 135. [CrossRef] [PubMed]
42. Celina, M.; Ottesen, D.K.; Gillen, K.T.; Clough, R.L. FTIR emission spectroscopy applied to polymer degradation. *Polym. Degrad. Stab.* **1997**, *58*, 15–31. [CrossRef]
43. Leroy, A.; Ribeiro, S.; Grossiord, C.; Alves, A.; Vestberg, R.H.; Salles, V.; Brunon, C.; Gritsch, K.; Grosgogeat, B.; Bayon, Y. FTIR microscopy contribution for comprehension of degradation mechanisms in PLA-based implantable medical devices. *J. Mater. Sci. Mater. Med.* **2017**, *28*, 87. [CrossRef] [PubMed]
44. Langueh, C.; Changotade, S.; Ramtani, S.; Lutomski, D.; Rohman, G. Combination of in Vitro Thermally-Accelerated Ageing and Fourier-Transform Infrared Spectroscopy to Predict Scaffold Lifetime. *Polym. Degrad. Stab.* **2021**, *183*, 109454. [CrossRef]

Disclaimer/Publisher's Note: The statements, opinions and data contained in all publications are solely those of the individual author(s) and contributor(s) and not of MDPI and/or the editor(s). MDPI and/or the editor(s) disclaim responsibility for any injury to people or property resulting from any ideas, methods, instructions or products referred to in the content.

Article

Characterization of Bioinks Prepared via Gelifying Extracellular Matrix from Decellularized Porcine Myocardia

Héctor Sanz-Fraile [1,†], Carolina Herranz-Diez [1,†], Anna Ulldemolins [1], Bryan Falcones [1], Isaac Almendros [1,2,3], Núria Gavara [1,4,5,6], Raimon Sunyer [1,6,7], Ramon Farré [1,2,3,6] and Jorge Otero [1,2,4,5,6,*]

1 Unitat de Biofísica i Bioenginyeria, Facultat de Medicina i Ciències de la Salut, Universitat de Barcelona, 08036 Barcelona, Spain; hector.sanz.fraile@hotmail.com (H.S.-F.); carolinaherranz@ub.edu (C.H.-D.); anna.ulldemolins@ub.edu (A.U.); bfalco86@gmail.com (B.F.); ialmendros@ub.edu (I.A.); ngavara@ub.edu (N.G.); rsunyer@ub.edu (R.S.); rfarre@ub.edu (R.F.)
2 CIBER de Enfermedades Respiratorias, 28029 Madrid, Spain
3 Institut d'Investigacions Biomèdiques August Pi i Sunyer, 08036 Barcelona, Spain
4 The Institute for Bioengineering of Catalonia (IBEC), 08028 Barcelona, Spain
5 The Barcelona Institute of Science and Technology (BIST), 08036 Barcelona, Spain
6 Institute of Nanoscience and Nanotechnology (IN2UB), Universitat de Barcelona, 08028 Barcelona, Spain
7 CIBER de Bioingeniería, Biomateriales y Nanomedicina, 28029 Madrid, Spain
* Correspondence: jorge.otero@ub.edu
† These authors contributed equally to this work.

Citation: Sanz-Fraile, H.; Herranz-Diez, C.; Ulldemolins, A.; Falcones, B.; Almendros, I.; Gavara, N.; Sunyer, R.; Farré, R.; Otero, J. Characterization of Bioinks Prepared via Gelifying Extracellular Matrix from Decellularized Porcine Myocardia. *Gels* 2023, 9, 745. https://doi.org/10.3390/gels9090745

Academic Editors: Enrique Aguilar and Helena Herrada-Manchón

Received: 1 August 2023
Revised: 1 September 2023
Accepted: 8 September 2023
Published: 13 September 2023

Copyright: © 2023 by the authors. Licensee MDPI, Basel, Switzerland. This article is an open access article distributed under the terms and conditions of the Creative Commons Attribution (CC BY) license (https://creativecommons.org/licenses/by/4.0/).

Abstract: Since the emergence of 3D bioprinting technology, both synthetic and natural materials have been used to develop bioinks for producing cell-laden cardiac grafts. To this end, extracellular-matrix (ECM)-derived hydrogels can be used to develop scaffolds that closely mimic the complex 3D environments for cell culture. This study presents a novel cardiac bioink based on hydrogels exclusively derived from decellularized porcine myocardium loaded with human-bone-marrow-derived mesenchymal stromal cells. Hence, the hydrogel can be used to develop cell-laden cardiac patches without the need to add other biomaterials or use additional crosslinkers. The scaffold ultrastructure and mechanical properties of the bioink were characterized to optimize its production, specifically focusing on the matrix enzymatic digestion time. The cells were cultured in 3D within the developed hydrogels to assess their response. The results indicate that the hydrogels fostered inter-cell and cell-matrix crosstalk after 1 week of culture. In conclusion, the bioink developed and presented in this study holds great potential for developing cell-laden customized patches for cardiac repair.

Keywords: 3D bioprinting; hydrogels; extracellular matrix; decellularized cardiac tissue; biomaterials; mesenchymal stromal cells

1. Introduction

There is a significant lack of organs for transplantation and replacements for implants and this problem will increase with the population aging, particularly in developed countries. Tissue engineering and regenerative medicine have emerged as potential solutions to overcome the progressive reduction of viable organ donors [1]. In the specific case of the heart, engineering implantable hearts built in the laboratory is still far away [2]. Hence, novel constructs based on decellularized tissue [3] or electroconductive scaffolds [4] have been recently developed to produce re-cellularized grafts (in the form of cardiac patches) to ameliorate the function of diseased hearts [5]. Several synthetic materials have been used to develop these scaffolds, but they still possess limitations related to immune responses and biodegradability [6,7]. Naturally derived materials, on the other hand, have been explored with the idea that they could better resemble the native environment of cardiac cells [8,9], especially those based on the extracellular matrix (ECM) obtained from the decellularization of cardiac tissues (dECM) [10–12]. An interesting feature of dECM patches is that, because of their biomimetic nature, they can be repopulated with human mesenchymal stromal cells (hMSCs) to enhance their regenerative potential [3,13]. Although such developed cardiac patches have shown good outcomes, they still present certain limitations, for instance, their inadequate mechanical properties (e.g., low stiffness) and the lack of patient-specific shapes [14,15]. Thus, there is a need for bioprintable cell-laden hydrogels (bioinks) to develop customized patches exhibiting enough mechanical strength.

In the present work, we describe an optimized protocol to produce cardiac hydrogels from the enzymatic digestion of decellularized porcine myocardium (cECM) that can be loaded with cells, and which are bioprintable. The bioink is generated in a way that can be fabricated and bioprinted without the need to mix the cECM with other biomaterials or incorporate external crosslinkers. The mechanical properties of the developed hydrogel were assessed by rheometry and its ultrastructure by imaging with a Scanning Electron Microscope. The cECM bioink (loaded with human-bone-marrow-derived mesenchymal stem cells -hBM-MSCs) was bioprinted by using a two-printhead strategy with a sacrificial material. The behavior of the cells when cultured in 3D within the developed scaffolds was then studied by immunostaining and contraction assays.

2. Results and Discussion

2.1. Macro- and Ultra-Structure of the Hydrogels Developed from the Digested cECM

The hydrogels showed a homogeneous structure with no macroscopically visible fiberboard clusters. Structures formed with the developed hydrogels showed enough strength to be manipulated with tweezers without breaking and with the capability of recovering the original shape after manipulation (Figure 1A). This improvement with respect to previously reported studies is achieved by working at higher powder concentration and by optimizing the digestion time (as compared with the common protocol originally published by Freytes and coworkers [16]). Thanks to such improvement in mechanical strength, the 3D-bioprinted structures can be manipulated with surgical tools so they could be implanted in vivo by using a bio-glue. This aspect represents an important advance in the case of patches developed from ECM-derived cardiac bioinks. Indeed, to have the required strength, previously reported materials presented weaker mechanical properties, so they should be mixed with other biomaterials [17] not naturally included in the native tissue or additionally crosslinked [18].

Regarding to ultrastructure, SEM images of the hydrogels showed a fibrillary structure as expected (Figure 1B). The average diameter of the fibers of the cECM hydrogel was 126 ± 8 nm. These data agree with the diameter of tropocollagen fibers [19], which is consistent with collagen I fibrils, reported to be in the 100 nm diameter range [20]. The structure observed in cECM hydrogels is closer to that of the native ECM compared with previous works [21], showing fibers with a diameter slightly below that of the collagen present in the ECM. As can be observed in Figure 1B, the structures obtained with the developed bioink presented a distribution of the fibers resembling that of the original cardiac ECM [22,23]. Moreover, thanks to the small diameter of the fibers, the hydrogel presented a high contact area [24], which has been reported to improve cell attachment and proliferation [25,26].

2.2. Rheological Properties of the cECM Hydrogels Depending on the Pepsin Digestion Time

The rheological properties of the cECM hydrogels depended on the digestion time (Figure 2). Gelation started gradually, being the hydrogel digested for 24 h the one showing the highest storage modulus ($G' = 23.9 \pm 10.6$ Pa), and the hydrogel digested for 16 h, the one with the lowest storage modulus ($G' = 4.8 \pm 1.4$ Pa). Of note, for digestion times above 24 h, the cECM did not reach gelation, in keeping with previous reports regarding other natural, organ-derived hydrogels, such as lung extracellular matrix or type I collagen hydrogels [27].

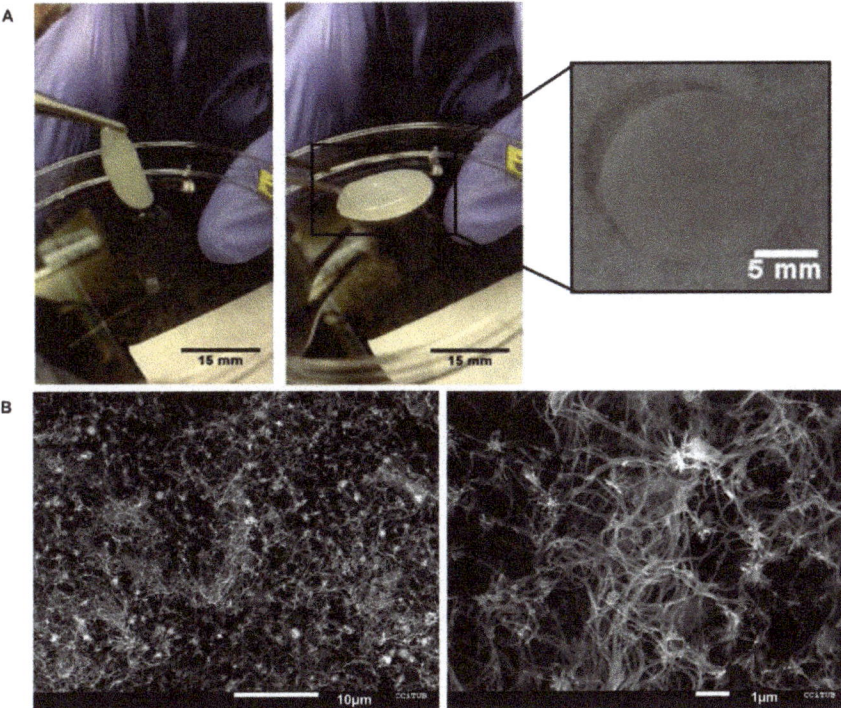

Figure 1. Structural characterization of the cECM hydrogels. Macro image of the cECM hydrogel, showing a homogeneous structure and its manipulation with tweezers (**A**). SEM images of the cECM hydrogel (**B**).

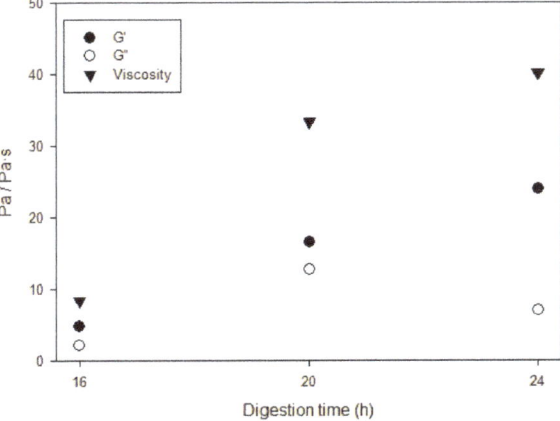

Figure 2. Rheological properties of the cECM hydrogels as a function of the pepsin digestion time ($N = 2$).

Thus, 24 h was the chosen digestion time for the bioinks used in the rest of the experiments in the present work. The required digestion time is highly dependent on the particle size obtained after decellularization and cryogenic milling. If the digestion time is not long enough, big particles cannot be fully solubilized, while if the time is higher than optimal, pepsin may start to degrade certain proteins that are crucial for hydrogel formation.

2.3. 3D Cultures of hBM-MSCs

When cultured in 3D, hBM-MSCs contracted the cECM hydrogels, as measured via the hydrogel contraction assay (Figure 3). As expected, the shrinkage of the structures increased with the culture time. After 1 day of 3D culture, the measured area of the cECM hydrogels with hBM-MSCs was reduced to 158.5 ± 8.3 mm^2 (12.9 ± 4.5 percent of contraction; $p = 0.076$). At day 4, the measured value was 111.2 ± 11.2 mm^2 (38.9 ± 6.2 percent of contraction; $p = 0.005$), and after 7 days in culture, the area was 85.7 ± 7.4 mm^2 (52.9 ± 4.1 percent of contraction; $p < 0.001$).

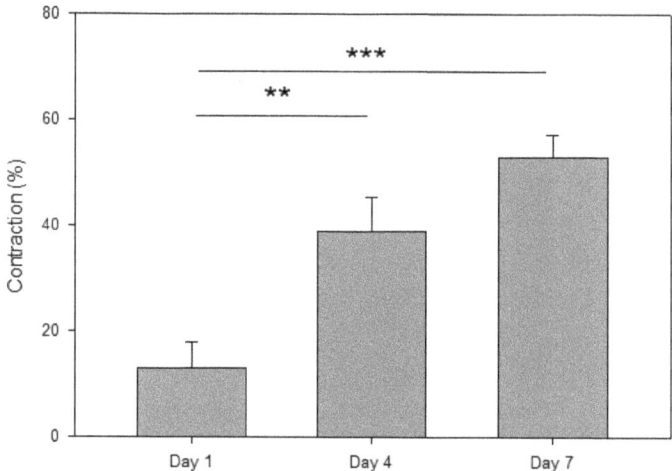

Figure 3. Contraction of the 3D-cultured cECM hydrogels (hBM-MSC) with respect to acellular ones (control) for 1, 4 and 7 days. **: $p < 0.01$; ***: $p < 0.001$.

The contraction observed in the cell-laden structures indicated an active crosstalk between the cECM hydrogel matrix and the cells. The concentration of cells used in the present study, which is in the lower range (2.5×10^5–5×10^5 cells/mL) recommended in the seminal paper by Freytes and coworkers for stem cell-laden cardiac hydrogels [16], seemed to be enough for this type of cells to interact with their surrounding ECM. Nevertheless, several effects could be overlapping, explaining the observed contraction, as cells can be effectively pulling the fibers of the structure while, at the same time, degrading matrix proteins by secreting metalloproteinases. Previous studies conducted with fibroblasts and stromal cells cultured in collagen matrices have shown that the combination of both factors strongly depends on structural mechanics and other stimuli that may alter cell contractility [28–30]. Although it is out of the scope of the present study, further work should be carried out to determine which of the mechanisms is dominating for this kind of cells in cECM hydrogels.

To better understand these cell–matrix interactions, the alteration of the mechanical properties of the structures due to the cell culture was studied via rheometry. As shown in Figure 4, cell-laden structures were softer when compared with the acellular ones. As expected, values for the shear modulus and viscosity decreased with the applied strain. Measured rheological properties were in the same range of values previously reported for other cECM hydrogels [15]. Interestingly, cell-laden hydrogels seemed to be softer than acellular ones as shown in the representative example in Figure 4. Nevertheless, these experiments have the limitation that the size of cell-laden structures varies from replicates due to contraction so an extensive analysis of this aspect would be extremely complex (due to physical limitations of the rheometer) and it is out of the scope of this work. Even in this case, an explanation for the observed softening of the structures could be that cells are generally softer than their surrounding ECM or by the fact that cells are degrading the ECM where they are cultured.

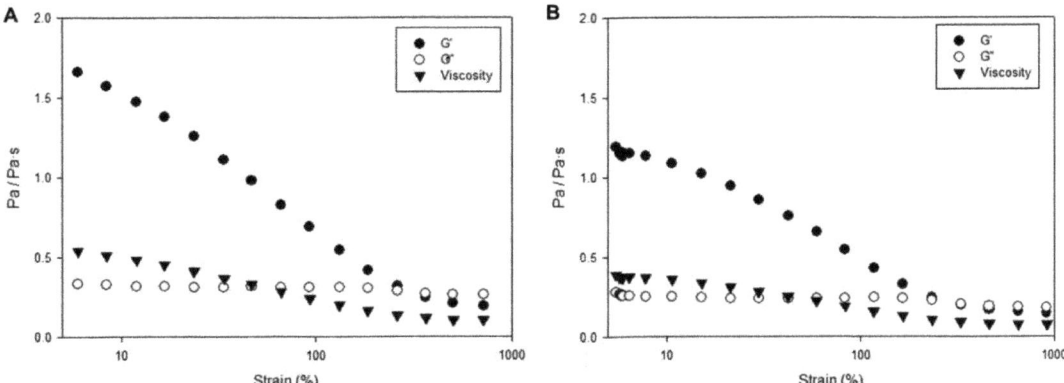

Figure 4. Comparison of the storage modulus (G', G" and viscosity) in acellular (**A**) and cellular hydrogels (**B**) in a representative example.

Immunofluorescence images of 3D cultured hBM-MSCs for 7 days can be observed in Figure 5. Cells were stained for Connexin 43 (Cx43) and α-Smooth muscle actin (α-SMA) in green. The actin filaments of the cytoskeleton and the nuclei were also stained. Together with high-magnification images to show single-cell morphology, low-magnification images reveal that there are several cells positive for the different stainings.

Even with the difficulties of imaging cells within hydrogels (cells are not as well aligned with the imaging plane as they are when imaged on top of plane surfaces), data from Figure 5 illustrate that cells presented a well-formed cytoskeleton with the characteristic spindle shape of MSCs. Additionally, immunostaining revealed that cells expressed Cx43 and α-SMA after 7 days of 3D culture within the developed scaffolds.

The observed results indicate that the cells were able to remain alive when cultured in the developed cardiac patches built with cECM bioinks, showing a cytoskeleton-spindle-like morphology typical of MSCs [31]. It has been previously proven that hMSCs have the ability to release paracrine factors and extracellular vesicles (EVs) exhibiting immunomodulatory, anti-inflammatory, and antimicrobial effects [32–36]. The proliferation and/or differentiation of the cells within the developed dECM hydrogels are not studied in the present work. Proliferation has not been generally observed in previous studies in short periods of culture (1 week), while differentiation was not expected to occur. Indeed, hBM-MSCs were maintained in culture by using manufacturer-recommended medium and supplements, which are designed to keep the stemness of the cultured cells. Although, the presented data show cell survival together with the positive expression of Cx43 and α-SMA (which indicated cell–cell and cell-matrix interactions, which have been shown to play an important role in the physiology of the heart [37,38] and its ECM remodeling [39,40]) it was shown that the developed bioinks at the cell concentration used are suitable for the development of hMSC-loaded cardiac patches (although it would be interesting to test lower and higher concentrations in future works).

Bioinks developed exclusively from the digestion of decellularized ECM should be printed in liquid phase, hence a support material is needed (F127 in our case), which should be removed after the crosslinking of the final structures. This aspect represents a limitation if complex shapes must be developed, but it is well suited for developing a customized cardiac patch. Moreover, as the bioink is printed in liquid phase, the pressure required is quite low (lower that the one exerted on the cells when manipulated with a standard laboratory pipette), thus fastening the test phase of new developments as they can be carried out with standard labware prior to testing in with the bioprinter.

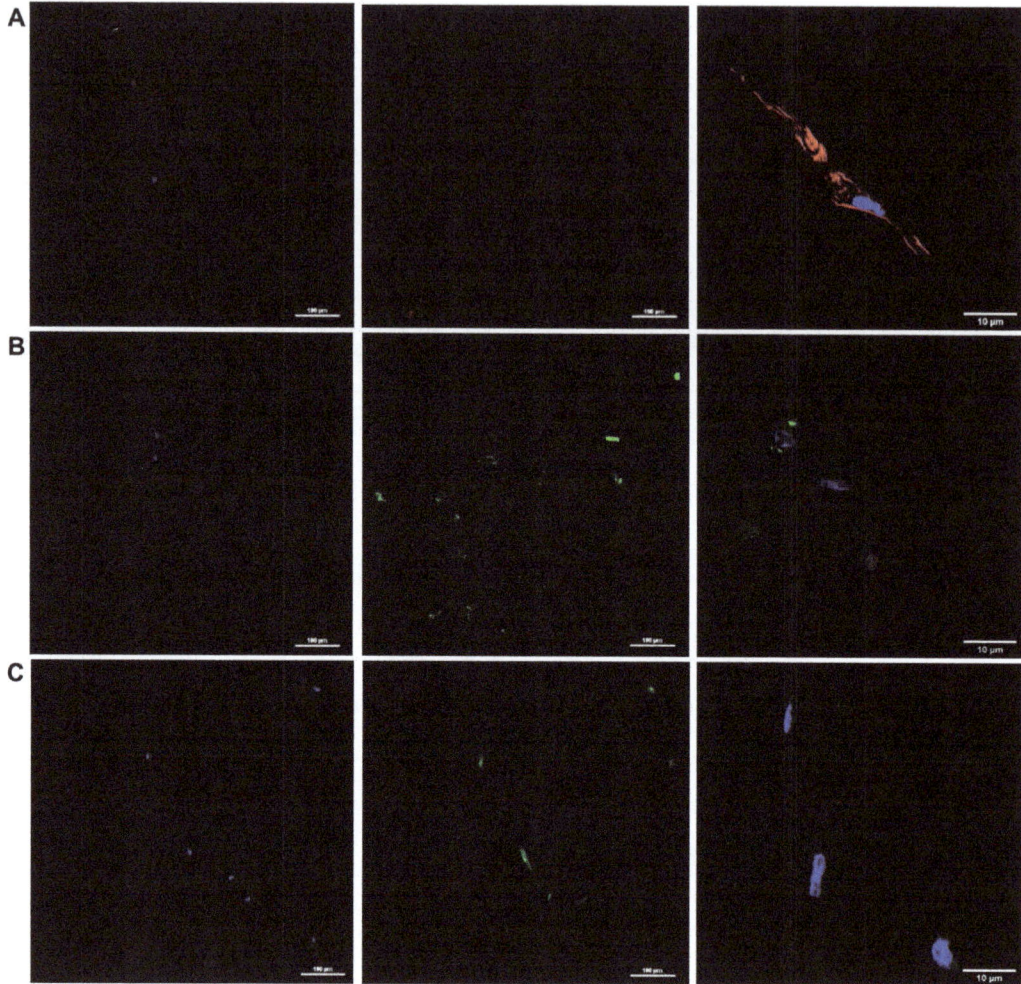

Figure 5. Immunostaining of hBM-MSC 3D-cultured in the cECM hydrogels for 7 days. (**A**) Phalloidin (red). (**B**) α-SMA (green). (**C**) Cx43 (green). Left and center images are low-magnification (20×) with split channels, while the nucleus is counter-stained in blue; right images are high-magnification (100×) images.

Previous studies have reported the use of pericardium dECM to engineer patches for the treatment of myocardial infarction [3]. Indeed, rat dECM patches have successfully replaced the right ventricular outflow tract defect in Lewis rat models, showing no differences between the area in contact with the patch and the healthy ventricles [41]. Moreover, patches have also shown signs of neovascularization and nerve sprouting in the infarcted area that was in contact with either human pericardial or porcine myocardial dECM scaffolds [13]. The results presented herein extend the potential of cardiac patches based on dECM, as they open the door to developing custom-shaped grafts using 3D bioprinting technology [42].

The idea of customizing the shape of the cardiac patches started when Pati et al. [43] printed three-dimensional cardiac tissue with cECM hydrogel made from porcine dECM and human-adipose-derived MSCs, achieving high cell viability. Jang et al. [44] added Vascular Endothelial Growth Factor (VEGF) and MSCs to a porcine dECM pregel to promote a rapid vascularization of cell-laden heart cECM bioink. In both cases, polycaprolactone (PCL, a synthetic polymer) was used to provide mechanical support to the structure, due to the weak mechanical properties of their cECM hydrogel alone. Other approaches used crosslinkers [45–47] and biomaterials from natural and synthetic origins (such as silk, genipin [18], or gelatin methacrylate-GelMA [14]) with the same objective of obtaining patches with enough strength to be manipulated with surgical tools and hence for in vivo application. The solution achieved in the present work by optimizing the digestion of cECM overcomes those previous limitations, as there is no need to use a complementary biomaterial or to crosslink the patches after 3D bioprinting.

3. Conclusions

The use of MSCs for the repair of the infarcted zones of the heart is a promising therapy, with the release of paracrine factors as the main mechanism of repair involved. The use of MSC-laden grafts in the form of myocardial patches to be implanted in the heart represents a major advance in this field, since the therapeutic effect of the cells can be sustained over time when using this strategy. Nevertheless, MSCs are highly sensitive to the physico-chemical properties of their ECM microenvironment. Therefore, it is important to mimic the characteristics of the ECM as much as possible when engineering a therapeutic graft. Among all possible procedures, the use of cardiac ECM obtained via native tissue decellularization as a biomaterial to develop the grafts is one the most promising nowadays, because the scaffold has components from the ECM exclusively. In this work, we have developed a novel cardiac ECM bioink (digested decellularized myocardial ECM mixed with hBM-MSCs and bioprinted in the pregel phase) in such a way that the obtained patches can be manipulated with surgical tools, and thus applied in vivo, not requiring the use of additional biomaterials or crosslinkers. Moreover, the printability (using a two-nozzle system with a supporting material) of the bioink allows for custom-shaped patch development. The morphology and expression of relevant markers of hMSCs cultured within the developed scaffolds showed effective cell–cell and cell–matrix interactions after a 1-week culture. However, further in vivo studies in animal models are required to ascertain the therapeutic effect of these cell-cultured scaffolds. Nevertheless, the preliminary results we show herein, together with results we previously reported for MSCs cultured within lung ECM hydrogels [24], are promising and suggest that the cECM bioink presented in this work could be an excellent material to develop customized patches for cell therapy in cardiac repair.

4. Materials and Methods

All the reagents employed were obtained from Sigma Aldrich (Saint Louis, MO, USA) unless otherwise specified.

4.1. Preparation of the cECM Bioinks from the Decellularization of Porcine Myocardia

Porcine hearts were obtained from a local slaughterhouse and then washed with deionized water before freezing them at −80 °C to promote cell lysis and storage until further processing. The decellularization method was adapted from a protocol developed for human hearts [46], with slight modifications. Briefly, hearts were thawed to room temperature (RT), and the left ventricular myocardium was sectioned into $1 \times 1 \times 1$ cm^3 cubes and frozen for further cryo-sectioning into 300 μm-thickness slices using a cryostat (HM 560, Thermo Fisher Scientific, Waltham, MA, USA). The resulting myocardial slices were then decellularized via sequential immersion in lysis solution (10 mM Tris, 0.1% w/v ethylenediaminetetraacetic acid (EDTA), pH 7.4 in dH$_2$O, 2 h at RT), 0.5% sodium dodecyl sulfate (SDS) (6 h at RT), and fetal bovine serum (FBS) (3 h, 37 °C) with intermediate washes in phosphate-buffered saline (PBS) 1x. At the end of the decellularization process, the slices were drained and stored at −80 °C. To produce the cECM powder, slices were freeze-dried for 48 h (Lyoquest-55 Plus, Telstar, Terrassa, Spain) and pulverized into a micrometric powder using a cryogenic miller (6775 Freezer/Mill, SPEX, Metuchen, NJ, USA).

The resulting cECM powder was digested at a concentration of 20 mg/mL in a 0.1 M HCl solution with pepsin from porcine gastric mucosa (1:10 concentration) under magnetic stirring at RT for different times (for experiments other than rheology, 24 h digestion was chosen as this time and resulted in the highest storage modulus). After digestion of the ECM (where pH never rose above 3), the pregel solution was stabilized to pH = 7.4 ± 0.4 using 1 M NaOH (1:10 v/v respect to the HCl volume) and PBS 10x (1:10 v/v respect to the neutralized volume). For 3D bioprinting, after loading the pregel with the cells (Figure 6A, Section 4.4 of the present work for more details) the dual-printhead method described in [27] was used as schematically represented in Figure 6C. Briefly, one printhead of the 3D bioprinter (3Ddiscovery, RegenHU, Switzerland, installed inside a safety laminar flow cabinet) was filled with the pregel and maintained at 4 °C to prevent gelification. A secondary printhead was filled with Pluronic F-127 gel at room temperature (RT). Structures were then bioprinted by alternatively printing an F127 layer (40% v/v in PBS 1x, printed at ~4.5 atm with a needle of 250 μm of diameter—Nordson EFD, Westlake, OH, USA), which served as a template, and a pregel layer to form the desired shape. As the pregel is maintained in liquid phase, the pressure exerted was just above atmospheric pressure. At the end of the bioprinting process, scaffolds were incubated at 37 °C for 45 min to crosslink the cell-laden cECM hydrogels. Finally, the Pluronic was dissolved by immersing the developed structures in culture media at 4 °C for 10 min.

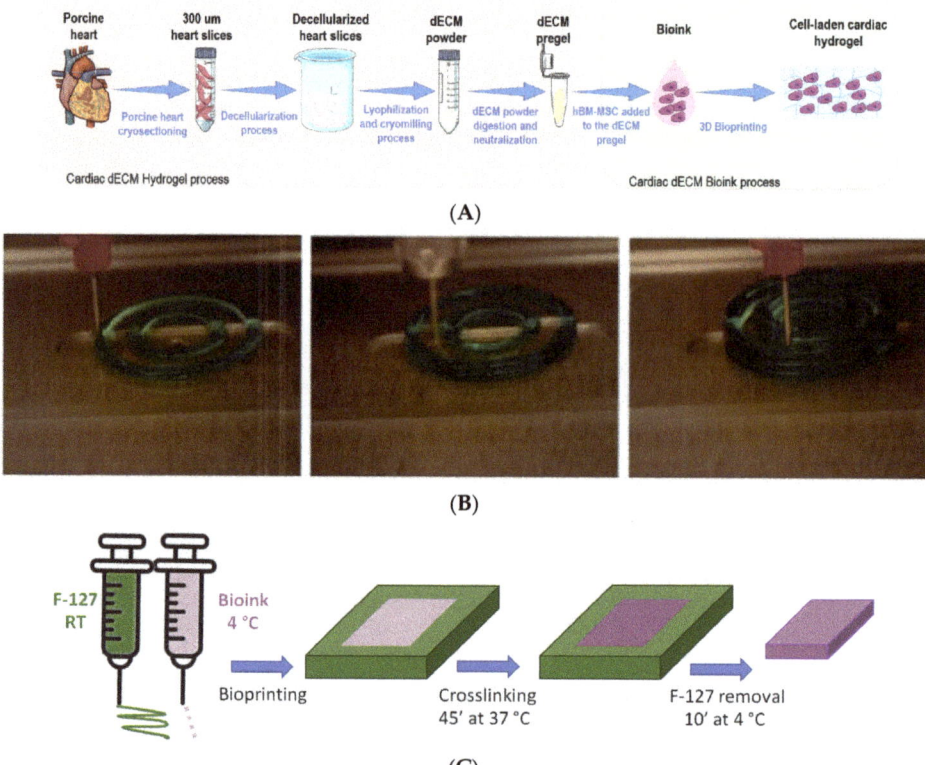

Figure 6. (**A**) Schematic description of the process followed to obtain cECM pregel and the bioink. (**B**) Bioprinting process using F-127 as supportive/sacrificial layer (in green), adapted from [24]. (**C**) Schematic of the technique used for the bioprinting the cell-laden pregel, crosslinking the structure and then removing the supportive layer of F-127.

4.2. Ultrastructural Characterization by Scanning Electron Microscopy

The ultrastructure of the cECM hydrogel scaffolds was visualized with a JSM-6510 (JEOL, Tokyo, Japan) scanning electron microscope (SEM). cECM scaffolds of $20 \times 9 \times 3$ mm^3 were produced via 3D bioprinting casting

The 3D scaffolds were fixed in 4% paraformaldehyde (PFA) in PBS for 48 h and then washed three times with 0.1 M phosphate buffer (PB). Next, the samples were incubated in 4% osmium tetroxide for 90 min and then rinsed with deionized water. Subsequently, samples were dehydrated by washing them with ethanol 80% ($\times 2$), 90% ($\times 3$), 96% ($\times 3$), and 100% ($\times 3$) and preserved in absolute ethanol at 4 °C until critical point drying (Autosamdri-815 critical point dryer, Tousimis, Rockville, MD, USA). Samples were then carbon coated and mounted using conductive adhesive tabs (TED PELLA, Redding, CA, USA). Imaging was performed by using an SEM (JSM-6510, JEOL, Tokyo, Japan) at 15 kV.

The diameter of the fibers was assessed following the method developed in [47]. Briefly, 10 fibers of three different zones of each sample were randomly selected, and their diameter was computed with ImageJ Software v1.53 (National Institute of Health, Bethesda, MD, USA).

4.3. Hydrogel Rheology

The rheology of the developed bioinks was measured using a Haake RheStress1 rheometer (Thermo Fisher, MA, USA) with a 35 mm serrated parallel plate. The storage (G') and loss (G'') moduli were measured, and the modulus of the complex viscosity (η^*) was calculated by using Equation (1) for a given frequency ω.

$$\eta^* = \eta' - i\eta'' = \frac{G''}{\omega} - \frac{iG'}{\omega} \qquad (1)$$

Hydrogels were digested for 16 h, 20 h, 24 h, and 28 h. The gelation kinetics of the acellular hydrogels was assessed by loading a pregel solution onto a Peltier plate at 4 °C for 10 min at 0.628 rad/s frequency (0.1 Hz). The temperature was kept constant for 10 min, subsequently increased to 37 °C, and kept constant for 15 min.

4.4. 3D Cell Culture of hBM-MSCs

Human bone marrow mesenchymal stromal cells (ATCC, Manassas, VA, USA) were expanded in tissue culture plates by following the manufacturer's instructions. Cells in passages 3–5 were used for conducting all the experiments in this study.

The 3D cultures were prepared by cooling the digested ECM solution to 4 °C in the fridge and then maintaining it in ice prior to neutralization by using ice-cold NaOH and PBS 10X as previously described. Once the pregel was at 4 °C and at physiological pH, it was mixed with the cells resuspended in ice-cold culture media (2.5×10^6 cells/mL) at a relation 10:1 v/v. To form the cell-laden hydrogel structures, the resulting mixture was then incubated at 37 °C for 45 min to form disk-shaped structures of 1.9 cm^2 of surface area. The final cell concentration in the scaffolds was 2.5×10^5 cells/mL. The cell-laden structures were cultured for 7 days, changing the medium every 3 days.

The diameter of cell-laden structures was measured on days 1, 4, and 7 after cell seeding. cECM hydrogels were imaged with a high-resolution camera fixed to a tripod after aspirating the culture media to avoid image errors due to the movement of the structure. The surface area of the 3D cultures was quantified from the acquired images by using ImageJ Software, and the contraction was calculated as the percentage of reduction with respect to the corresponding acellular controls [27].

The influence of the cells on the rheological properties once the structures were gellified was assessed for the cultures at 7 days. Then, rheometry on the cell-laden hydrogels (24 h of pepsin digestion) was characterized at 37 °C by using an amplitude sweep from 5% (which showed to be the deformation for the cell-laden hydrogels when applying the lowest value for tension that the equipment allowed) to 1000% at a frequency of 3.14 rad/s (0.5 Hz).

For the immunohistochemical analysis of the cells cultured in 3D within the cardiac bioinks, cell-laden hydrogels at day 7 were immersed in Optimal Cutting Temperature (OCT) and frozen at −80 °C. Thin hydrogel slices (≈12 μm) were obtained via cryosectioning (HM 560, Thermo Fisher Scientific, MA, USA) and placed on top of positively charged glass slides. OCT was removed by thawing and washing the samples in PBS 1X solution at room temperature. Subsequently, cells were fixed with 4% PFA for 15 min. Primary and secondary antibodies were incubated overnight and for 2 h at 37 °C. Nuclei were stained with Hoechst 33342 (Thermo Fisher Scientific, MA, USA) for 15 min. Primary antibodies employed were anti-Cx43 (ab11370, abcam, Cambridge, UK) and anti-αSMA (ab32575, abcam, Cambridge, UK). Secondary antibodies used were Alexa 488 goat anti-rabbit (ab150081, abcam, Cambridge, UK) and Alexa 488 goat anti-mouse antibody (ab150117, abcam, Cambridge, UK). Images were acquired with a Nikon D-Eclipse Ci confocal microscope with a 20× and 100× Plan Apo objectives (Nikon, Tokyo, Japan). Usual controls for immunostaining (e.g., without the primary antibody or without the secondary antibody) were performed to ensure non-specific staining in the experiments.

4.5. Statistical Analysis

Data are expressed as mean ± SE. For the digestion time and contraction analyses, one-way analysis of variance (ANOVA), followed by the Holm–Sidak's post hoc test, was conducted, while paired t-test was performed for rheological characterization. Statistical significance was considered at p-values < 0.05.

Author Contributions: Conceptualization, H.S.-F., C.H.-D., R.F. and J.O.; Formal analysis, H.S.-F., C.H.-D., A.U., R.S., N.G., R.F., I.A. and J.O.; Funding acquisition, R.F., I.A., N.G. and J.O.; Investigation, H.S.-F., C.H.-D., B.F., N.G., I.A., R.S., R.F. and J.O.; Methodology, H.S.-F., C.H.-D., A.U., B.F., R.F. and J.O.; Project administration, R.F. and J.O.; Resources, N.G., R.S., R.F., N.G., I.A. and J.O.; Software, H.S.-F., C.H.-D., A.U. and J.O.; Supervision, R.F. and J.O.; Writing of original draft, H.S.-F., C.H.-D. and J.O. All authors have read and agreed to the published version of the manuscript.

Funding: This research was funded by the Spanish Ministry of Science, Innovation and Universities, grants numbers PID2020-113910RB-I00-AEI/10.13039/501100011033, PGC2018-097323-A-I00 and PID2021-128674OB-I00.

Data Availability Statement: The datasets generated for this study are available on justified request from the corresponding author.

Acknowledgments: The authors wish to thank Miguel A. Rodríguez and Elisabet Urrea (Unit of Biophysics and Bioengineering) for their excellent technical assistance, and Josep Rebled and Eva Prats (Scientific Services of the University of Barcelona) for their aid with the scanning electron microscope.

Conflicts of Interest: The authors declare no conflict of interest.

References

1. Muzzio, N.; Moya, S.; Romero, G. Multifunctional Scaffolds and Synergistic Strategies in Tissue Engineering and Regenerative Medicine. *Pharmaceutics* **2021**, *13*, 792. [CrossRef] [PubMed]
2. Yadid, M.; Oved, H.; Silberman, E.; Dvir, T. Bioengineering Approaches to Treat the Failing Heart: From Cell Biology to 3D Printing. *Nat. Rev. Cardiol.* **2022**, *19*, 83–99. [CrossRef] [PubMed]
3. Prat-Vidal, C.; Rodríguez-Gómez, L.; Aylagas, M.; Nieto-Nicolau, N.; Gastelurrutia, P.; Agustí, E.; Gálvez-Montón, C.; Jorba, I.; Teis, A.; Monguió-Tortajada, M.; et al. First-in-Human PeriCord Cardiac Bioimplant: Scalability and GMP Manufacturing of an Allogeneic Engineered Tissue Graft. *eBioMedicine* **2020**, *54*, 102729. [CrossRef] [PubMed]
4. Esmaeili, H.; Patino-Guerrero, A.; Hasany, M.; Ansari, M.O.; Memic, A.; Dolatshahi-Pirouz, A.; Nikkhah, M. Electroconductive Biomaterials for Cardiac Tissue Engineering. *Acta Biomater.* **2022**, *139*, 118–140. [CrossRef] [PubMed]
5. Kc, P.; Hong, Y.; Zhang, G. Cardiac Tissue-Derived Extracellular Matrix Scaffolds for Myocardial Repair: Advantages and Challenges. *Regen. Biomater.* **2019**, *6*, 185–199. [CrossRef]
6. Peña, B.; Laughter, M.; Jett, S.; Rowland, T.J.; Taylor, M.R.G.; Mestroni, L.; Park, D. Injectable Hydrogels for Cardiac Tissue Engineering. *Macromol. Biosci.* **2018**, *18*, 1800079. [CrossRef]
7. Sadtler, K.; Wolf, M.T.; Ganguly, S.; Moad, C.A.; Chung, L.; Majumdar, S.; Housseau, F.; Pardoll, D.M.; Elisseeff, J.H. Divergent Immune Responses to Synthetic and Biological Scaffolds. *Biomaterials* **2019**, *192*, 405–415. [CrossRef]

8. Sanz-Fraile, H.; Amoros, S.; Mendizabal, I.; Galvez-Monton, C.; Prat-Vidal, C.; Bayes-Genis, A.; Navajas, D.; Farre, R.; Otero, J. Silk-Reinforced Collagen Hydrogels with Raised Multiscale Stiffness for Mesenchymal Cells 3D Culture. *Tissue Eng. Part A* **2020**, *26*, 358–370. [CrossRef]
9. Elkhoury, K.; Morsink, M.; Sanchez-Gonzalez, L.; Kahn, C.; Tamayol, A.; Arab-Tehrany, E. Biofabrication of Natural Hydrogels for Cardiac, Neural, and Bone Tissue Engineering Applications. *Bioact. Mater.* **2021**, *6*, 3904–3923. [CrossRef]
10. Taylor, D.A.; Sampaio, L.C.; Ferdous, Z.; Gobin, A.S.; Taite, L.J. Decellularized Matrices in Regenerative Medicine. *Acta Biomater.* **2018**, *74*, 74–89. [CrossRef]
11. Bejleri, D.; Davis, M.E. Decellularized Extracellular Matrix Materials for Cardiac Repair and Regeneration. *Adv. Healthc. Mater.* **2019**, *8*, 1801217. [CrossRef] [PubMed]
12. Saldin, L.T.; Cramer, M.C.; Velankar, S.S.; White, L.J.; Badylak, S.F. Extracellular Matrix Hydrogels from Decellularized Tissues: Structure and Function. *Acta Biomater.* **2017**, *49*, 1–15. [CrossRef] [PubMed]
13. Perea-Gil, I.; Gálvez-Montón, C.; Prat-Vidal, C.; Jorba, I.; Segú-Vergés, C.; Roura, S.; Soler-Botija, C.; Iborra-Egea, O.; Revuelta-López, E.; Fernández, M.A.; et al. Head-to-Head Comparison of Two Engineered Cardiac Grafts for Myocardial Repair: From Scaffold Characterization to Pre-Clinical Testing. *Sci. Rep.* **2018**, *8*, 6708. [CrossRef] [PubMed]
14. Bejleri, D.; Streeter, B.W.; Nachlas, A.L.Y.; Brown, M.E.; Gaetani, R.; Christman, K.L.; Davis, M.E. A Bioprinted Cardiac Patch Composed of Cardiac-Specific Extracellular Matrix and Progenitor Cells for Heart Repair. *Adv. Healthc. Mater.* **2018**, *7*, 1800672. [CrossRef]
15. Johnson, T.D.; Lin, S.Y.; Christman, K.L. Tailoring Material Properties of a Nanofibrous Extracellular Matrix Derived Hydrogel. *Nanotechnology* **2011**, *22*, 494015. [CrossRef]
16. Freytes, D.O.; O'Neill, J.D.; Duan-Arnold, Y.; Wrona, E.A.; Vunjak-Novakovic, G. Natural Cardiac Extracellular Matrix Hydrogels for Cultivation of Human Stem Cell-Derived Cardiomyocytes. *Methods Mol. Biol.* **2014**, *1181*, 69–81. [CrossRef]
17. Basara, G.; Gulberk Ozcebe, S.; Ellis, B.W.; Zorlutuna, P. Tunable Human Myocardium Derived Decellularized Extracellular Matrix for 3d Bioprinting and Cardiac Tissue Engineering. *Gels* **2021**, *7*, 70. [CrossRef]
18. Efraim, Y.; Sarig, H.; Cohen Anavy, N.; Sarig, U.; de Berardinis, E.; Chaw, S.Y.; Krishnamoorthi, M.; Kalifa, J.; Bogireddi, H.; Duc, T.V.; et al. Biohybrid Cardiac ECM-Based Hydrogels Improve Long Term Cardiac Function Post Myocardial Infarction. *Acta Biomater.* **2017**, *50*, 220–233. [CrossRef]
19. Buehler, M.J. Nature Designs Tough Collagen: Explaining the Nanostructure of Collagen Fibrils. *Proc. Natl. Acad. Sci. USA* **2006**, *103*, 12285–12290. [CrossRef]
20. Vesentini, S.; Redaelli, A.; Gautieri, A. Nanomechanics of Collagen Microfibrils. *Muscles Ligaments Tendons J.* **2013**, *3*, 23–34. [CrossRef]
21. Pouliot, R.A.; Link, P.A.; Mikhaiel, N.S.; Schneck, M.B.; Valentine, M.S.; Gninzeko, F.J.K.; Herbert, J.A.; Sakagami, M.; Heise, R.L. Development and Characterization of a Naturally Derived Lung Extracellular Matrix Hydrogel. *J. Biomed. Mater. Res. Part A* **2016**, *104*, 1922–1935. [CrossRef] [PubMed]
22. Oropeza, B.P.; Adams, J.R.; Furth, M.E.; Chessa, J.; Boland, T. Bioprinting of Decellularized Porcine Cardiac Tissue for Large-Scale Aortic Models. *Front. Bioeng. Biotechnol.* **2022**, *10*, 855186. [CrossRef] [PubMed]
23. Reid, J.A.; Callanan, A. Hybrid Cardiovascular Sourced Extracellular Matrix Scaffolds as Possible Platforms for Vascular Tissue Engineering. *J. Biomed. Mater. Res. B Appl. Biomater.* **2020**, *108*, 910–924. [CrossRef] [PubMed]
24. Eichhorn, S.J.; Sampson, W.W. Relationships between Specific Surface Area and Pore Size in Electrospun Polymer Fibre Networks. *J. R. Soc. Interface* **2010**, *7*, 641–649. [CrossRef]
25. Wang, B.; Borazjani, A.; Tahai, M.; De Jongh Curry, A.L.; Simionescu, D.T.; Guan, J.; To, F.; Elder, S.H.; Liao, J. Fabrication of Cardiac Patch with Decellularized Porcine Myocardial Scaffold and Bone Marrow Mononuclear Cells. *J. Biomed. Mater. Res. A* **2010**, *94*, 1100–1110. [CrossRef]
26. Singelyn, J.M.; DeQuach, J.A.; Seif-Naraghi, S.B.; Littlefield, R.B.; Schup-Magoffin, P.J.; Christman, K.L. Naturally Derived Myocardial Matrix as an Injectable Scaffold for Cardiac Tissue Engineering. *Biomaterials* **2009**, *30*, 5409–5416. [CrossRef]
27. Falcones, B.; Sanz-Fraile, H.; Marhuenda, E.; Mendizábal, I.; Cabrera-Aguilera, I.; Malandain, N.; Uriarte, J.J.; Almendros, I.; Navajas, D.; Weiss, D.J.; et al. Bioprintable Lung Extracellular Matrix Hydrogel Scaffolds for 3D Culture of Mesenchymal Stromal Cells. *Polymers* **2021**, *13*, 2350. [CrossRef]
28. Kobayashi, N.; Yasu, T.; Ueba, H.; Sata, M.; Hashimoto, S.; Kuroki, M.; Saito, M.; Kawakami, M. Mechanical stress promotes the expression of smooth muscle-like properties in marrow stromal cells. *Exp. Hematol.* **2004**, *32*, 1238–1245. [CrossRef]
29. Rouabhia, M.; Park, H.; Meng, S.; Derbali, H.; Zhang, Z. Electrical Stimulation Promotes Wound Healing by Enhancing Dermal Fibroblast Activity and Promoting Myofibroblast Transdifferentiation. *PLoS ONE* **2013**, *8*, e0071660. [CrossRef]
30. Grinnell, F. Fibroblast Biology in Three-Dimensional Collagen Matrices. *Trends Cell Biol.* **2003**, *13*, 264–269. [CrossRef]
31. Ichinose, S.; Tagami, M.; Muneta, T.; Sekiya, I. Morphological Examination during in Vitro Cartilage Formation by Human Mesenchymal Stem Cells. *Cell Tissue Res.* **2005**, *322*, 217–226. [CrossRef] [PubMed]
32. Lee, J.W.; Krasnodembskaya, A.; McKenna, D.H.; Song, Y.; Abbott, J.; Matthay, M.A. Therapeutic Effects of Human Mesenchymal Stem Cells in Ex Vivo Human Lungs Injured with Live Bacteria. *Am. J. Respir. Crit. Care Med.* **2013**, *187*, 751–760. [CrossRef] [PubMed]
33. Gore, A.V.; Bible, L.E.; Song, K.; Livingston, D.H.; Mohr, A.M.; Sifri, Z.C. Mesenchymal Stem Cells Increase T-Regulatory Cells and Improve Healing Following Trauma and Hemorrhagic Shock. *J. Trauma Acute Care Surg.* **2015**, *79*, 48–52. [CrossRef] [PubMed]

34. Curley, G.F.; Hayes, M.; Ansari, B.; Shaw, G.; Ryan, A.; Barry, F.; O'Brien, T.; O'Toole, D.; Laffey, J.G. Mesenchymal Stem Cells Enhance Recovery and Repair Following Ventilator-Induced Lung Injury in the Rat. *Thorax* **2012**, *67*, 496–501. [CrossRef]
35. Devaney, J.; Horie, S.; Masterson, C.; Elliman, S.; Barry, F.; O'brien, T.; Curley, G.F.; O'toole, D.; Laffey, J.G. Human Mesenchymal Stromal Cells Decrease the Severity of Acute Lung Injury Induced by E. Coli in the Rat. *Thorax* **2015**, *70*, 625–635. [CrossRef]
36. Laffey, J.G.; Matthay, M.A. Cell-Based Therapy for Acute Respiratory Distress Syndrome: Biology and Potential Therapeutic Value. *Am. J. Respir. Crit. Care Med.* **2017**, *196*, 266–273. [CrossRef]
37. Camci-Unal, G.; Annabi, N.; Dokmeci, M.R.; Liao, R.; Khademhosseini, A. Hydrogels for Cardiac Tissue Engineering. *NPG Asia Mater.* **2014**, *6*, 740–761. [CrossRef]
38. Kotini, M.; Barriga, E.H.; Leslie, J.; Gentzel, M.; Rauschenberger, V.; Schambon, A.; Mayor, R. Gap Junction Protein Connexin-43 Is a Direct Transcriptional Regulator of N-Cadherin in Vivo. *Nat. Commun.* **2018**, *9*, 3846. [CrossRef]
39. Shinde, A.V.; Humeres, C.; Frangogiannis, N.G. The Role of α-Smooth Muscle Actin in Fibroblast-Mediated Matrix Contraction and Remodeling. *Biochim. Biophys. Acta Mol. Basis Dis.* **2017**, *1863*, 298–309. [CrossRef]
40. Dye, B.K.; Butler, C.; Lincoln, J. Smooth Muscle α-Actin Expression in Mitral Valve Interstitial Cells Is Important for Mediating Extracellular Matrix Remodeling. *J. Cardiovasc. Dev. Dis.* **2020**, *7*, 32. [CrossRef]
41. Wainwright, J.M.; Hashizume, R.; Fujimoto, K.L.; Remlinger, N.T.; Pesyna, C.; Wagner, W.R.; Tobita, K.; Gilbert, T.W.; Badylak, S.F. Right Ventricular Outflow Tract Repair with a Cardiac Biologic Scaffold. *Cells Tissues Organs* **2011**, *195*, 159–170. [CrossRef] [PubMed]
42. Cui, H.; Miao, S.; Esworthy, T.; Zhou, X.; Lee, S.J.; Liu, C.; Yu, Z.X.; Fisher, J.P.; Mohiuddin, M.; Zhang, L.G. 3D Bioprinting for Cardiovascular Regeneration and Pharmacology. *Adv. Drug Deliv. Rev.* **2018**, *132*, 252–269. [CrossRef] [PubMed]
43. Pati, F.; Cho, D.-W. Bioprinting of 3D Tissue Models Using Decellularized Extracellular Matrix Bioink. In *3D Cell Culture: Methods and Protocols*; Koledova, Z., Ed.; Springer: New York, NY, USA, 2017; pp. 381–390. ISBN 978-1-4939-7021-6.
44. Jang, J.; Park, H.J.; Kim, S.W.; Kim, H.; Park, J.Y.; Na, S.J.; Kim, H.J.; Park, M.N.; Choi, S.H.; Park, S.H.; et al. 3D Printed Complex Tissue Construct Using Stem Cell-Laden Decellularized Extracellular Matrix Bioinks for Cardiac Repair. *Biomaterials* **2017**, *112*, 264–274. [CrossRef] [PubMed]
45. Gu, L.; Shan, T.; Ma, Y.X.; Tay, F.R.; Niu, L. Novel Biomedical Applications of Crosslinked Collagen. *Trends Biotechnol.* **2019**, *37*, 464–491. [CrossRef]
46. Becker, M.; Maring, J.A.; Oberwallner, B.; Kappler, B.; Klein, O.; Falk, V.; Stamm, C. Processing of Human Cardiac Tissue toward Extracellular Matrix Self-Assembling Hydrogel for in Vitro and in Vivo Applications. *J. Vis. Exp.* **2017**, *2017*, e56419. [CrossRef]
47. Baker, S.C.; Atkin, N.; Gunning, P.A.; Granville, N.; Wilson, K.; Wilson, D.; Southgate, J. Characterisation of Electrospun Polystyrene Scaffolds for Three-Dimensional in Vitro Biological Studies. *Biomaterials* **2006**, *27*, 3136–3146. [CrossRef]

Disclaimer/Publisher's Note: The statements, opinions and data contained in all publications are solely those of the individual author(s) and contributor(s) and not of MDPI and/or the editor(s). MDPI and/or the editor(s) disclaim responsibility for any injury to people or property resulting from any ideas, methods, instructions or products referred to in the content.

Article

Role of pH and Crosslinking Ions on Cell Viability and Metabolic Activity in Alginate–Gelatin 3D Prints

Andrea Souza [1], Matthew Parnell [1], Brian J. Rodriguez [2] and Emmanuel G. Reynaud [1,*]

1. School of Biomolecular and Biomedical Science, University College Dublin, D04 V1W8 Dublin, Ireland; andrea.souza@ucdconnect.ie (A.S.); matthew.parnell@ucdconnect.ie (M.P.)
2. School of Physics, University College Dublin, D04 V1W8 Dublin, Ireland; brian.rodriguez@ucd.ie
* Correspondence: emmanuel.reynaud@ucd.ie

Abstract: Alginate–gelatin hydrogels are extensively used in bioengineering. However, despite different formulations being used to grow different cell types in vitro, their pH and its effect, together with the crosslinking ions of these formulations, are still infrequently assessed. In this work, we study how these elements can affect hydrogel stability and printability and influence cell viability and metabolism on the resulting 3D prints. Our results show that both the buffer pH and crosslinking ion (Ca^{2+} or Ba^{2+}) influence the swelling and degradation rates of prints. Moreover, buffer pH influenced the printability of hydrogel in the air but did not when printed directly in a fluid-phase $CaCl_2$ or $BaCl_2$ crosslinking bath. In addition, both U2OS and NIH/3T3 cells showed greater cell metabolic activity on one-layer prints crosslinked with Ca^{2+}. In addition, Ba^{2+} increased the cell death of NIH/3T3 cells while having no effect on U2OS cell viability. The pH of the buffer also had an important impact on the cell behavior. U2OS cells showed a 2.25-fold cell metabolism increase on one-layer prints prepared at pH 8.0 in comparison to those prepared at pH 5.5, whereas NIH/3T3 cells showed greater metabolism on one-layer prints with pH 7.0. Finally, we observed a difference in the cell arrangement of U2OS cells growing on prints prepared from hydrogels with an acidic buffer in comparison to cells growing on those prepared using a neutral or basic buffer. These results show that both pH and the crosslinking ion influence hydrogel strength and cell behavior.

Keywords: 3D printing; alginate–gelatin hydrogel; pH; $CaCl_2$; $BaCl_2$; U2OS; NIH/3T3; fluid phase

1. Introduction

Recently, 3D (bio)printing technology has shown its potential to replace and complement existing methods in basic research, drug delivery, screening, and medical procedures. For instance, it allows the production of a bioprinted full-skin equivalent, which can support bone and capillary formation and breast fat tissue growth, among other biological structures created in (bio)printed scaffolds in vitro [1–4]. However, even though some important advances have been made to date, this technology is still evolving and requires more research on the basic requirements needed to create optimal matrices to grow tissues in vitro. To construct a successful environment, the biological, physical, mechanical, and chemical aspects of these scaffolds should be established specifically for each tissue and/or purpose of the study, as pathologic tissues behave differently from healthy ones.

Hydrogels are largely used in the bioengineering field due to their ability to mimic the extracellular environment. Alginate hydrogel is the second most used natural bioink in this field because of its biocompatibility, biomimetic, tunability, good printability, and crosslink characteristics [5]. However, alginate's highly swollen polymer crosslinked network can lead to instability and degradation in just a few days under cell culture conditions [6]. Therefore, tailoring the alginate into a more biomimetic matrix can make it a more attractive biomaterial. The addition of gelatin can improve the mechanical properties of the hydrogel, increasing its elasticity and decreasing its stiffness [7,8]. Moreover, Alsberg et al. (2001)

have shown that the incorporation of RGD molecules into alginate gel transforms the hydrogel into a suitable matrix to promote bone formation [9]. Finally, an additional way to tailor its mechanical properties is the degree of crosslinking, which can be determined by the crosslinker type, concentration, temperature, time of exposure, etc. [10,11].

Alginate can be chemically or physically crosslinked. Chemical crosslinking involves the formation of irreversible covalent bonds between alginate chains and leads, in general, to better stability under cell culture conditions [12–15]. Physical crosslinking involves hydrogen bonding, hydrophobic bonding, ionic bonding, electrostatic interactions, etc., to create reversible three-dimensional structures [13–16]. Alginate is usually ionically crosslinked in the bioengineering field using divalent cations, including calcium. However, Ca^{2+} molecules from the crosslinked hydrogel can be released into the medium, causing inflammatory responses in vitro and in vivo [17]. Another possibility for the ionic crosslinking of alginate substrates for cell-based therapies is the use of Ba^{2+} molecules. Barium has a better affinity to alginate, therefore providing hydrogels with stronger mechanical properties and increased Young's modulus in comparison to calcium [18]. Yet, Ba^{2+} has shown diverse effects on cell viability and proliferation depending on the cell type. For instance, Luca et al. (2007) reported that encapsulated Sertoli cells in alginate microbeads showed greater viability after barium crosslinking in comparison to calcium [19]. However, Mores et al. (2015) showed a decrease in the mononuclear phagocyte viability caused by apoptosis due to barium crosslinking [20]. Therefore, as few studies regarding barium crosslinking are available in the literature, alginate crosslinking with barium should be studied carefully for each cell type and condition.

Another important aspect during the preparation of hydrogel scaffolds for bioengineering is the pH of the solution. FitzSimons et al. (2022) described how pH influences the polymer bonding kinetics, the mechanical properties, and the protein release of PEG hydrogels crosslinked via reversible thia-Michael addition [21]. In addition, pH is involved in regulating the solubility of alginate/gelatin in the aqueous phase [22]. Bouhadir et al. (2001) showed that increasing the pH leads to an increase in the alginate hydrogel's degradability [23]. In addition, the alginate hydrogel viscosity increased with decreasing pH, reaching a maximum viscosity at pH 3.0–3.5 once the carboxylate groups were present in the alginate chain protonate and formed hydrogen bonds [15]. Recently, the effect of pH on hydrogels has been largely studied, especially regarding the production of responsive hydrogels used to deliver drugs or optimize cell growth and differentiation. Tailoring hydrogel's mechanical properties, degradation rate, and affinity to proteins via pH changes is a costless and important tool for the bioengineering field. Therefore, understanding the effect of hydrogel's intrinsic pH as well as the surrounding pH on the hydrogel's mechanics and kinetics is fundamental to accurately control hydrogel behavior.

In this work, we used a hydrogel composed of 6% alginate and 2% gelatin (w/v) to investigate the impact of different pH values (5.5, 6.5, 7.0, 8.0) and crosslinking ions ($CaCl_2$, $BaCl_2$) on hydrogel stability under cell culture conditions incubated with either DMEM or RPMI 1640. These two media are widely used in cell culture and present differences in their composition regarding calcium (RPMI: 0.8 mM, DMEM: 1.8 mM) and phosphate (RPMI: 5 mM, DMEM: 1 mM) concentrations [24]. Both media present a physiological pH of 7.0–7.4. The printability of the alginate–gelatin hydrogel with different pH values was tested in the air and under fluid phase using a support bath containing either 100 mM $CaCl_2$ or 100 mM $BaCl_2$. Air printing allowed the printability of high-viscosity biomaterials and bioinks, while fluid-phase printing allowed the printability of low-viscosity biomaterials and bioinks [25]. Finally, to broaden our understanding, different cell types (the human osteosarcoma cell line U2OS and the murine fibroblast cell line NIH/3T3) were used to study the effect of the substrate's pH (as modified by buffer) and crosslinking on cell viability and metabolism.

2. Results and Discussion

2.1. Results

Alginate–gelatin hydrogels were prepared using an MES buffer with a pH of 5.5, 6.5, 7.0, or 8.0 and crosslinked with either 100 mM $CaCl_2$ or 100 mM $BaCl_2$ to study the role of the buffer pH and crosslinking on hydrogel stability over time. With all other factors remaining the same, the addition of a different pH buffer was expected to change the pH of the hydrogel, hereafter the substrate's pH. Hydrogels were molded into 10 mm discs and kept under cell culture conditions incubated with either complete RMPI 1640 + 10% FBS or complete DMEM + 10% FBS for 25 days to measure their swelling and degradation rates. As a result, we observed that all the conditions presented a weight loss of approximately 50% during overnight crosslinking with either calcium or barium. However, they regained their initial weight at different speeds. Hydrogels crosslinked with calcium and incubated with RPMI were the least stable under cell culture conditions, showing, in general, the fastest degradation rate among all the conditions tested in this work. They regained their initial weight within 24 h, and the swelling of all the hydrogels was between 150 and 200% before degrading (Figure 1A). Crosslinking with calcium and incubation with DMEM showed better stability in comparison to the previous conditions. The swelling of the samples was around 135% before degradation. However, all the substrates, regardless of their pH, degraded within 25 days under cell culture conditions (Figure 1B). Crosslinking with barium provided overall good stability to the alginate–gelatin hydrogel. Incubation with RPMI 1640 promoted the higher swelling of samples (~150%) in comparison to their incubation with DMEM (~120%) (Figure 1C,D). Regarding the hydrogel pH, the hydrogel pH 8.0 presented the weakest stability under cell culture conditions in RPMI regardless of the crosslinking and in DMEM after calcium crosslinking (Figure 1).

Afterward, alginate–gelatin hydrogels with different pH values had their printability tested under the following three different conditions: air printing, fluid-phase embedding printing with 100 mM of a $CaCl_2$ support bath, and fluid-phase embedding printing with 100 mM of a $BaCl_2$ support bath. We reported in previous findings that to produce printings of the alginate–gelatin hydrogel, the fluid phase is an efficient way to increase printing fidelity and resolution [26]. To generate these prints, a needle size of 0.43 mm ID was used, and the diameter of the strand spread was measured just after printing for the three conditions. First, we observed that the spread of the air printing strand was around 200%, while both fluid-phase embedding printing restrained the spread of the printed strand as the hydrogel was immediately crosslinked during printing. In addition, the two crosslinker solutions were used as support baths and showed no significant difference in the spread of the strand between them (Figure 2A). The pH of the alginate–gelatin hydrogel had a significant influence on air printing, increasing the spread of the printed strand with the increase in pH. By contrast, fluid-phase embedding printings only showed a slight increase in the spread of the strand with a hydrogel pH of 8.0, which was not significant (Figure 2B). The spread of the strand is important for the resolution of the (bio)print and should be calculated and adjusted to obtain the desired printing design. Knowing how the pH of the substrate can influence this aspect and how to overcome it is an important parameter for the biofabrication field. Finally, our results showed that both crosslinker (Ca^{2+} and Ba^{2+})-containing support baths increased the printing fidelity, which is particularly important at the corners of the prints, as seen in Figure 2C.

Next, we tested the effect of crosslinking on the cell viability and metabolic activity of the human osteosarcoma U2OS and the murine fibroblast NIH/3T3. Both cell types are well established in the literature and present high-cell proliferative rates, making them good models for this study. U2OS cells did not show significant a difference in the cell death rate of cells growing on alginate–gelatin substrates crosslinked with calcium or barium either on day 1 or day 7 of cell culture (Figure 3A). However, the substrate crosslinked with barium led to a 3.62-fold decrease in U2OS cell metabolic activity in comparison to the substrate crosslinked with calcium (Figure 3B). NIH/3T3 cells presented a high-cell death rate on both day 1 and day 7 of cell culture on alginate–gelatin hydrogel crosslinked with barium at 45% and 60%, respectively, in comparison to 10% and 20%, respectively, of the cell death rate on the same hydrogel crosslinked with calcium (Figure 3C). Moreover, NIH/3T3 cells also showed less metabolic activity on the substrate crosslinked with barium (83.91%) in comparison to the one crosslinked with calcium (100%) (Figure 3D). Even though crosslinking with barium showed better stability of the alginate–gelatin hydrogel, the two cell types studied in this work did not show good cell viability and/or cell metabolic activity on substrates crosslinked with Ba^{2+}.

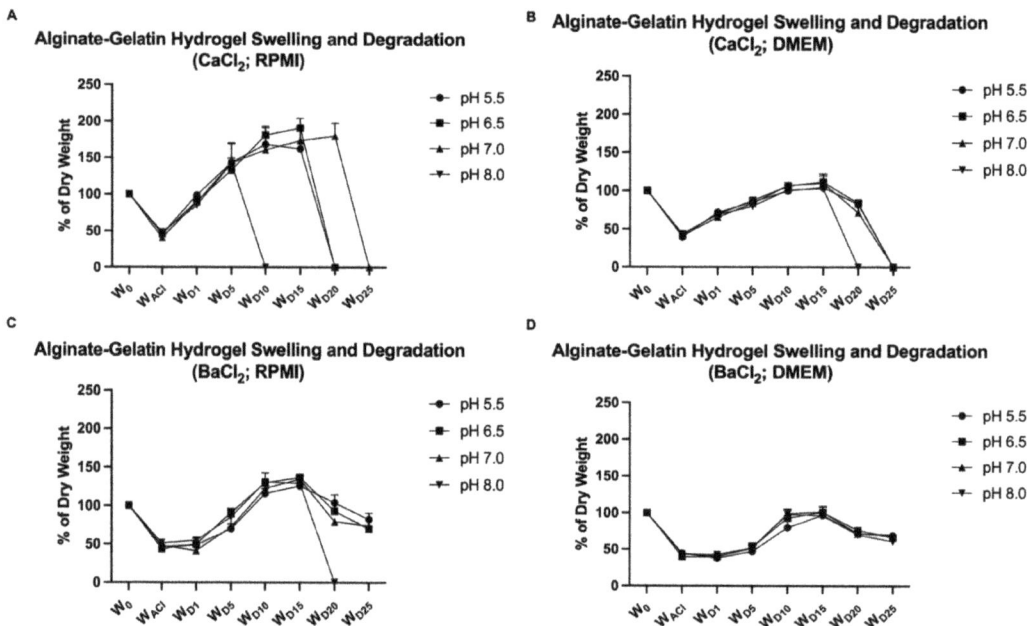

Figure 1. Alginate–gelatin hydrogel swelling and degradation curve. 6% alginate + 2% gelatin hydrogel prepared with 0.1 M MES buffer pH 5.5, 6.5, 7.0 or 8.0 and crosslinked with either 100 mM $CaCl_2$ or 100 mM $BaCl_2$. The samples were kept under cell culture conditions at 37 °C with saturating humidity with either RPMI 1640 + 10% FBS or DMEM + 10% FBS. Hydrogels were dried and weighed after being molded into 10 mm diameter discs (W_0 = 100%), after crosslinking (W_{Acl}), and on day 1 (W_{D1}), day 5 (W_{D5}), day 10 (W_{D10}), day 15 (W_{D15}), day 20 (W_{D20}), and day 25 (W_{D25}). (**A**) Hydrogels crosslinked with 100 mM $CaCl_2$ and incubated with RPMI 1640 + 10% FBS. (**B**) Hydrogels crosslinked with 100 mM $CaCl_2$ and incubated with DMEM + 10% FBS. (**C**) Hydrogels crosslinked with 100 mM $BaCl_2$ and incubated with RPMI 1640 + 10% FBS. (**D**) Hydrogels crosslinked with 100 mM $BaCl_2$ and incubated with DMEM + 10% FBS. (n = 4, graph of mean ± SD).

To study the effect of the substrate's pH on cell viability and metabolic activity, U2OS and NIH/3T3 were seeded onto alginate–gelatin prepared with the MES buffer of pH 5.5, 6.5, 7.0, and 8.0. Cells were kept in culture for 7 days incubated with complete RPMI +10% FBS (U2OS) or complete DMEM + 10% FBS (NHI/3T3). Our results show that U2OS cells growing on alginate–gelatin hydrogels prepared with different pH values present good cell viability of approximately 90% of viable cells on day 1 and day 7 of cell culture. Some cell protective effects were observed by the substrate at pH 7.0; however, no significant difference in cell viability was caused by the substrate pH (Figure 4A). On the contrary, the substrate presenting different pH values led to a significant difference in cell metabolic activity. Cells growing on substrates prepared with pH 7.0 and 8.0 showed an important increase in cell metabolism in comparison to cells growing on acidic substrates after 7 days of cell culture. An increase of 2.25-fold was observed in cell metabolism on the substrate at pH 8.0 in comparison to cells on the substrate at pH 5.5. All the cells were grown in an RPMI 1640 cell culture medium under physiological pH. The results show that it is possible to grow cells over the influence of the pH of interest by preparing a substrate with a specific pH independent of the cell culture medium utilized. In the same way, NIH/3T3 cells also showed good viability of cells regardless of substrate pH values on day 1 and day 7 of cell culture (Figure 4C). However, in the same fashion as U2OS cells, NIH/3T3 also showed a significant difference in cell metabolism due to the substrate pH. NIH/3T3 cells growing on substrate at pH 7.0 presented higher metabolic activity in comparison to cells growing on substrates at pH 6.5 and 8.0 and even greater cell metabolic activity in comparison to cells growing on substrates at pH 5.5 (Figure 4D).

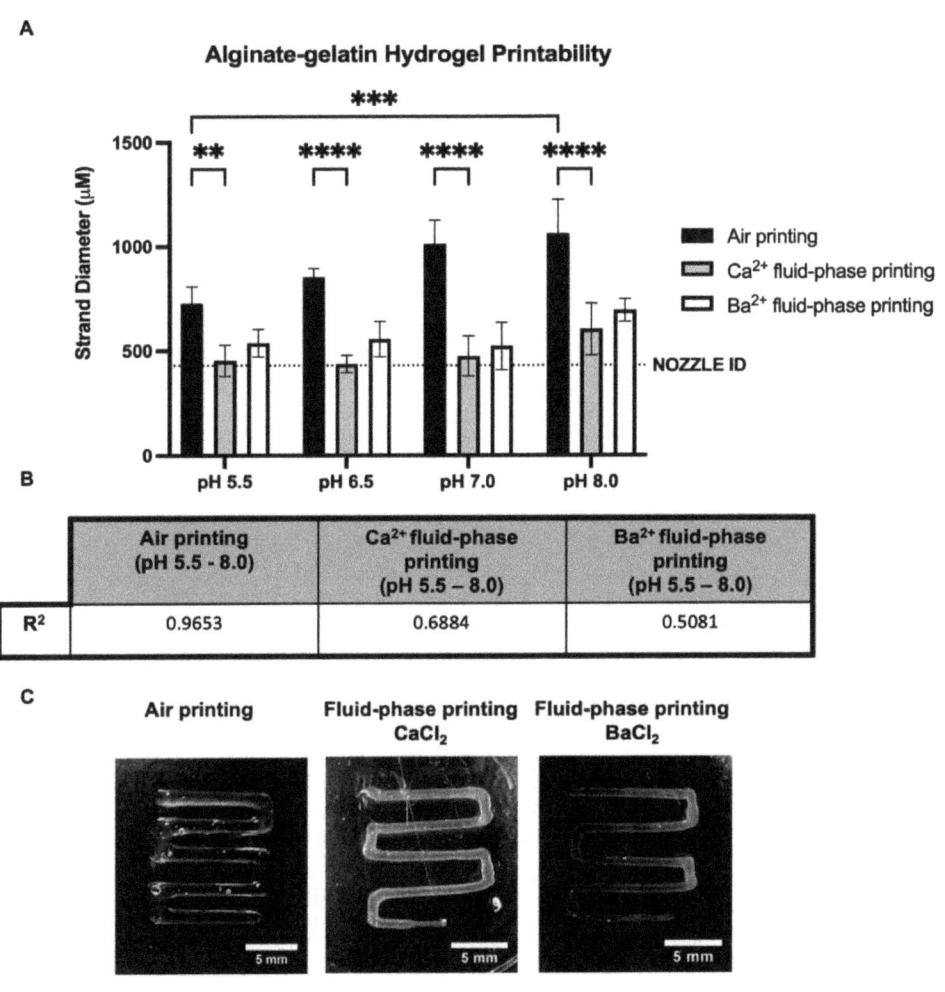

Figure 2. Alginate–gelatin hydrogel printability. (A) Quantification of the alginate–gelatin hydrogel at pH 5.5, 6.5, 7.0, or 8.0 for the diameter of the strand printed in the air and the fluid-phase embedding support bath containing either 100 mM $CaCl_2$ or 100 mM $BaCl_2$. (n = 4, Two-way ANOVA, ** $p < 0.01$, *** $p < 0.001$, **** $p < 0.0001$, ns: not significant). (B) R-squared of the effect of increasing pH of alginate–gelatin hydrogel on the spread of the printed line (strand). (C) Alginate–gelatin hydrogel pH 7.0 square wave printing obtained under 3 bars and 1 mm/s in the air and fluid-phase embedding support bath containing either 100 mM $CaCl_2$ or 100 mM $BaCl_2$.

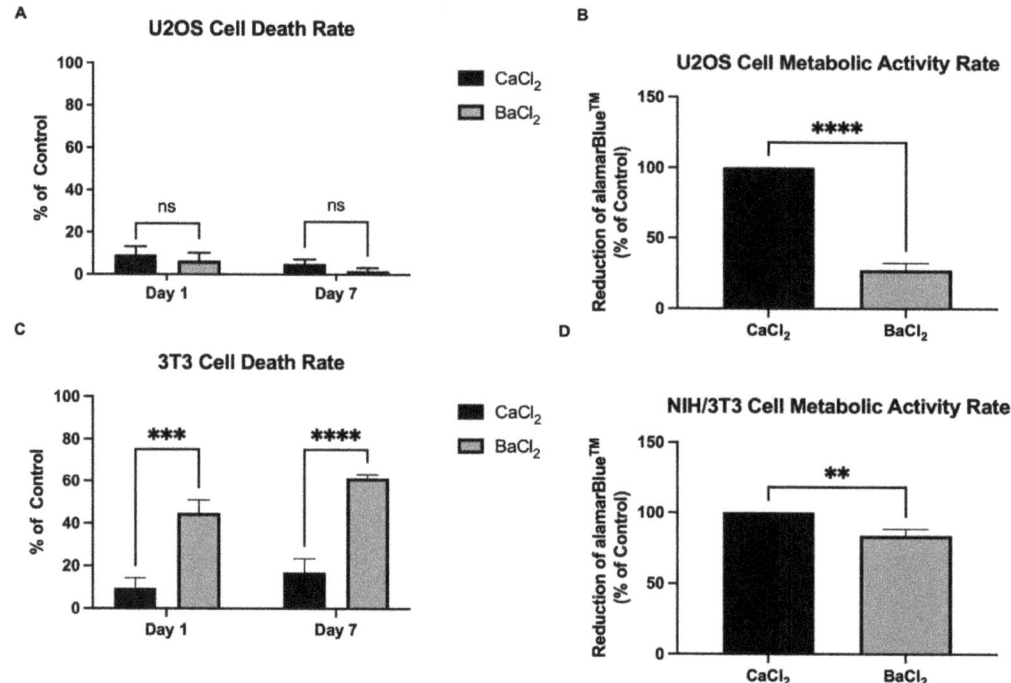

Figure 3. (**A**) LDH cytotoxicity assay of U2OS cells on alginate–gelatin hydrogel at pH 7.0 crosslinked with either 100 mM CaCl$_2$ or 100 mM BaCl$_2$ on days 1 and 7 of cell culture (day 1 n = 4 and day 7 n = 3, Two-way ANOVA, ns: not significant). (**B**) U2OS metabolic activity rate quantified by the % of alamarBlue^{TM} reduction on day 7 of cell culture on alginate–gelatin hydrogel at pH 7.0 crosslinked with either 100 mM CaCl$_2$ or 100 mM BaCl$_2$ (n = 4, Two-way ANOVA, **** $p < 0.0001$). (**C**) LDH cytotoxicity assay of NIH/3T3 cells on alginate–gelatin hydrogel at pH 7.0 crosslinked with either 100 mM CaCl$_2$ or 100 mM BaCl$_2$ on days 1 and 7 of cell culture (n = 3, Two-way ANOVA, *** $p < 0.001$, **** $p < 0.0001$). (**D**) NIH/3T3 metabolic activity rate quantified by the % of alamarBlue^{TM} reduction on day 7 of cell culture on alginate–gelatin hydrogel at pH 7.0 crosslinked with either 100 mM CaCl$_2$ or 100 mM BaCl$_2$ (n = 4, Two-way ANOVA, ** $p < 0.01$).

Finally, we observed that the pH of the substrate also influenced the cell–cell and cell–matrix interactions, with cells favoring interactions with each other below pH 7 and the matrix at pH 7 and above. Alginate–gelatin substrates at pH 5.5 and 6.5 led to the formation of cell aggregates, whereas substrates at pH 7.0 and 8.0 showed cells that grew in a more spread-out layout after 7 days of cell culture (Figure 5).

2.2. Discussion

Alginate hydrogels can be crosslinked with many divalent cations. In the present study, we tested the swelling and degradation rates of alginate–gelatin hydrogels with different pH values after crosslinking with either 100 mM CaCl$_2$ or 100 mM BaCl$_2$. Both cations are widely used in the biofabrication field to crosslink the alginate for different purposes, such as drug encapsulation and delivery, wound dressing, tissue formation, etc. In addition, the choice of the crosslinker and its concentration can tailor some of the hydrogel's mechanical properties like elasticity, strength, stiffness, swelling, etc. [27]. Haper and Barbut (2014) showed that alginate films, when crosslinked with BaCl$_2$, had the highest tensile strength and Young's modulus among Ba^{2+}, Ca^{2+}, Mg^{2+}, Sr^{2+}, and Zn^{2+} while films crosslinked with CaCl$_2$ had the highest puncture strength [18]. Alginate's affinity for divalent cations was shown to decrease in the following order: Pb^{2+} > Cu^{2+} > Cd^{2+} > Ba^{2+} > Sr^{2+} > Ca^{2+} >

Co^{2+}, Ni^{2+}, Zn^{2+} > Mn^{2+} [28,29]. Gel strength decreased with decreasing affinity. Our study is in accordance with the literature, as barium crosslinking provided better stability for all the tested hydrogels in comparison to calcium crosslinking. Additionally, we observed a shrinkage of approximately 50% on all hydrogels incubated with both crosslinker solutions ($CaCl_2$ and $BaCl_2$) for 24 h. Saitoh et al. described a similar effect of alginate hydrogel when crosslinked with Ca^{2+} and incubated with increasing concentrations of $CaCl_2$, which showed an increasing shrinkage rate. Increasing the binding between Ca^{2+} and alginate residual carboxylate groups led to an increase in the crosslinking degree, which facilitated gel shrinkage [30]. In the present study, we show that Ba^{2+} presents a similar effect to Ca^{2+} regarding the alginate hydrogel shrinkage extent after overnight incubation. In addition to the crosslinker effect on the hydrogel, we also showed that the cell culture medium in which the hydrogel was kept played an important role as hydrogels incubated with complete DMEM + 10% FBS presented overall better stability over time in comparison to the same hydrogels incubated with complete RPMI 1640 + 10% FBS, even though both media had the same physiological pH. In addition, among the hydrogels presenting different pH values, the hydrogel at pH 8.0 showed the weakest strength under cell culture conditions in general, which degraded faster than the hydrogels at pH 5.5, 6.5, and 7.0. Anionic hydrogels such as alginate swell at a high pH and shrank at a lower pH. The deprotonation of carboxylic groups of the alginate molecules at high pH decreased the strength of the hydrogel as the negatively charged ions repelled each other, leading to hydrogel swelling and fast degradation. In the opposite fashion, acidic media can lead to the protonation of alginate carboxylic groups, decreasing repulsion and causing shrinkage because of water loss [31–33]. In this study, we showed that we can also tailor the alginate hydrogel swelling and degradation/dissociation rates by changing the pH of the solvent used to produce the hydrogel independently of the medium's pH.

This pH effect was also seen in the printability of the hydrogel, as the spread of the strand significantly increased with the increase in the hydrogel's pH during printing in the air. However, fluid-phase embedding printing using either 100 mM $CaCl_2$ or 100 mM $BaCl_2$ as support baths was shown to be efficient in preventing spread from occurring. The slight increase in the spreading of the line of hydrogel at pH 8.0 printed in the fluid phase was not significant, as the immediate crosslink after printing the hydrogel was sufficient to keep its shape. No significant difference was seen between calcium and barium in relation to the printability of the hydrogel with different pH values. The barium support bath showed a lower correlation between the spread of the strand and the increasing pH of the hydrogel in comparison to calcium. Jui-Jung et al. (2017) described that barium crosslinking baths with different pH values do not influence the shape of alginate particles; however, crosslinking with calcium at lower pH values does exert an influence [34].

However, even though the crosslinking with barium provided more stability and strength to the alginate–gelatin hydrogel, both U2OS and NIH/3T3 cells showed significantly higher cell metabolic activity on hydrogels crosslinked with calcium in comparison to hydrogels crosslinked with barium. It was described that both crosslinkers, Ca^{2+} and Ba^{2+}, presented a rate of release of molecules to the medium due to their relatively weak ionic interaction and competition with other cations present in the medium. Chan and Mooney (2013) described how alginate crosslinked with calcium releases around 43% of the Ca^{2+} incorporated in its meshes within the first 10 h of incubation with a cell culture medium and keeps releasing Ca^{2+} molecules at a lower rate thereafter as Ca^{2+} is slowly exchanged by sodium cations present in the cell culture medium [17]. Ba^{2+} also presented a high release rate from alginate in vitro and in vivo, which might be a safety concern [35]. In addition, it is widely known that Ca^{2+} is one of the most important intracellular second messengers participating in an extensive number of cell signaling pathways in relation to cell adhesion, proliferation, metabolism, apoptosis, etc. [36]. There is the possibility that Ba^{2+} ions released from alginate can compete with Ca^{2+} in important metabolic pathways, decreasing the metabolic rate of cells seeded onto the substrate crosslinked with Ba^{2+} [37–39]. However, there are few studies on the effect of barium crosslinking on cell activity and metabolism, and further investigations should be conducted for better understanding. We also observed that Ba^{2+} crosslinking did not increase cell death in comparison to Ca^{2+} crosslinking on U2OS cells, whereas Ba^{2+} crosslinking increased the NIH/3T3 cell death rate significantly.

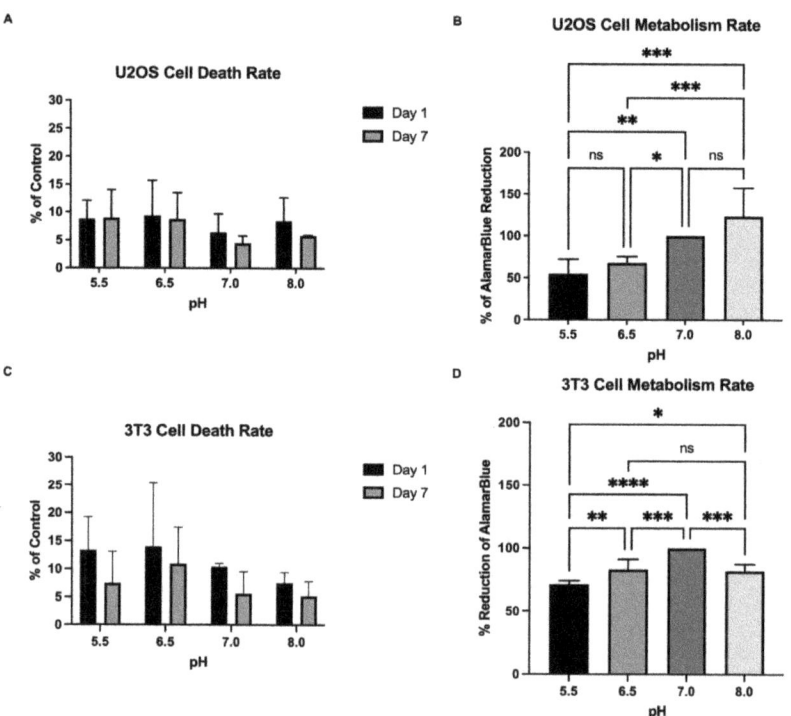

Figure 4. (**A**) LDH cytotoxicity assay of U2OS cells on alginate–gelatin hydrogel at pH 5.5, 6.5, 7.0, and 8.0 on days 1 and 7 of cell culture (day 1 n = 4 and day 7 n = 3, Two-way ANOVA, not significant). (**B**) U2OS metabolic activity rate quantified by the % of alamarBlue^{TM} reduction on day 7 of cell culture on alginate–gelatin hydrogel at pH 5.5, 6.5, 7.0, and 8.0 (n = 6, Two-way ANOVA, * $p < 0.05$, ** $p < 0.01$, *** $p < 0.001$, ns: not significant). (**C**) LDH Cytotoxicity assay of NIH/3T3 cells on alginate–gelatin hydrogel at pH 5.5, 6.5, 7.0, and 8.0 on days 1 and 7 of cell culture (day 1 n = 3 and day 7 n = 3, Two-way ANOVA, not significant). (**D**) NIH/3T3 metabolic activity rate quantified by the % of alamarBlue^{TM} reduction on day 7 of cell culture on alginate–gelatin hydrogel at pH 5.5, 6.5, 7.0, and 8.0 (n = 6, Two-way ANOVA, * $p < 0.05$, ** $p < 0.01$, *** $p < 0.001$, **** $p < 0.0001$, ns: not significant).

To study the influence of substrate pH on cell viability and metabolism, U2OS and NIH/3T3 cells were seeded onto substrates with different pH values. Bone cells responded to even slight differences in the environment's pH with higher osteoclast activity in acidic pH, while osteoblasts have higher activity in basic pH [40–42]. This is one of the homeostatic mechanisms to keep the systemic acid-base balance. Matsubara et al. (2013) showed that U2OS cells modulate their proliferation rate in response to extracellular pH, increasing proliferation in higher pH values [43]. In this work, we used the same cell type to investigate if changing the substrate pH could change cell behavior. Our results showed that a different substrate pH led to a significant difference in cell metabolism and no influence on cell viability. Corroborating the literature, osteoblasts growing on basic substrates showed greater cell metabolic activity in comparison to cells growing on acidic substrates. Our results, however, show that independently of the cell culture medium pH, the substrate's pH played a crucial role in cell behavior. In addition, we observed that the U2OS cells can modulate the cell arrangement in response to the substrate pH, forming aggregates on acidic printings and spreading out on basic printings. Cells start to grow in 3D aggregates when the environment is unfavorable to cell–substrate interactions, leading cells to perform cell–cell and cell-ECM interactions instead [44].

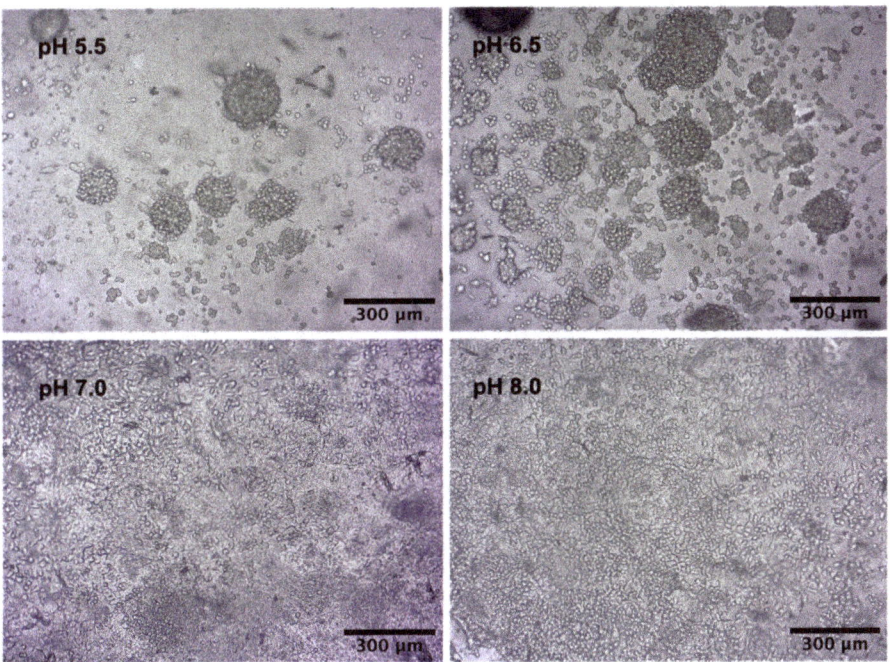

Figure 5. Light microscopy images of U2OS cells cultured on alginate–gelatin hydrogel at pH 5.5, 6.5, 7.0, and 8.0 on day 7 of cell culture. Magnification of 10×.

Studying the interstitial pH of each tissue is a difficult task. There is no available data for human bone tissue. However, mice embryos were reported to have no impairment in their development in pH between 7.17 and 7.37 [45]. In this work, we used NHI/3T3 cells, which are murine embryonic fibroblasts. It is not reported from which tissue it originated, but Dastagir et al. (2014) showed that this cell type can differentiate into adipogenic, chondrogenic, and osteogenic lineages [46]. Our results showed that NHI/3T3 cells had a higher metabolic activity on substrates prepared with a buffer of pH 7.0 in comparison to substrates prepared with pH 5.5, 6.5, and 8.0, even though all cells were kept in a cell culture medium under physiological pH. In addition, the pH of the buffer did not influence NHI/3T3 cell viability. Future work in this direction should attempt to directly measure substrate pH as a function of time.

3. Conclusions

The bioengineering field is a very complex area of science because it involves several steps, which are all very specific to each cell type and circumstance. In this study, we aimed to understand the importance of carefully studying the pH of 3D prints. In addition, we also aimed to understand how crosslinking affects not only the mechanical properties of hydrogels but also influences cell viability, growth, and behavior on the prints. In essence, our findings show that the pH of the microenvironment in which the cells are in direct contact to grow influences cell behavior independently of the pH of the cell culture medium in the range studied. Moreover, barium crosslinking provided better stability for alginate–gelatin hydrogels independently of the substrate pH. However, U2OS and NIH/3T3 cells showed significantly lower metabolic activity when grown on substrates crosslinked with Ba^{2+} in comparison to Ca^{2+} crosslinking. In addition, Ba^{2+} crosslinking increased NIH/3T3 cell death. Therefore, Ca^{2+} has been shown to be better at crosslinking for the growth of both U2OS and NIH/3T3 cells on alginate–gelatin hydrogels. This illustrates that each cell type can respond differently to different substrate pH values and crosslinking protocols, indicating the importance of adjusting the conditions in 3D printing to achieve desired results.

4. Materials and Methods

4.1. Materials

Alginic acid sodium salt type 1, sodium chloride, calcium chloride, and EDC (1-ethyl-3-(3-dimethylaminopropyl) carbodiimide hydrochloride) were purchased from Thermo Fisher Scientific, Dublin, Ireland; for gelatin type B, bovine skin and Triton X 100 were purchased from Sigma-Aldrich Ireland Limited, Arklow, Co. Wicklow, Ireland; the MES buffer at pH 5.5, 6.5, 7.0, and 8.0, Polyetherimide (PEI), Barium Chloride, and NHS (N-hydroxysuccinimide) were purchased from Alfa Aesar, Ward Hill, MA, USA; MVG GRGDSP (RGD) was purchased from Novatech, Hattiesburg, MS, USA; RPMI 1640, DMEM, fetal bovine serum (FBS), and trypsin/EDTA were purchased from Gibco Life Technologies at Thermo Fisher Scientific, Dublin, Ireland. The U2OS human osteosarcoma cell line (ATCC® HTB-96™) and NIH/3T3 murine fibroblast cell line (ATCC® CRL-1658™) were used as cell models and purchased from ATCC, Manassas, VA, USA.

4.2. Methods

4.2.1. Cell Culture

U2OS cells and NIH/3T3 cells were cultured in 75 cm^2 flasks in complete RPMI 1640 and DMEM, respectively, both supplemented with 10% FBS under a saturated humidified atmosphere at 37 °C and 5% CO_2. Subconfluent cultures were passaged at a ratio of 1:5 (U2OS cells) or 1:3 (NIH/3T3 cells) using a 0.05% trypsin/EDTA solution. High-density cells, $0.5 \times 10^6/cm^2$, were seeded onto alginate–gelatin printings coated with 120 μM/mL of RGD + 5% EDC + 2.5% NHS was diluted in 100 mM $CaCl_2$ for 1 h [47,48].

Cells growing on alginate–gelatin hydrogel printings were monitored under an inverted phase contrast microscope (Olympus CKX53, Olympus Corporation, Tokyo, Japan), and photomicrographs were taken by phase contrast using the 10× objective. Images were analyzed by the software ImageJ/FIJI 2.9.0 (https://imagej.net/software/fiji/ [26], accessed on 12 September 2023).

4.2.2. Alginate–Gelatin Hydrogel Preparation and Crosslinking

Adapted from [49], 2% gelatin (w/v) was added to a sterile 0.1 M MES buffer at pH 5.5, 6.5, 7.0 or 8.0 + 0.3 M NaCl at 50 °C and stirred for 10 min. Sodium alginate was then added at a final concentration of 6% (w/v) and stirred thoroughly at 50 °C until the hydrogel was homogeneous. Different hydrogel pH values were obtained with the use of an MES buffer with different pH values as a solvent. The hydrogel pH was not measured throughout the experiments. Hydrogels were either molded or printed in the air or fluid phase. Hydrogels were then crosslinked overnight with either 5 mL of 100 mM calcium chloride or 5 mL of 100 mM barium chloride, both dissolved in deionized water.

4.2.3. Swelling and Degradation Measurement

In total, 6% alginate + 2% gelatin pH 5.5, 6.5, 7.0, or 8.0 were molded into 10 mm diameter discs and crosslinked with 5 mL of either 100 mM calcium chloride or 100 mM barium chloride. Hydrogels were weighed before crosslinking (W_0) as a control. To measure the swelling and degradation rates, hydrogels were also weighed after crosslinking (W_{ACl}), and on day 1 (W_{D1}), day 5 (W_{D5}), day 10 (W_{D10}), day 15 (W_{D15}), day 20 (W_{D20}) and day 25 (W_{D25}) under cell culture conditions, they were incubated with either complete RPMI 1640 or complete DMEM both supplemented with 10% FBS at 37 °C with saturating humidity. Hydrogels were air-dried for 30 min before having their weight measured. Calculations were made for the percentage of W_0, and experiments were finished when 2/3 of the samples disintegrated.

4.2.4. 3D Printing and Printing Resolution

The extrusion-based NAIAD 3D bioprinter, developed by the 3D Bioprinting Group from the Conway Institute at the University College Dublin, was used to print the alginate–gelatin hydrogel in the air and fluid phase. Briefly, hydrogels were warmed at 37 °C for 30 min in the bioprinter before the printing process. Alginate–gelatin printings were obtained using a 0.43 mm ID nozzle gauge 23 under 3 bars of pressure and 1mm/s of speed in the air or fluid phase containing either 100 mM $CaCl_2$ or 100 mM $BaCl_2$. Prior to the fluid-phase printing, six-well plates were treated with 0.1% PEI overnight for hydrogel attachment purposes and were washed 3 times with PBS to remove excess PEI. The width of the strands was measured immediately after printing and was analyzed via the software ImageJ/FIJI 2.9.0 (https://imagej.net/software/fiji/ [26], accessed on 12 September 2023).

4.2.5. Cell Viability Assay

An LDH Cytotoxicity Detection Kit plus (Roche Diagnostics, Basel, Switzerland) was used to quantify U2OS and NIH/3T3 cell death on day 1 and day 7 of cell culture. Briefly, cells were incubated for 24 h with either RPMI 1640 or DMEM, each supplemented with 1% FBS, centrifuged, and the amount of LDH on the cell medium was measured at 490 nm by spectrophotometer (SpectraMax M3). Positive and negative controls were conducted on conventional 2D cell culture on plastic plates. The positive control was treated with 2% Triton X-100. Calculations were given as a percentage of the control.

4.2.6. Adapted alamarBlue™ Assay

To quantify the metabolic activity of U2OS and NIH/3T3 cells, 10% of the alamarBlue™ cell viability reagent (Invitrogen) was added to the DMEM + 10% FBS (NHI/3T3) or RPMI 1640 + 10% FBS (U2OS) media on day 7 of cell culture and incubated for 4 hours. Next, the cell media containing reduced alamarBlue™ were centrifuged, and 100 µL was transferred to a 96-well plate. The absorbance of the reduced alamarBlue™ was measured at 570 nm and 600 nm. Calculations were performed according to the manufacturer's instructions and given as a percentage of the control. For calculation purposes, the metabolic activity of cells growing alginate–gelatin hydrogel at pH 7.0 crosslinked with 100 mM of $CaCl_2$ was considered a control (100%).

4.2.7. Statistical Analysis

All statistical data processing was performed using multiple comparison tests on either One-way ANOVA or Two-way ANOVA on GraphPad Prism 9 software. Differences between the groups were considered reliable if $p < 0.05$ and values were expressed as the mean ± SD of at least 3 independent experiments.

Author Contributions: Conceptualization, B.J.R. and E.G.R.; methodology, A.S. and M.P.; software, A.S.; validation, A.S. and M.P.; formal analysis, A.S.; investigation, A.S. and M.P.; resources, B.J.R. and E.G.R.; data curation, A.S., B.J.R. and E.G.R.; writing—original draft preparation, A.S., B.J.R. and E.G.R.; writing—review and editing, A.S., B.J.R. and E.G.R.; visualization, A.S.; supervision, B.J.R. and E.G.R.; project administration, B.J.R. and E.G.R.; funding acquisition, E.G.R. All authors have read and agreed to the published version of the manuscript.

Funding: A. Souza has received a PhD scholarship under the demonstrator's program from the School of Biomolecular and Biomedical Sciences, University College Dublin, Belfield, Dublin 4, Belfield.

Data Availability Statement: The data presented in this study are openly available in article.

Conflicts of Interest: The authors declare no conflict of interest.

References

1. Derr, K.; Zou, J.; Luo, K.; Song, M.J.; Sittampalam, G.S.; Zhou, C.; Michael, S.; Ferrer, M.; Derr, P. Fully Three-Dimensional Bioprinted Skin Equivalent Constructs with Validated Morphology and Barrier Function. *Tissue Eng. Part C Methods* **2019**, *25*, 334–343. [CrossRef]
2. Diloksumpan, P.; de Ruijter, M.; Castilho, M.; Gbureck, U.; Vermonden, T.; van Weeren, P.R.; Malda, J.; Levato, R. Combining multi-scale 3D printing technologies to engineer reinforced hydrogel-ceramic interfaces. *Biofabrication* **2020**, *12*, 025014. [CrossRef]
3. Lv, S.; Nie, J.; Gao, Q.; Xie, C.; Zhou, L.-Y.; Qiu, J.; Fu, J.; Zhao, X.; He, Y. Micro/nanofabrication of brittle hydrogels using 3D printed soft ultrafine fiber molds for damage-free demolding. *Biofabrication* **2019**, *12*, 025015. [CrossRef]
4. Weisgrab, G.; Guillaume, O.; Guo, Z.; Heimel, P.; Slezak, P.; Poot, A.; Grijpma, D.; Ovsianikov, A. 3D Printing of large-scale and highly porous biodegradable tissue engineering scaffolds from poly(trimethylene-carbonate) using two-photon-polymerization. *Biofabrication* **2020**, *12*, 045036. [CrossRef]
5. Huang, N.F.; Li, S. Regulation of the Matrix microenvironment for stem cell engineering and regenerative medicine. *Ann. Biomed. Eng.* **2011**, *39*, 1201–1214. [CrossRef]
6. Di Giuseppe, M.; Law, N.; Webb, B.; Macrae, A.R.; Liew, J.L.; Sercombe, B.T.; Dilley, J.R.; Doyle, J.B. Mechanical behaviour of alginate-gelatin hydrogels for 3D bioprinting. *J. Mech. Behav. Biomed. Mater.* **2018**, *79*, 150–157. [CrossRef]
7. Jiang, T.; Munguia-Lopez, G.J.; Gu, K.; Bavoux, M.M.; Flores-Torres, S.; Kort-Mascort, J.; Grant, J.; Vijayakumar, S.; Leon-Rodriguez, D.A.; Ehrlicher, J.A.; et al. Engineering bioprintable alginate/gelatin composite hydrogels with tunable mechanical and cell adhesive properties to modulate tumor spheroid growth kinetics. *Biofabrication* **2019**, *12*, 015024. [CrossRef]
8. Mondal, A.; Gebeyehu, A.; Miranda, M.; Bahadur, D.; Patel, N.; Ramakrishnan, S.; Rishi, A.K.; Singh, M. Characterization and printability of sodium alginate-gelatin hydrogel for bioprinting NSCLC co-culture. *Sci. Rep.* **2019**, *9*, 19914. [CrossRef]
9. Alsberg, E.; Anderson, K.; Albeiruti, A.; Franceschi, R.; Mooney, D. Cell interactive alginate hydrogels for bone tissue engineering. *J. Dent. Res.* **2001**, *80*, 2025–2029. [CrossRef]
10. Pan, T.; Song, W.; Cao, X.; Wang, Y. 3D bioplotting of gelatin/alginate scaffolds for tissue engineering: Influence of crosslinking degree and pore architecture on physicochemical properties. *J. Mater. Sci. Technol.* **2016**, *32*, 889–900. [CrossRef]
11. Lee, K.Y.; Rowley, J.A.; Eiselt, P.; Moy, E.M.; Bouhadir, K.H.; Mooney, D.J. Controlling mechanical and swelling properties of alginate hydrogels independently by cross-linker type and crosslinking density. *Macromolecules* **2000**, *33*, 4291–4294. [CrossRef]
12. Liu, J.; Su, C.; Chen, Y.; Tian, S.; Lu, C.; Huang, W.; Lv, Q. Current Understanding of the Applications of Photocrosslinked Hydrogels in Biomedical Engineering. *Gels* **2022**, *8*, 216. [CrossRef] [PubMed]
13. Hennink, W.; van Nostrum, C. Novel crosslinking methods to design hydrogels. *Adv. Drug Deliv. Rev.* **2012**, *64*, 223–236. [CrossRef]
14. Zhao, X.H.; Huebsch, N.; Mooney, D.J.; Suo, Z.G. Stress-relaxation behavior in gels with ionic and covalent crosslinks. *J. Appl. Phys.* **2010**, *107*, 063509. [CrossRef] [PubMed]
15. Lee, Y.K.; Mooney, J.D. Alginate: Properties and biomedical applications. *Prog. Polym. Sci.* **2012**, *37*, 106–126. [CrossRef] [PubMed]
16. Grant, G.T.; Morris, E.R.; Rees, D.A.; Smith, P.J.; Thom, D. Biological interactions between polysaccharides and divalent cations—Egg-box model. *FEBS Lett.* **1973**, *32*, 195–198. [CrossRef]
17. Chan, G.; Mooney, D.J. Ca^{2+} released from calcium alginate gels can promote inflammatory responses in vitro and in vivo. *Acta Biomater.* **2013**, *9*, 9281–9291. [CrossRef] [PubMed]
18. Harper, B.A.; Barbut, S.; Lim, L.T.; Marcone, M.F. Effect of Various Gelling Cations on the Physical Properties of "Wet" Alginate Films. *J. Food Sci.* **2014**, *79*, E562–E567. [CrossRef]
19. Luca, G.; Calviti, M.; Nastruzzi, C.; Bilancetti, L.; Becchetti, E.; Angeletti, G.; Mancuso, F.; Calafiore, R. Encapsulation, in vitro characterization, and in vivo biocompatibility of Sertoli cells in alginate-based microcapsules. *Tissue Eng.* **2007**, *13*, 641–648. [CrossRef]
20. Mores, L.; França, E.L.; Silva, N.A.; Suchara, E.A.; Honorio-França, A.C. Nanoparticles of barium induce apoptosis in human phagocytes. *Int. J. Nanomed.* **2015**, *10*, 6021–6026. [CrossRef]
21. FitzSimons, T.M.; Anslyn, E.V.; Rosales, A.M. Effect of pH on the Properties of Hydrogels Cross-Linked via Dynamic Thia-Michael Addition Bonds. *ACS Polym. Au* **2021**, *2*, 129–136. [CrossRef] [PubMed]
22. Kolotova, D.S.; Borovinskaya, E.V.; Bordiyan, V.V.; Zuev, Y.F.; Salnikov, V.V.; Zueva, O.S.; Derkach, S.R. Phase Behavior of Aqueous Mixtures of Sodium Alginate with Fish Gelatin: Effects of pH and Ionic Strength. *Polymers* **2023**, *15*, 2253. [CrossRef] [PubMed]
23. Bouhadir, K.H.; Lee, K.Y.; Alsberg, E.; Damm, K.L.; Anderson, K.W.; Mooney, D.J. Degradation of partially oxidized alginate and its potential application for tissue engineering. *Biotechnol. Prog.* **2001**, *17*, 945–950. [CrossRef] [PubMed]
24. Wu, X.; Lin, M.; Li, Y.; Zhao, X.; Yan, F. Effects of DMEM and RPMI 1640 on the biological behavior of dog periosteum-derived cells. *Cytotechnology* **2009**, *59*, 103–111. [CrossRef]
25. Thomas, J.H.; Quentin, J.; Rachelle, N.P.; Joon, H.P.; Martin, S.G.; Hao-Jan, S.; Mohamed, H.R.; Andrew, R.H.; Adam, W.F. Three-dimensional printing of complex biological structures by freeform reversible embedding of suspended hydrogels. *Sci. Adv.* **2015**, *1*, e1500758. [CrossRef]
26. Schindelin, J.; Arganda-Carreras, I.; Frise, E.; Kaynig, V.; Longair, M.; Pietzsch, T.; Preibisch, S.; Rueden, C.; Saalfeld, S.; Schmid, B.; et al. Fiji: An open-source platform for biological-image analysis. *Nat. Methods* **2012**, *9*, 676–682. [CrossRef]
27. Tan, J.; Luo, Y.; Guo, Y.; Zhou, Y.; Liao, X.; Li, D.; Lai, X.; Liu, Y. Development of alginate-based hydrogels: Crosslinking strategies and biomedical applications. *Int. J. Biol. Macromol.* **2023**, *239*, 124275. [CrossRef]

28. Haug, A. The affinity of some divalent metals to different types of alginates. *Acta Chem. Scand.* **1961**, *15*, 1794–1795. [CrossRef]
29. Haug, A.; Smidsrød, O. Selectivity of some anionic polymers for divalent metal ions. *Acta Chem. Scand.* **1970**, *24*, 843–854. [CrossRef]
30. Saitoh, S.; Araki, Y.; Kon, R.; Katsura, H.; Taira, M. Swelling/deswelling mechanism of calcium alginate gel in aqueous solutions. *Dent. Mater. J.* **2000**, *19*, 396–404. [CrossRef]
31. Shi, X.; Zheng, Y.; Wang, G.; Lin, Q.; Fan, J. pH- and electro-response characteristics of bacterial cellulose nanofiber/sodium alginate hybrid hydrogels for dual controlled drug delivery. *RSC Adv.* **2014**, *4*, 47056–47065. [CrossRef]
32. George, M.; Abraham, T.E. pH sensitive alginate-guar gum hydrogel for the controlled delivery of protein drugs. *Int. J. Pharm.* **2007**, *335*, 123–129. [CrossRef] [PubMed]
33. Ramdhan, T.; Ching, S.H.; Prakash, S.; Bhandari, B. Time dependent gelling properties of cuboid alginate gels made by external gelation method: Effects of alginate-CaCl2 solution ratios and pH. *Food Hydrocoll.* **2018**, *90*, 232–240. [CrossRef]
34. Chuang, J.J.; Huang, Y.Y.; Lo, S.H.; Hsu, T.F.; Huang, W.Y.; Huang, S.L.; Lin, Y.S. Effects of pH on the Shape of Alginate Particles and Its Release Behavior. *Int. J. Polym. Sci.* **2017**, *2017*, 3902704. [CrossRef]
35. Mørch, Y.A.; Qi, M.; Gundersen, P.O.M.; Formo, K.; Lacik, I.; Skjåk-Braek, G.; Oberholzer, J.; Strand, B.L. Binding and leakage of barium in alginate microbeads. *J. Biomed. Mater. Res. Part A* **2012**, *100A*, 2939–2947. [CrossRef] [PubMed]
36. Clapham, D.E. Calcium signaling. *Cell* **2007**, *131*, 1047–1058. [CrossRef]
37. Przywara, D.A.; Chowdhury, P.S.; Bhave, S.V.; Wakade, T.D.; Wakade, A.R. Barium-induced exocytosis is due to internal calcium release and block of calcium efflux. *Proc. Natl. Acad. Sci. USA* **1993**, *90*, 557–561. [CrossRef]
38. Heldman, E.; Levine, M.; Raveh, L.; Pollard, H.B. Barium ions enter chromaffin cells via voltage-dependent calcium channels and induce secretion by a mechanism independent of calcium. *J. Biol. Chem.* **1989**, *264*, 7914–7920. [CrossRef]
39. Gilmore, R.S.C.; Bullock, C.G.; Sanderson, G.; Wallace, W.F. Ultrastructural evidence in rabbit ear arteries that barium enters smooth muscle cells through calcium channels. *Q. J. Exp. Physiol.* **1986**, *71*, 417–422. [CrossRef]
40. Sprague, S.M.; Krieger, N.S.; Bushinsky, D.A. Greater inhibition of in vitro bone mineralization with metabolic than respiratory acidosis. *Kidney Int.* **1994**, *46*, 1199–1206. [CrossRef]
41. Brandao-Burch, A.; Utting, J.C.; Orriss, I.R.; Arnett, T.R. Acidosis inhibits bone formation by osteoblasts in vitro by preventing mineralization. *Calcif. Tissue Int.* **2005**, *77*, 167–174. [CrossRef] [PubMed]
42. Arnett, T.R. Extracellular pH regulates bone cell function. *J. Nutr.* **2008**, *138*, 415S–418S. [CrossRef] [PubMed]
43. Matsubara, T.; DiResta, G.R.; Kakunaga, S.; Li, D.; Healey, J.H. Additive Influence of Extracellular pH, Oxygen Tension, and Pressure on Invasiveness and Survival of Human Osteosarcoma Cells. *Front. Oncol.* **2013**, *3*, 199. [CrossRef] [PubMed]
44. Białkowska, K.; Komorowski, P.; Bryszewska, M.; Miłowska, K. Spheroids as a Type of Three-Dimensional Cell Cultures-Examples of Methods of Preparation and the Most Important Application. *Int. J. Mol. Sci.* **2020**, *21*, 6225. [CrossRef] [PubMed]
45. John, D.P.; Kiessling, A.A. Improved pronuclear mouse embryo development over an extended pH range in Ham's F-10 medium without protein. *Fertil. Steril.* **1988**, *49*, 150–155. [CrossRef]
46. Dastagir, K.; Reimers, K.; Lazaridis, A.; Jahn, S.; Maurer, V.; Strauß, S.; Dastagir, N.; Radtke, C.; Kampmann, A.; Bucan, V.; et al. Murine embryonic fibroblast cell lines differentiate into three mesenchymal lineages to different extents: New models to investigate differentiation processes. *Cell. Reprogram.* **2014**, *16*, 241–252. [CrossRef]
47. Rowley, J.A.; Madlambayan, G.; Mooney, D.J. Alginate hydrogels as synthetic extracellular matrix materials. *Biomaterials* **1999**, *20*, 45–53. [CrossRef]
48. Souza, A.; McCarthy, K.; Rodriguez, B.J.; Reynaud, E.G. The Use of Fluid-phase 3D Printing to Pattern Alginate-gelatin Hydrogel Properties to Guide Cell Growth and Behaviour In Vitro. *bioRxiv* **2023**. [CrossRef]
49. Alruwaili, M.; Lopez, J.A.; McCarthy, K.; Reynaud, E.G.; Rodriguez, B.J. Liquid-phase 3D bioprinting of gelatin alginate hydrogels: Influence of printing parameters on hydrogel line width and layer height. *Bio-Des. Manuf.* **2019**, *2*, 172–180. [CrossRef]

Disclaimer/Publisher's Note: The statements, opinions and data contained in all publications are solely those of the individual author(s) and contributor(s) and not of MDPI and/or the editor(s). MDPI and/or the editor(s) disclaim responsibility for any injury to people or property resulting from any ideas, methods, instructions or products referred to in the content.

Article

Polydopamine Blending Increases Human Cell Proliferation in Gelatin–Xanthan Gum 3D-Printed Hydrogel

Preetham Yerra [1], Mario Migliario [2], Sarah Gino [1], Maurizio Sabbatini [3], Monica Bignotto [4], Marco Invernizzi [1] and Filippo Renò [4,*]

[1] Health Sciences Department, Università del Piemonte Orientale, Via Solaroli n.17, 28100 Novara, Italy; 20035225@studenti.uniupo.it (P.Y.); sarah.gino@uniupo.it (S.G.); marco.invernizzi@uniupo.it (M.I.)

[2] Traslational Medicine Department, Università del Piemonte Orientale, Via Solaroli n.17, 28100 Novara, Italy; mario.migliario@med.uniupo.it

[3] Department of Sciences and Innovative Technology, Università del Piemonte Orientale, Viale T. Michel 11, 15121 Alessandria, Italy; muarizio.sabbatini@uniupo.it

[4] Department of Health Sciences, Università degli Studi di Milano, Via A. di Rudini n.8, 20142 Milano, Italy; monica.bignotto@unimi.it

* Correspondence: filippo.reno@unimi.it

Abstract: Background: Gelatin–xanthan gum (Gel–Xnt) hydrogel has been previously modified to improve its printability; now, to increase its ability for use as cell-laden 3D scaffolds (bioink), polydopamine (PDA), a biocompatible, antibacterial, adhesive, and antioxidant mussel-inspired biopolymer, has been added (1–3% v/v) to hydrogel. Methods: Control (CT) and PDA-blended hydrogels were used to print 1 cm^2 grids. The hydrogels' printability, moisture, swelling, hydrolysis, and porosity were tested after glutaraldehyde (GTA) crosslinking, while biocompatibility was tested using primary human-derived skin fibroblasts and spontaneously immortalized human keratinocytes (HaCaT). Keratinocyte or fibroblast suspension (100 μL, 2.5×10^5 cells) was combined with an uncrosslinked CT and PDA blended hydrogel to fabricate cylinders (0.5 cm high, 1 cm wide). These cylinders were then cross-linked and incubated for 1, 3, 7, 14, and 21 days. The presence of cells within various hydrogels was assessed using optical microscopy. Results and discussion: PDA blending did not modify the hydrogel printability or physiochemical characteristics, suggesting that PDA did not interfere with GTA crosslinking. On the other hand, PDA presence strongly accelerated and increased both fibroblast and keratinocyte growth inside. This effect seemed to be linked to the adhesive abilities of PDA, which improve cell adhesion and, in turn, proliferation. Conclusions: The simple PDA blending method described could help in obtaining a new bioink for the development of innovative 3D-printed wound dressings.

Keywords: polydopamine; bioprinting; hydrogel; fibroblast; keratinocyte; cell proliferation; skin wound healing

Citation: Yerra, P.; Migliario, M.; Gino, S.; Sabbatini, M.; Bignotto, M.; Invernizzi, M.; Renò, F. Polydopamine Blending Increases Human Cell Proliferation in Gelatin–Xanthan Gum 3D-Printed Hydrogel. Gels 2024, 10, 145. https://doi.org/10.3390/gels10020145

Academic Editors: Enrique Aguilar and Helena Herrada-Manchón

Received: 9 January 2024
Revised: 7 February 2024
Accepted: 11 February 2024
Published: 14 February 2024

Copyright: © 2024 by the authors. Licensee MDPI, Basel, Switzerland. This article is an open access article distributed under the terms and conditions of the Creative Commons Attribution (CC BY) license (https:// creativecommons.org/licenses/by/ 4.0/).

1. Introduction

Polydopamine (PDA) stands out as a remarkable bioinspired biopolymer, drawing inspiration from mussel adhesive proteins and exhibiting a multifaceted array of attributes that render it highly applicable across diverse biomedical domains [1]. Its remarkable biocompatibility, coupled with potent antioxidant, antibacterial, and adhesive properties, positions PDA as a material of choice in diverse applications in the biomedical field [1]. The versatility of PDA is exemplified by its ability to form layers on nearly all types of organic and inorganic substrates, facilitated by the self-polymerization of dopamine (DA) under alkaline conditions [2]. This characteristic enables PDA to serve diverse functions, including tissue adhesion, sealing, surface coating, and biomolecule immobilization, amplifying its utility in biomedical contexts [2]. However, the efficacy of the oxidative polymerization process of DA hinges upon the meticulous control of parameters such as temperature, the pH of Tris-HCl, and the initial dopamine concentration [3]. Dopamine (DA), as the precursor to PDA, emerges as a pivotal material due to its versatility and cost-effectiveness, laying the foundation for the exploration of advanced applications of PDA in biomedical fields [4]. In recent years, PDA-enriched mussel-inspired hydrogels have garnered significant attention for their superior properties compared to conventional hydrogels [5]. These hydrogels, enriched with dynamic catechol-based bonds, exhibit exceptional toughness, extensibility, and a rapid self-healing capacity, marking significant strides in biomaterials engineering [5,6]. Furthermore, PDA modifications in hydrogels allow for increased applications not only in regenerative medicine (increasing cell and tissue adhesion capacity), but also in drug delivery systems [6–11]. Notably, PDA-modified hydrogels hold promise as wound dressings, accelerating the healing of chronic wounds [12–14]. Bioprinting emerges as a transformative technology in tissue engineering, facilitating the fabrication of intricate, cell-laden 3D scaffolds that mimic native tissue architecture [15,16]. Hydrogel-based bioinks, particularly those incorporating PDA, have emerged as a cornerstone in 3D bioprinting, enabling the precise deposition of cells and biomaterials to generate complex tissue constructs [17–20]. While the application of 3D bioprinted PDA-modified hydrogels has primarily focused on bone tissue engineering, their versatility suggests potential applications across various tissue types [17,21,22]. In our prior work, we introduced a novel 3D-printable bioink comprising gelatin and xanthan gum, which was successfully employed for printing structures laden with human skin cells [23]. Herein, our investigation revolves around evaluating the impact of incorporating polydopamine (PDA) into the gelatin–xanthan gum hydrogel blend. We aim to ascertain whether this modification enhances the hydrogel's capacity to facilitate skin cell proliferation while preserving its original physicochemical properties. The ultimate goal is to harness this innovative mussel-inspired hydrogel for dual purposes: serving as a potential wound dressing and as a scaffold for constructing intricate 3D skin-like models.

2. Results and Discussion

2.1. PDA Does Not Alter Hydrogel Printability and Moisture

The incorporation of Polydopamine (PDA) into Gel–Xnt hydrogel resulted in a dose-dependent darkening of the hydrogel, as depicted in Figure 1A, consistent with expectations [24]. Despite this darkening effect, all PDA-blended hydrogels retained their printability, as evidenced by Figure 1A, indicating that the presence of PDA did not significantly impact the viscosity of the hydrogel. In previous studies, it has been shown that soaking substrates in a dilute aqueous solution of dopamine, buffered to a pH of 8.5, results in the spontaneous deposition of a thin adherent polymer film [25]. This polymer film typically attains a thickness ranging from 10 to 50 nm after 24 h, with its formation appearing independent of the substrate used [25]. In our experimental setup, we introduced a PDA solution into the Gel–Xnt mixture before crosslinking with GTA. Remarkably, this introduction of PDA did not appear to affect the viscosity or crosslinking behavior of the hydrogel, indicating the compatibility of PDA with both Gel and Xnt components.

The main goal of our research on Gel–Xnt printable hydrogels was to use 3D printing technology to fabricate cell-seeded hydrogels for therapeutic purposes (e.g., skin grafts [26], restoring tissue layers [27] or to promote wound healing). Specifically, hydrogels that are employed in wound healing ought to contain a significant water content. This attribute is essential as it enhances cellular interaction and facilitates the diffusion of molecules [28], all while preventing tissue dehydration.

In CT hydrogels (3% Gel–1.2% Xnt) the percentage of water measured, using the method described by Shawan et al. [28], was $91 \pm 0.2\%$ and the blending with different % of PDA did not modify hydrogel water content (Figure 2). In fact, the water percent measured for PDA 1%, 2%, and 3% hydrogels was, respectively, $90.9 \pm 0.2\%$, $91.1 \pm 0.3\%$, and $90.6 \pm 1.1\%$ (Figure 2). PDA is an excellent adhesive material with superhydrophilic properties [29], but, in this case, the hydrophilicity of the starting hydrogel was so high that the PDA blending was not able to further increase it.

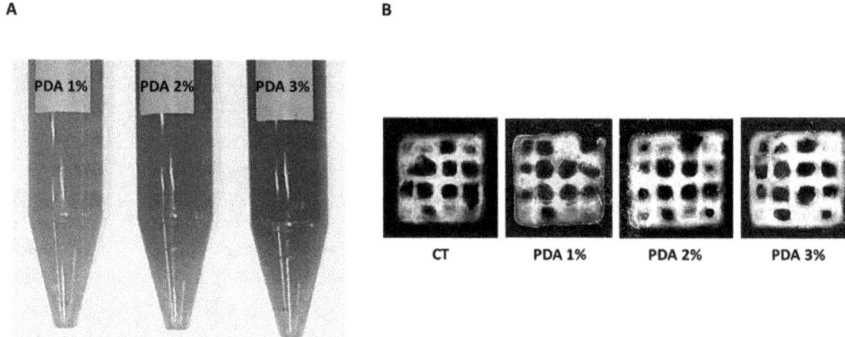

Figure 1. (**A**) Dose-dependent darkening of PDA-blended hydrogels. (**B**) Morphological appearance of CT and PDA-blended hydrogels printed with a pressure of 8 kPa to obtain 1 cm^2 squares samples crosslinked for 20 min using a solution of 0.3 v/v% GTA.

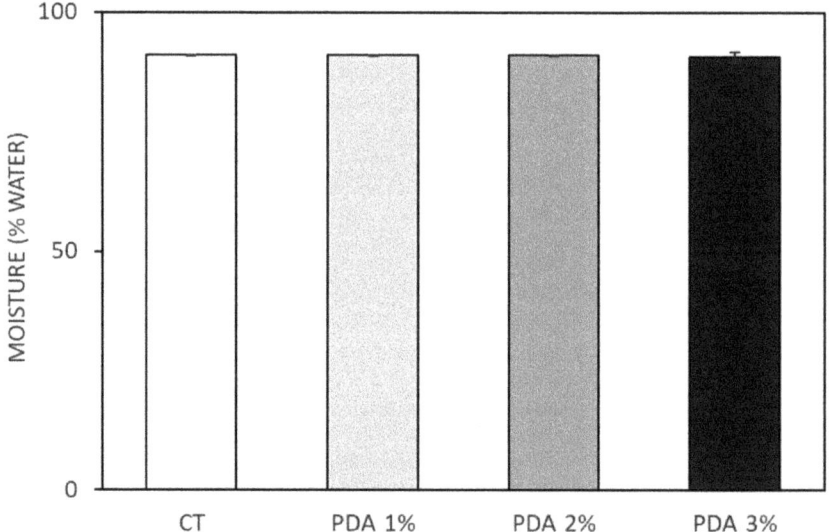

Figure 2. Moisture percentage of CT (Gel3%-Xnt 1.2%) and PDA-blended hydrogels after hydration and drying process at 37 °C for 2 days.

2.2. PDA Blending Effect on Hydrogel Swelling

Evaluating swelling is pivotal in the assessment of hydrogel properties due to its direct correlation with the water absorption capacity, resulting in increased weight and volume [30]. Hydrogels prone to higher degrees of swelling may undergo shape changes and fractures, particularly in moist environments. The swelling behavior of hydrogels can vary based on factors such as gelatin concentration and the presence of glutaraldehyde in their formulation. Glutaraldehyde facilitates the crosslinking of gelatin by reacting with non-protonated ε-amino groups (-NH2) in lysine or hydroxylysine within the gelatin structure, leading to the formation of amide linkages [31]. These crosslinking interactions significantly influence the hydrogel's water retention capacity [30]. In our CT bioink, gel functional groups effectively bind with all aldehyde groups of the crosslinker, limiting their interaction with water molecules and thereby positively impacting the swelling rate [24]. Figure 3 illustrates that PDA 2% and 3% hydrogels exhibited increased swelling after 1 h (65.5 ± 11.4% and 60.9 ± 7%, respectively) compared to the CT samples (41.4 ± 4.3%,

$p < 0.05$), which was attributed to the presence of hydrophilic PDA enhancing the water binding ability.

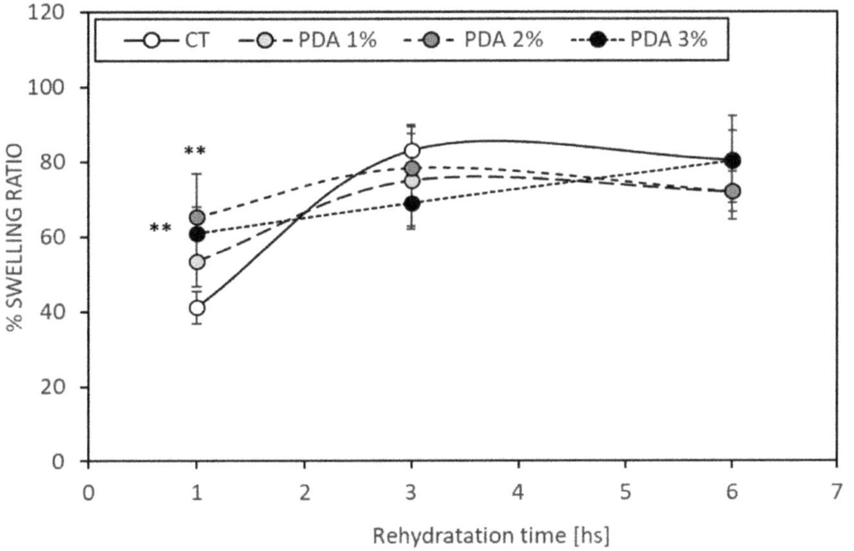

Figure 3. Swelling ratio of rehydrated CT and PDA-blended hydrogels. Dried hydrogels were (1) weighed, (2) rehydrated by immersion in deionized water, and (3) weighed, after the residual water was removed by capillary action of filter paper, at different time points (time 1, 3, 6 hs). ** $p < 0.05$ compared to CT samples.

However, this increased water binding ability demonstrated by PDA-blended hydrogels diminished after 3 h, as the CT samples showed a swelling ratio of 83.1 ± 6.8%, while PDA 3% exhibited a ratio of 69.0 ± 7%. Similar trends were observed after 6 h (Figure 3). Notably, PDA did not appear to interfere with GTA crosslinking abilities in forming bonds between Gel and Xnt. This suggests a complex interplay between hydrogel components influencing swelling behavior over time. Further investigations are warranted to elucidate the long-term effects and optimize hydrogel formulations for specific applications.

2.3. PDA Effect on Hydrogel Hydrolysis

Hydrolysis values were determined following 7 and 14 days of submersion at 37 °C for both CT and PDA-blended printed hydrogels. Despite being predominantly composed of water, hydrogels are susceptible to degradation via hydrolysis, a process where water molecules interact with the hydrogel structure over time, especially under elevated temperatures. The CT hydrogel exhibited a significant percentage of hydrolysis after 7 days (86.1 ± 1.1%), which decreased notably after 14 days (61.4 ± 3.8%) (refer to Figure 4). Interestingly, the incorporation of PDA did not appear to influence the hydrolysis rate of the hydrogel, indicating that PDA does not interfere with the crosslinking process, mediated by GTA, between Gel and Xnt. This observation aligns with our previous findings [23], where we suggested that the sensitivity of CT hydrogel to hydrolysis might stem from the robust crosslinking induced by GTA, resulting in numerous aldehyde groups linked to gelatin via an amide bond, rendering it less stable and more prone to hydrolytic degradation.

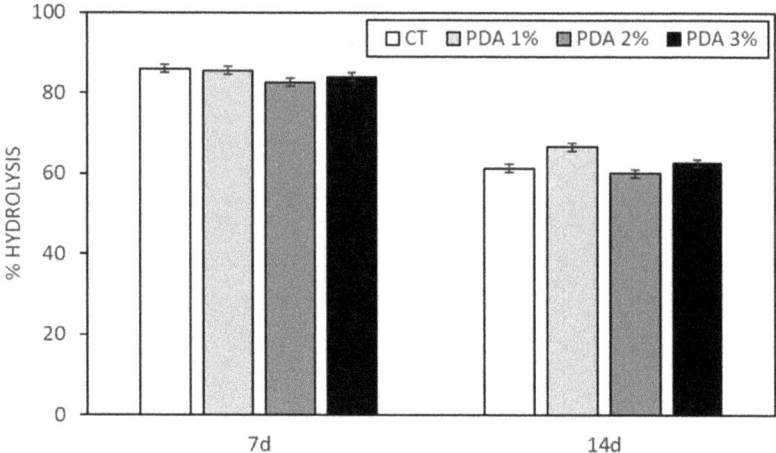

Figure 4. Hydrolysis percentage of CT and PDA-blended hydrogels after soaking in deionized water at 37 °C for 7 and 14 days.

2.4. PDA Effect on Hydrogel Porosity

Since the ability of cells to proliferate in a 3D structure depends on the pore size and the biodegradability of the structure, it is important to consider the size and number of pores, as well as their geometry and connectivity [32]. Pore sizes should be less than 500 µm to allow for vascularization and tissue formation, as larger pore sizes could decrease cell–cell interactions and, thus, their proliferation [32,33]. In a previously published paper from our lab [23], the porosity of CT hydrogel (3% Gel–1.2% Xnt) was measured using both morphological analysis and a liquid displacement method. The average number of pores measured in an area of 25 µm^2 was 12,100 ± 561, while their average diameter was 0.723 ± 0.38 µm [23]. The porosity of the CT gel was, therefore, in a range that allows cell survival and communication. Moreover, the percentage of porosity measured using the liquid displacement method for the CT hydrogel was 30.82 ± 4.93% [23], comparable with that measured in the current work (44.1 ± 11.2%,); even the real porosity (average number and size of pores) is reasonable and comparable with that observed in that paper. As shown in Figure 5, PDA blending did not alter the percentage of porosity values which were 43.1 ± 8.2%, 45.3 ± 10.2%, and 46.2 ± 9.4%, respectively, for PDA 1%, 2%, and 3%.

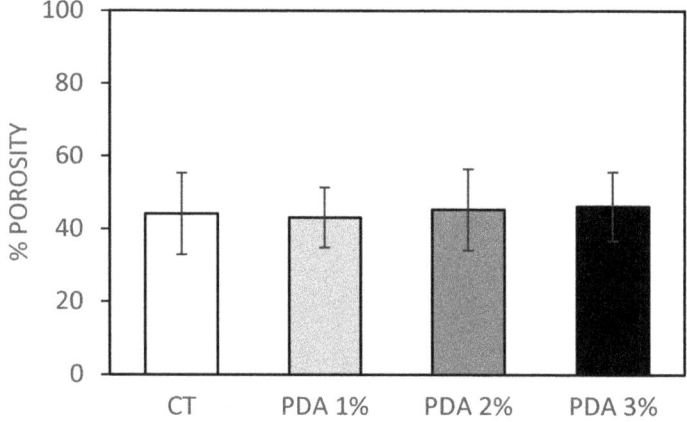

Figure 5. Porosity analysis of CT and PDA-blended hydrogels carried out by liquid displacement method.

2.5. PDA Effect on Cell Proliferation in Cell-Laden Hydrogel

The important indicators of the cytocompatibility of a biomaterial are cell adhesion and proliferation. We already demonstrated that Gel–Xnt hydrogels were nontoxic and cytocompatible; however, to evaluate the effects of hydrogel functionalization with various concentrations of PDA, human keratinocytes and fibroblasts were mixed with CT and PDA-blended. Cell-laden hydrogels were used to print small cylinders crosslinked using TGA, as described in the Materials and Methods section. Cell-laden cylinders were then washed using PBS and, finally, were cultured for 21 days. Cell growth was monitored b counting the cells present in the hydrogel at different time points. As shown in Figure 6A,B, both keratinocytes and fibroblasts grew slowly until day 7, when, for both cell populations, a significant acceleration in growth in the PDA-blended hydrogel was seen. In fact, on day 7, keratinocyte growth in the CT samples was 120.1 ± 6.9% of T0 (0d), similar to the value scored in PDA 1% samples (123.1 ± 5.7%), while, in PDA 2% and 3%, keratinocyte growth significantly increased, respectively, to 138.2 ± 6.1% and 142.4 ± 3.9% that of CT ($p < 0.05$) (Figure 6A). Keratinocyte growth further increased at day 14, reaching a plateau at day 21 with an ever significant positive difference in PDA-blended hydrogels. In fact, at day 21, keratinocyte growth in the CT samples was 131.8 ± 2.5% of T0 (0 d), while, in PDA 1%, 2%, and 3%, it was, respectively, 141.0 ± 2.1%, 149.2 ± 2.5%, and 156.4 ± 4.7% that of CT ($p < 0.05$) (Figure 6A,B).

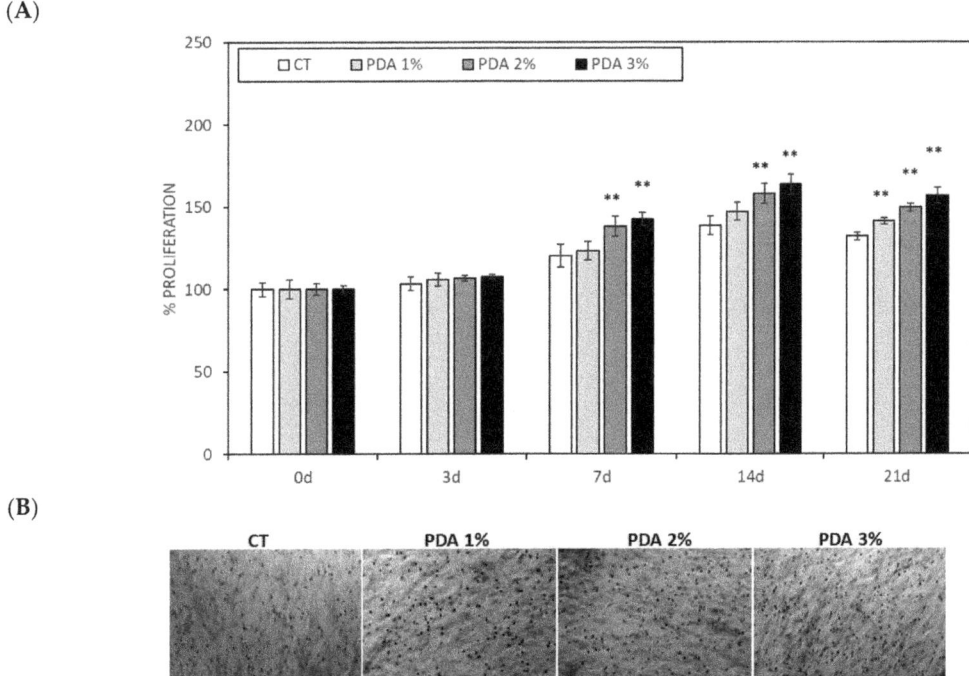

Figure 6. (**A**) Human keratinocytes spontaneously immortalized (HaCaT) proliferation measured in keratinocyte-laden hydrogel printed cylinders. (**B**) Representative images of cell-laden hydrogels after 21 days. Bar = 100 μ. ** $p < 0.05$ as compared to CT samples.

A similar cellular growth pattern was observed for human fibroblasts (Figure 7A,B), which, however, grew faster than keratinocytes and reached their maximum growth at day 7. In fact, at day 7, fibroblast growth in the CT samples was 139.5 ± 7.2% of T0 (0 d), while, in PDA 1%, 2%, and 3%, hydrogels fibroblast growth significantly increased, respectively, to 159.7 ± 9.1%, 176.4 ± 9.9%, and 196.1 ± 12.1% that of CT ($p < 0.05$) (Figure 7A). These growth values remained substantially unchanged at day 14 and 21, indicating that fibroblasts reached their maximum space of proliferation (Figure 7A,B)

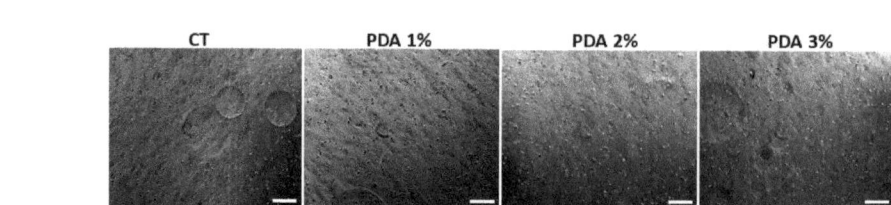

Figure 7. (A) Primary human-derived skin fibroblast proliferation measured in cell-laden hydrogel printed cylinders. (B) Representative images of fibroblast-laden hydrogels after 21 days. Bar = 100 µ. ** $p < 0.05$ as compared to CT samples.

The faster and higher growth observed for both keratinocytes and fibroblasts in PDA-blended hydrogels compared to the CT samples, even with some important differences due to cell characteristics (epithelial cells vs. connective tissue cells), is believed to be due to the strong adhesion of the cells to the surface of the PDA-mixed hydrogels and the resulting cell survival and proliferation [8]. Of course, a more detailed cell morphological and proteomic analysis is necessary to fully understand the PDA positive effect on cell growth observed in this kind of printable hydrogel. In addition, it may be interesting to evaluate whether a different percentage of PDA or a different quantity of PDA nanoparticles present in a hydrogel [34] could modulate the growth of the different cell population, such as endothelial cells or osteoblasts, to be able, for example, to create a PDA gradient that can act as an "organizer" of small models of complex organs.

Two potential mechanisms of action for PDA in promoting cell proliferation can be postulated. The first, an indirect mechanism, may be linked to the pronounced hydrophilicity imparted by the PDA coating, attributed to its catechol/quinone polar units, amines, and imines [11]. This increased hydrophilicity could play a crucial role in modulating the adsorption of serum proteins on the hydrogel substrate, thereby indirectly influencing cell adhesion and cytoskeletal reorganization, as observed in other scaffold types [35]. Notably, studies have indicated that PDA facilitates the adsorption of proteins while preserving their native configuration, which is pivotal for effective cell adhesion and spreading [36].

However, despite the anticipated increase in hydrogel swelling properties due to the enhanced hydrophilicity induced by PDA, our study revealed no significant rise in hydrogel swelling upon PDA addition. This result could be due to the high hydrogel wettability; consequently, the hydrophilic impact of PDA in our experimental setup may not be paramount. Hence, an alternative direct mechanism of action for PDA in augmenting cell adhesion and proliferation can be proposed. Specifically, this involves the capacity of the PDA coat to elevate the expression of paxillin, a key protein within the focal adhesion complex facilitating cell–substrate anchoring [37,38], as well as inducing the overexpression of the marker MYO, which is responsible for encoding myosin expression in the cytoskeleton. The upregulation of MYO induced by PDA could facilitate the rearrangement of the actin–myosin filaments crucial for establishing robust cell–matrix interactions [38].

3. Conclusions

Three-dimensional (3D) bioprinting has emerged as a transformative technology in modern medicine, particularly heralded for its applications in tissue engineering and the treatment of challenging wound conditions. This innovative approach enables the fabrication of complex 3D biological structures that closely mimic the microenvironment of native tissues, fostering crucial interactions between cells and their extracellular matrix. In previous studies, we successfully adapted a Gel–Xnt hydrogel, demonstrating its efficacy as a wound dressing material through rat skin burn experiments [28], and subsequently developed it into a printable bioink [23]. In our latest endeavor, we sought to further enhance the biocompatibility and bioactivity of this hydrogel by incorporating Polydopamine (PDA) during the crosslinking process. PDA, renowned for its natural bioadhesive properties and versatility as a polymer, holds immense promise across various biomedical applications owing to its exceptional chemical adaptability and inherent antioxidant and antibacterial characteristics. Moreover, PDA exhibits the remarkable ability to modulate cell adhesion dynamics and promote cell proliferation at the interface of biomaterials. These attributes position PDA as a compelling biomaterial for facilitating wound healing and skin regeneration. Importantly, our incorporation of PDA into the original Gel–Xnt bioink did not compromise its printability or alter its physicochemical properties, including its moisture content, swelling behavior, hydrolysis rate, and porosity. However, the addition of PDA significantly augmented the growth of human keratinocytes and fibroblasts, suggesting a pronounced enhancement in bioactivity. This positive effect can primarily be attributed to the unique adhesive properties of PDA, which facilitate superior cell adhesion and proliferation within the hydrogel matrix.

Thus, the straightforward method of blending PDA described herein holds promise for the development of a novel bioink with enhanced antibacterial and antioxidant properties, courtesy of PDA's presence. Such advancements could prove invaluable in the fabrication of next-generation printed wound dressings, offering innovative solutions for improved wound care and tissue regeneration.

4. Materials and Methods

4.1. Preparation of Polydopamine

Dopamine hydrochloride (2 mg/mL) (Sigma-Aldrich, St. Louis, MO, USA) was polymerized to polydopamine (PDA) in Tris-HCl buffer (pH 8.5) [25]. The prepared dark PDA solution was stored at 4 °C for a maximum period of a month [12].

4.2. Preparation of Hydrogel

Shawan et al. [12] laid the groundwork for the preparation of hydrogels targeting wound healing, employing various formulations of gelatin (Gel) and xanthan gum (Xnt). Building upon this research, we developed two novel printable Gel–Xnt hybrid composite hydrogels [23]. For the hydrogel discussed in this paper, we utilized a formulation consisting of 3% Gel and 1.2% Xnt. The synthesis process involved gradually adding bovine Gel (0.6 g) (Sigma-Aldrich, St. Louis, MO, USA) to 20 mL of deionized sterilized water at 70 °C under magnetic stirring to prevent clumping. Subsequently, Xnt (0.12 g) (Sigma-Aldrich, St. Louis, MO, USA) was introduced into the gelatin solution, with a temperature range of 60–70 °C, until water evaporation occurred. The resulting hydrogel (3% Gel-1.2% Xnt) was then stored at 4 °C until use, typically within a week. To incorporate polydopamine (PDA) into the hydrogels, a PDA solution was added drop by drop during mixing to achieve concentrations of 1%, 2%, or 3% (v/v). Following preparation, the hydrogels were refrigerated for 3 to 10 days and brought to room temperature prior to printing. This meticulous synthesis procedure ensures the stability and consistency of the hydrogel formulations for subsequent applications.

4.3. Hydrogel Bioprinting and Crosslinking

CellInk Inkredible 3D printer (CellInk, Gothenburg, Sweden) was used for hydrogels printing. Bio-ink syringes (3 mL) were filled with the different hydrogels and then loaded into the 3D printers (3 mL). Hydrogels were printed with a pressure of 8 kPa to obtain 1 cm^2 samples. Hydrogels crosslinking was then performed by using 0.3% glutaraldehyde (GTA) (v/v%), prepared from a 25 v/v% stock solution. Samples were submerged in GTA solution for 20 min then washed twice with phosphate buffer (PBS) pH 7.4.

4.4. Characterization of 3D-Printed Hydrogel

4.4.1. Moisture

The moisture percentage was calculated according to Shawan's method [13]. On the third day of stabilization, the hydrated prints were weighed (WH), stored at 37 °C for 2 days to allow the drying process to complete, and then weighed again (WD). The percentage of moisture (water) in a printed hydrogel was calculated by using the following equation:

$$\text{Moisture (\%)} = [(\text{WH} - \text{WD})/\text{WH}] \times 100$$

where WH is the original weight of the sample before drying and WD is the weight of the sample after drying.

All the experiments were replicated three times, with at least three samples for each condition.

4.4.2. Swelling

Swelling test was performed using crosslinked printed hydrogel dried at 37 °C for 48 h. Dried hydrogels were weighed, rehydrated by soaking in deionized water, and weighed again after the residual water was removed by the capillary action of filter paper, at different time points (1, 3, and 6 hs). The swelling ratio (S) was calculated according to Zheng [13].

All the experiments were replicated three times, with at least three samples for each condition.

4.4.3. Hydrolysis

To assess hydrolysis, the cross-linked printed hydrogels were weighed (time 0) and immersed in deionized water at 37 °C. After a period of 4–14 days, the weight was calculated by removing the deionized water and lightly blotting the cross-linked prints with filter paper. The percentage of hydrolysis was obtained with the following equation:

$$\text{Hydrolysis (\%)} = [(\text{WI} - \text{WF})/\text{WI}] \times 100$$

where WI is the weight before soaking and WF is the weight remaining after soaking and removal of deionized water. All experiments were replicated three times, with at least three samples for each condition.

4.4.4. Porosity

The porosity of the printed hydrogels was assessed by the liquid displacement method. Absolute ethanol, which causes neither swelling nor shrinking of gelatin [14], was used to immerse the prints. After 5 min of submersion in a known amount of absolute ethanol, the samples were weighed. The porosity of the hydrogel was calculated as follows:

$$\text{Porosity (\%)} = [(W1 - W3)/(W2 - W3)] \times 100$$

where W1 is the initial weight of pure ethanol, W2 is the total weight combining the weight of the hydrogel with that of the ethanol, and W3 is the final weight of ethanol without hydrogel.

4.5. Cell Culture

Primary human-derived skin fibroblasts (human fibroblasts were kindly donated by Prof. Marco De Andrea (University of Turin, Italy) and spontaneously immortalized human keratinocytes (HaCaT) (HaCaT were purchased from Cell Lines Service GmbH (Eppelheim, Germany)) were used to test the biocompatibility of the printable hydrogels [23]; Fibroblasts were grown in Petri dishes in RPMI 1640 and keratinocytes in culture flasks in DMEM, both supplemented with 10% heat-inactivated fetal bovine serum (FBS) and 1% Penicillin–Streptomycin and L-Glutamine (Immunological Science, Milan, Italy) in a cell incubator with a humidified atmosphere containing 5% CO_2 at 37 °C. Small volumes of 100 µL cell suspension containing 2.5×10^5 cells (fibroblast or keratinocytes) were gently mixed, with two syringes connected under a sterile hood. Cylinders, 0.5 cm high and 1 cm wide, were printed in 12-well plates using 12-well model tissue G-code. The cylinders were then crosslinked using GTA. The cell-laden cylinders were washed three times with sterile PBS and incubated for 1, 3, 7, 14, and 21 days in DMEM 10% FBS at 37 °C in a 5% CO_2 atmosphere. Then, the culture medium was removed and the hydrogels containing the cells were washed 3 times with PBS. Hydrogel images (3 different fields for samples) were digitally acquired using a Zeiss Axiovert 40 cfl (Carl Zeiss Microscopy, LLC, New York, NY, USA) at an original magnification of 20×. The images were analyzed by two different operators blind about the nature of images and the cell number was scored using Image J 1.53m software. Cell number was expressed as % cells present in CT samples at time 0.

4.6. Statistical Analysis

Data were presented as mean ± standard deviation (SD). Statistical analyses were performed with GraphPad PRISM software. The Kolgomorov–Smirnoff test was applied in order to understand the normal distribution of data. One-way ANOVA was used for multiple comparisons, while the *t*-Test was performed to compare between the two groups. *p*-values < 0.05 were considered statistically significant.

Author Contributions: Conceptualization, F.R. and P.Y.; methodology, F.R. and P.Y.; software, F.R.; validation, M.M. and S.G.; formal analysis, S.G. and M.B.; investigation, F.R. and P.Y.; resources, F.R. and M.M.; data curation, P.Y. and M.S.; writing—original draft preparation, F.R., P.Y. and M.S.; writing—review and editing, F.R., M.I., S.G. and M.I.; visualization, M.B.; supervision, F.R. and M.I.; project administration, F.R., M.B. and M.I.; funding acquisition, M.I. and M.S. All authors have read and agreed to the published version of the manuscript.

Funding: This research was funded by the Health Sciences Department research fund, Università del Piemonte Orientale, grant number: RIVinvernizzimRL16_BCRLrehab17_19, and by the Department of Science and Technology Innovation, Università del Piemonte Orientale, grant number: RIVsabbatinimFAR17_astrociti_19-21.

Institutional Review Board Statement: Not applicable.

Informed Consent Statement: Not applicable.

Data Availability Statement: The datasets presented in this article are not readily available because the data are part of an ongoing study. Requests to access the datasets should be directed to filippo.reno@unimi.it.

Acknowledgments: The authors thank Marco De Andrea from the University of Turin for the kind gift of the human fibroblasts.

Conflicts of Interest: The authors declare no conflict of interest.

References

1. Li, H.; Yin, D.; Li, W.; Tang, Q.; Zou, L.; Peng, Q. Polydopamine-based nanomaterials and their potentials in advanced drug delivery and therapy. *Colloids Surf. B Biointerfaces* **2021**, *199*, 111502. [CrossRef] [PubMed]
2. Lu, M.; Yu, J. Mussel-Inspired Biomaterials for Cell and Tissue Engineering. *Adv. Exp. Med. Biol.* **2018**, *1077*, 451–474. [PubMed]
3. Sarkari, S.; Khajehmohammadi, M.; Davari, N.; Li, D.; Yu, B. The effects of process parameters on polydopamine coatings employed in tissue engineering applications. *Front. Bioeng. Biotechnol.* **2022**, *10*, 1005413. [CrossRef] [PubMed]
4. Beckwith, K.M.; Sikorski, P. Patterned cell arrays and patterned co-cultures on polydopamine-modified poly(vinyl alcohol) hydrogels. *Biofabrication* **2013**, *5*, 045009. [CrossRef] [PubMed]
5. Zhang, C.; Wu, B.; Zhou, Y.; Zhou, F.; Liu, W.; Wang, Z. Mussel-inspired hydrogels: From design principles to promising applications. *Chem. Soc. Rev.* **2020**, *49*, 3605–3637. [CrossRef]
6. Hu, Y.; Dan, W.; Xiong, S.; Kang, Y.; Dhinakar, A.; Wu, J.; Gu, Z. Development of collagen/polydopamine complexed matrix as mechanically enhanced and highly biocompatible semi-natural tissue engineering scaffold. *Acta Biomater.* **2017**, *47*, 135–148. [CrossRef]
7. Yazdi, M.K.; Zare, M.; Khodadadi, A.; Seidi, F.; Sajadi, S.M.; Zarrintaj, P.; Arefi, A.; Saeb, M.R.; Mozafari, M. Polydopamine Biomaterials for Skin Regeneration. *ACS Biomater. Sci. Eng.* **2022**, *13*, 2196–2219. [CrossRef]
8. Pacelli, S.; Paolicelli, P.; Petralito, S.; Subham, S.; Gilmore, D.; Varani, G.; Yang, G.; Lin, D.; Casadei, M.A.; Paul, A. Investigating the role of polydopamine to modulate stem cell adhesion and proliferation on gellan gum-based hydrogels. *ACS Appl. Bio Mater.* **2020**, *3*, 945–951. [CrossRef]
9. Singh, N.; Sallem, F.; Mirjolet, C.; Nury, T.; Sahoo, S.K.; Millot, N.; Kumar, R. Polydopamine modified Superparamagnetic Iron oxide nanoparticles as multifunctional nanocarrier for targeted prostate cancer treatment. *Nanomaterials* **2019**, *9*, 138. [CrossRef] [PubMed]
10. Zhang, M.; Zou, Y.; Zuo, C.; Ao, H.; Guo, Y.; Wang, X.; Han, M. Targeted antitumor comparison study between dopamine self-polymerization and traditional synthesis for nanoparticle surface modification in drug delivery. *Nanotechnology* **2021**, *32*, 30. [CrossRef] [PubMed]
11. Jia, L.; Han, F.; Wang, H.; Zhu, C.; Guo, Q.; Li, J.; Zhao, Z.; Zhang, Q.; Zhu, X.; Li, B. Polydopamine-assisted surface modification for orthopaedic implants. *J. Orthop. Transl.* **2019**, *17*, 82–95. [CrossRef] [PubMed]
12. Fan, F.; Saha, S.; Hanjaya-Putra, D. Biomimetic hydrogels to promote wound healing. *Front. Bioeng. Biotechnol.* **2021**, *9*, 718377. [CrossRef]
13. Zheng, D.; Huang, C.; Zhu, X.; Huang, H.; Xu, C. Performance of polydopamine complex and mechanisms in wound healing. *Int. J. Mol. Sci.* **2021**, *22*, 10563. [CrossRef]
14. Zhang, S.; Ge, G.; Qin, Y.; Li, W.; Dong, J.; Mei, J.; Ma, R.; Zhang, X.; Bai, J.; Zhu, C.; et al. Recent advances in responsive hydrogels for diabetic wound healing. *Mater. Today Bio* **2022**, *18*, 100508. [CrossRef]
15. Chi, M.; Li, N.; Cui, J.; Karlin, S.; Rohr, N.; Sharma, N.; Thieringer, F.M. Biomimetic, mussel-inspired surface modification of 3D-printed biodegradable polylactic acid scaffolds with nano-hydroxyapatite for bone tissue engineering. *Front. Bioeng. Biotechnol.* **2022**, *10*, 989729. [CrossRef]
16. Ghorai, S.K.; Dutta, A.; Roy, T.; Guha Ray, P.; Ganguly, D.; Ashokkumar, M.; Dhara, S.; Chattopadhyay, S. Metal ion augmented mussel inspired polydopamine immobilized 3D printed osteoconductive scaffolds for accelerated bone tissue regeneration. *ACS Appl. Mater. Interfaces* **2022**, *14*, 28455–28475. [CrossRef] [PubMed]
17. Yang, Z.; Yi, P.; Liu, Z.; Zhang, W.; Mei, L.; Feng, C.; Tu, C.; Li, Z. Stem cell-laden hydrogel-based 3D bioprinting for bone and cartilage tissue engineering. *Front. Bioeng. Biotechnol.* **2022**, *10*, 865770. [CrossRef] [PubMed]
18. Teixeir, M.C.; Lameirinhas, N.S.; Carvalho, J.P.F.; Silvestre, A.J.D.; Vilela, C.; Freire, C.S.R. A guide to polysaccharide-based hydrogel bioinks for 3D bioprinting applications. *Int. J. Mol. Sci.* **2022**, *23*, 6564. [CrossRef]
19. Gao, Q.; Kim, B.S.; Gao, G. Advanced strategies for 3D bioprinting of tissue and organ analogs using alginate hydrogel bioinks. *Mar. Drugs* **2021**, *19*, 708. [CrossRef]
20. O'Connor, N.A.; Syed, A.; Wong, M.; Hicks, J.; Nunez, G.; Jitianu, A.; Siler, Z.; Peterson, M. Polydopamine antioxidant hydrogels for wound healing applications. *Gels* **2020**, *6*, 39. [CrossRef]
21. Im, S.; Choe, G.; Seok, J.M.; Yeo, S.J.; Lee, J.H.; Kim, W.D.; Lee, J.Y.; Park, S.A. An osteogenic bioink composed of alginate, cellulose nanofibrils, and polydopamine nanoparticles for 3D bioprinting and bone tissue engineering. *Int. J. Biol. Macromol.* **2022**, *205*, 520–529. [CrossRef]
22. Chen, Y.W.; Fang, H.Y.; Shie, M.Y.; Shen, Y.F. The mussel-inspired assisted apatite mineralized on PolyJet material for artificial bone scaffold. *Int. J. Bioprint* **2019**, *5*, 197. [CrossRef]

23. Piola, B.; Sabbatini, M.; Gino, S.; Invernizzi, M.; Renò, F. 3D Bioprinting of gelatin-xanthan gum composite hydrogels for growth of human skin cells. *Int. J. Mol. Sci.* **2022**, *23*, 539. [CrossRef] [PubMed]
24. Liebscher, J.; Mrówczyński, R.; Scheidt, H.A.; Filip, C.; Hădade, N.D.; Turcu, R.; Bende, A.; Beck, S. Structure of polydopamine: A never-ending story? *Langmuir* **2013**, *29*, 10539–10548. [CrossRef]
25. Lee, H.; Dellatore, S.M.; Miller, W.M.; Messersmith, P.B. Mussel-inspired surface chemistry for multifunctional coatings. *Science* **2007**, *318*, 426–430. [CrossRef] [PubMed]
26. Nilforoushzadeh, M.A.; Sisakht, M.M.; Amirkhani, M.A.; Seifalian, A.M.; Banafshe, H.R.; Verdi, J.; Nouradini, M. Engineered skin graft with stromal vascular fraction cells encapsulated in fibrin–collagen hydrogel: A clinical study for diabetic wound healing. *J. Tissue Eng. Regen. Med.* **2020**, *14*, 424–440. [CrossRef] [PubMed]
27. Wu, Y.; Heikal, L.; Ferns, G.; Ghezzi, P.; Nokhodchi, A.; Maniruzzaman, M. 3D bioprinting of novel biocompatible scaffolds for endothelial cell repair. *Polymers* **2019**, *11*, 1924. [CrossRef]
28. Shawan, M.M.A.K.; Islam, N.; Aziz, S.; Khatun, N.; Sarker, S.R.; Hossain, M.; Hossan, T.; Morshed, M.; Sarkar, M.; Shakil, M.S. Fabrication of xanthan gum: Gelatin (Xnt: Gel) hybrid composite hydrogels for evaluating skin wound healing efficacy. *Mod. Appl. Sci.* **2019**, *13*, 101. [CrossRef]
29. Sanbhal, N.; Saitaer, X.; Peerzada, M.; Habboush, A.; Wang, F.; Wang, L. One-step surface functionalized hydrophilic polypropylene meshes for hernia repair using bio-inspired polydopamine. *Fibers* **2019**, *7*, 6. [CrossRef]
30. Poursamar, S.A.; Lehner, A.N.; Azami, M.; Ebrahimi-Barough, S.; Samadikuchaksaraei, A.; Antunes, A.P.M. The effects of crosslinkers on physical, mechanical, and cytotoxic properties of gelatin sponge prepared via in-situ gas foaming method as a tissue engineering scaffold. *Mater. Sci. Eng. C* **2016**, *63*, 1–9. [CrossRef]
31. Farris, S.; Song, J.; Huang, Q. Alternative reaction mechanism for the cross-linking of gelatin with glutaraldehyde. *J. Agric. Food Chem.* **2010**, *58*, 998–1003. [CrossRef] [PubMed]
32. Bose, S.; Vahabzadeh, S.; Bandyopadhyay, A. Bone tissue engineering using 3D printing. *Mater. Today* **2013**, *16*, 496–504. [CrossRef]
33. An, J.; Teoh, J.E.M.; Suntornnond, R.; Chua, C.K. Design and 3D printing of scaffolds and tissues. *Engineering* **2015**, *1*, 261–268. [CrossRef]
34. Chen, L.; Huang, C.; Zhong, Y.; Chen, Y.; Zhang, H.; Zheng, Z.; Jiang, Z.; Wei, X.; Peng, Y.; Huang, L.; et al. Multifunctional sponge scaffold loaded with concentrated growth factors for promoting wound healing. *Iscience* **2022**, *26*, 105835. [CrossRef] [PubMed]
35. Yang, K.; Lee, J.S.; Kim, J.; Lee, Y.B.; Shin, H.; Um, S.H.; Kim, J.B.; Park, K.I.; Lee, H.; Cho, S.W. Polydopamine-mediated surface modification of scaffold materials for human neural stem cell engineering. *Biomaterials* **2012**, *33*, 6952–6964. [CrossRef]
36. Lv, H.; Li, L.; Sun, M.; Zhang, Y.; Chen, L.; Rong, Y.; Li, Y. Mechanism of regulation of stem cell differentiation by matrix stiffness. *Stem Cell Res. Ther.* **2015**, *6*, 103. [CrossRef]
37. Fraley, S.I.; Feng, Y.; Krishnamurthy, R.; Kim, D.-H.; Celedon, A.; Longmore, G.D.; Wirtz, D. A distinctive role for focal adhesion proteins in three-dimensional cell motility. *Nat. Cell Biol.* **2010**, *12*, 598. [CrossRef]
38. Geiger, B.; Spatz, J.P.; Bershadsky, A.D. Environmental sensing through focal adhesions. *Nat. Rev. Mol. Cell Biol.* **2009**, *10*, 21–33. [CrossRef] [PubMed]

Disclaimer/Publisher's Note: The statements, opinions and data contained in all publications are solely those of the individual author(s) and contributor(s) and not of MDPI and/or the editor(s). MDPI and/or the editor(s) disclaim responsibility for any injury to people or property resulting from any ideas, methods, instructions or products referred to in the content.

MDPI
St. Alban-Anlage 66
4052 Basel
Switzerland
www.mdpi.com

Gels Editorial Office
E-mail: gels@mdpi.com
www.mdpi.com/journal/gels

Disclaimer/Publisher's Note: The statements, opinions and data contained in all publications are solely those of the individual author(s) and contributor(s) and not of MDPI and/or the editor(s). MDPI and/or the editor(s) disclaim responsibility for any injury to people or property resulting from any ideas, methods, instructions or products referred to in the content.